# GeoPlanet: Earth and Planetary Sciences

More information about this series at http://www.springer.com/series/8821

Krzysztof A. Mizerski

# Foundations of Convection with Density Stratification

Springer

Krzysztof A. Mizerski
Department of Magnetism
Institute of Geophysics
Warsaw, Poland

The GeoPlanet: Earth and Planetary Sciences Book Series is in part a continuation of Monographic Volumes of Publications of the Institute of Geophysics, Polish Academy of Sciences, the journal published since 1962 (http://pub.igf.edu.pl/index.php).

ISSN 2190-5193 ISSN 2190-5207 (electronic)
GeoPlanet: Earth and Planetary Sciences
ISBN 978-3-030-63056-0 ISBN 978-3-030-63054-6 (eBook)
https://doi.org/10.1007/978-3-030-63054-6



This Springer imprint is published by the registered company Springer Nature Switzerland AG
The registered company address is: Gewerbestrasse 11, 6330 Cham, Switzerland

*To my wife,*
*my daughter,*
*and my son*

# Preface

The phenomenon of thermal and compositional (chemical) convection is very common in nature and therefore of great importance from the point of view of understanding of many fundamental aspects of the environment and universe. It has attracted a great deal of attention over the last half a century and a significant progress in a detailed quantitative description of this phenomenon has been made, in various physical contexts. A number of books have been written on the topic, such as the seminal work of Chandrasekhar (1961) on *Hydrodynamic and Hydromagnetic Stability*, a large portion of which is devoted to the convective instability near its onset or the outstanding book of Getling (1998) where systematization of the knowledge on convection has been continued with a thorough description of the weakly nonlinear stages. Most of the works, however, considered weakly stratified, that is the so-called Boussinesq convection.

It is the aim of this book to continue the process of systematization. It seems important to put the current knowledge on weakly and strongly stratified convection in order and provide a comprehensive description of the marginal, weakly nonlinear and fully developed stages of convective flow in both cases. To that end the book provides a short compendium of knowledge on the linear and weakly nonlinear limits of the Boussinesq convection, as a useful reference for a reader and than proceeds with a review of the theory on fully developed, weakly stratified convection. The entire third chapter is devoted to a detailed derivation and a study of the three aforementioned stages of stratified (*anelastic*) convection. The description of stratified convection requires extreme care, since many aspects have to be considered simultaneously for full consistency. Detailed and systematic explanations are therefore provided. It is not the aim to deliver a comprehensive review of findings on convection, but rather to pinpoint precisely to the relevant works (even pages and equations) on particular aspects of the dynamics of convective flows. This book is meant as a textbook for courses on hydrodynamics and convective flows, for the use of lecturers and students, however, it may also be of use for the entire scientific community as a practical reference.

# Contents

# Notation and Definitions

| Symbol | Mathematical definition | Explanation |
|---|---|---|
| $L$ | - | vertical span of the fluid layer |
| $L_x$, $L_y$ | - | horizontal periods of the domain |
| $\langle f\rangle, \bar{f}$ | $\frac{1}{L_xL_yL}\int_{-L_x/2}^{L_x/2}\int_{-L_y/2}^{L_y/2}\int_{-L/2}^{L/2} f\mathrm{d}x\mathrm{d}y\mathrm{d}z$ | spatial average |
| $\langle f\rangle_h$ | $\frac{1}{L_xL_y}\int_{-L_x/2}^{L_x/2}\int_{-L_y/2}^{L_y/2} f\mathrm{d}x\mathrm{d}y$ | average over a horizontal plane |
| $\tilde{f}$ | - | variable in the hydrostatic reference state |
| $\approx f$ | $\tilde{f}-\bar{\tilde{f}}$ | variation above the mean in hydrostatic state |
| $f_{ad}$ | - | variable in the hydrostatic adiabatic state |
| $f^{\#}$ | - | non-dimensional variable $f$ |
| $f_B, f_T$ | $f(z=0), f(z=L)$ | bottom and top values of the field $f$ |
| $\left(\frac{\partial f}{\partial A}\right)_{B,C}$ | - | derivative with respect to $A$ at constant $B$, $C$ |
| $\mathbf{u}$ | - | velocity field |
| $\zeta$ | $\nabla\times\mathbf{u}$ | vorticity field |
| $\rho$ | - | density |
| $V$ | - | volume of the fluid |
| $p$ | - | pressure |
| $s$ | - | entropy per unit mass |
| $S$ | - | entropy |
| $T$ | - | temperature |
| $\xi$ | $\frac{m_lN^{(l)}}{m_lN^{(l)}+m_hN^{(h)}}$ | mass fraction of a light constituent in a binary alloy |
| $N$ | - | number of particles in the fluid |

(continued)

(continued)

| Symbol | Mathematical definition | Explanation |
|---|---|---|
| $N^{(l)}$ | - | number of particles of light constituent in alloy |
| $N^{(h)}$ | - | number of particles of heavy constituent in alloy |
| $\mu_c$ | $\frac{\mu_l}{m_l} - \frac{\mu_h}{m_h}$ | chemical potential for a binary alloy |
| $\mu_l$ | - | chemical potential of light constituent in alloy |
| $\mu_h$ | - | chemical potential of heavy constituent in alloy |
| $\varepsilon$ | - | internal energy per unit mass |
| $\mathcal{E}$ | - | internal energy |
| $\psi$ | - | potential energy per unit mass from external forcing |
| $\mathbf{F}$ | $-\nabla\psi$ | external force per unit mass |
| $e$ | $e = \frac{1}{2}\mathbf{u}^2 + \psi + \varepsilon$ | total energy per unit mass |
| $m_m$ | - | molecular mass of fluid particles |
| $m_l$ | - | molecular mass of light constituent in alloy |
| $m_h$ | - | molecular mass of heavy constituent in alloy |
| $r_m$ | $\frac{m_h}{m_l}$ | molecular mass ratio |
| $\alpha$ | $-\frac{1}{\rho}\left(\frac{\partial\rho}{\partial T}\right)_{p,\xi}$ | coefficient of thermal expansion |
| $\alpha_a$ | (2.128) | radiation absorption coefficient per unit volume |
| $\beta$ | $\frac{1}{\rho}\left(\frac{\partial\rho}{\partial p}\right)_{T,\xi}$ | coefficient of isothermal compressibility |
| $c_{p,\xi}$ | $T\left(\frac{\partial s}{\partial T}\right)_{p,\xi}$ | specific heat at constant pressure |
| $c_{v,\xi}$ | $T\left(\frac{\partial s}{\partial T}\right)_{\rho,\xi}$ | specific heat at constant volume |
| $\gamma$ | $\frac{c_p}{c_v} > 1$ | specific heat ratio |
| $C$ | $\sqrt{\left(\frac{\partial p}{\partial\rho}\right)_{s,\xi}}$ | speed of sound |
| $C_T$ | $\sqrt{\left(\frac{\partial p}{\partial\rho}\right)_{T,\xi}}$ | isothermal speed of sound |
| $\Gamma$ | $\frac{T_B}{T_T}$ | temperature bottom to top ratio |
| $D_\rho$ | $\left\lvert\frac{1}{\rho}\frac{\mathrm{d}\rho}{\mathrm{d}z}\right\rvert^{-1}$ | density scale height |
| $D_p$ | $\left\lvert\frac{1}{p}\frac{\mathrm{d}p}{\mathrm{d}z}\right\rvert^{-1}$ | pressure scale height |
| $D_T$ | $\left\lvert\frac{1}{T}\frac{\mathrm{d}T}{\mathrm{d}z}\right\rvert^{-1}$ | temperature scale height |

(continued)

(continued)

| Symbol | Mathematical definition | Explanation |
|---|---|---|
| $\Delta_S$ | $-\left(\frac{\partial T}{\partial z}+\frac{g\alpha T}{c_{p,\xi}}\right)$ | superadiabatic excess |
| $\Delta\tilde{s}$ | $\tilde{s}_B-\tilde{s}_T$ | basic entropy jump across the fluid layer |
| $\Delta T$ | $T_B-T_T$ | temperature jump across the fluid layer |
| $\Delta\tilde{T}$ | $\tilde{T}_B-\tilde{T}_T$ | basic temperature jump across the fluid layer |
| $(\Delta f)_B$ | $\langle f\rangle_h(z=\delta_{th,B})-\langle f\rangle_h(z=0)$ | jump in value of $\langle f\rangle_h$ across bottom boundary layer |
| $(\Delta f)_T$ | $\langle f\rangle_h(z=L)-\langle f\rangle_h(z=L-\delta_{th,T})$ | jump in value of $\langle f\rangle_h$ across top boundary layer |
| $(\Delta T)_{bulk}$ | $\langle T\rangle_h(z=L-\delta_{th,T})-\langle T\rangle_h(z=\delta_{th,B})>0$ | mean temperature jump across the bulk |
| $(\Delta T')_{bulk}$ | $\langle T'\rangle_h(z=\delta_{th,B})-\langle T'\rangle_h(z=L-\delta_{th,T})>0$ | mean temperature fluctuation jump across the bulk |
| $(\Delta T)_{vel}$ | (3.258) | convective correction to bulk temperature jump |
| $\delta$ | $\left\langle\frac{L}{T}\Delta_S\right\rangle$ | non-dimensional superadiabatic excess |
| $\delta_{th}$ | $Nu^{-1}$ | thermal boundary layer thickness |
| $\delta_\nu$ | $Re^{-1/2}$ | viscous boundary layer thickness |
| $\epsilon$ | $\frac{\Delta\tilde{\tilde{\rho}}}{\tilde{\rho}}\ll 1$ | small density stratification (Boussinesq) |
| $\epsilon_a$ | $\frac{L}{\tilde{T}_B}\left(\frac{\Delta\tilde{T}}{L}-\frac{g}{c_p}\right)\ll 1$ | small departure from adiabaticity (anelastic) |
| $F_{total}$ | (3.147) | total, horizontally averaged heat flux |
| $F_{conv.}$ | (3.149a) | convective, horizontally averaged heat flux |
| $F_S$ | (3.150) | superadiabatic, horizontally averaged heat flux |
| $G_{total}$ | (4.151) | total, horizontally averaged material flux |
| $\mathbf{G}$ | $\frac{\partial u_i}{\partial x_j}$ | velocity gradient tensor |
| $G$ | $6.67\times10^{-11}\ m^3/s^2kg$ | gravitational constant |
| $\mathbf{g}$ | - | acceleration of gravity |
| $\eta$ | (2.165) | departure from threshold of convection |
| $h_{p,T}$ | $T\left(\frac{\partial s}{\partial\xi}\right)_{p,T}$ | heat of reaction |
| $\theta$ | $\frac{\Delta T}{T_B}$ | temperature stratification parameter |

(continued)

(continued)

| Symbol | Mathematical definition | Explanation |
|---|---|---|
| $\mathbf{I}$ | $\delta_{ij}$ | unitary matrix |
| $\mathbf{j}_A$ | - | flux of quantity $A$ |
| $\mathbf{j}_{A,\mathrm{mol}}$ | - | molecular flux of quantity $A$ |
| $k$ | - | coefficient of heat conduction |
| $\kappa$ | $\frac{k}{\rho c_{p,\xi}}$ | coefficient of thermal diffusion |
| $K$ | - | coefficient of material conductivity |
| $D$ | $\frac{K}{\rho}$ | coefficient of material diffusion |
| $k_T$ | (4.54) | Soret coefficient |
| $k_p$ | (4.54) | pressure gradient coefficient in material flux |
| $k_B$ | $1.38 \times 10^{-23}\ J/K$ | Boltzmann constant |
| $\Lambda$ | (4.54) | Dufour coefficient |
| $m$ | (3.68c) | polytropic index |
| $\mu$ | - | dynamic shear viscosity |
| $\mu_b$ | - | dynamic bulk viscosity |
| $\nu$ | $\frac{\mu}{\rho}$ | kinematic shear viscosity |
| $\nu_b$ | $\frac{\mu_b}{\rho}$ | kinematic bulk viscosity |
| $Q$ | - | heat sources other than viscous friction |
| $q_{tot}$ | - | heat delivered to fluid parcel from surroundings |
| $r_\delta$ | $\frac{\delta_{th,T}}{\delta_{th,B}}$ | top to bottom thermal boundary layer thickness ratio |
| $r_s$ | $\frac{(\Delta s)_T}{(\Delta s)_B}$ | ratio of jumps of $\langle s_h \rangle$ across boundary layers |
| $r_T$ | $\frac{(\Delta T')_T}{(\Delta T')_B}$ | ratio of jumps of $\langle T'_h \rangle$ across boundary layers |
| $r_U$ | $\frac{U_T}{U_B}$ | top to bottom thermal wind magnitude ratio |
| $R$ | $\frac{k_B}{m_m}$ | specific gas constant |
| $\sigma_A$ | - | volume sources of quantity $A$ |
| $\sigma$ | - | growth rate of convective instability |
| $\sigma_{rad}$ | $5.67 \times 10^{-8}\ W/m^2K^4$ | Stefan-Boltzmann constant |
| $\tau$ | (1.8) | stress tensor |
| $\mathscr{T}$ | (2.17), (3.15) | time scale |
| $\mathscr{U}$ | (2.17), (3.15) | velocity scale |
| $\Upsilon$ | $\left(\frac{\partial \mu_c}{\partial \xi}\right)_{p,T}$ | compositional derivative of chemical potential |

(continued)

(continued)

| Symbol | Mathematical definition | Explanation |
|---|---|---|
| $\chi$ | $-\frac{1}{\rho}\left(\frac{\partial\rho}{\partial\xi}\right)_{p,s}$ | compositional expansion cofficient |
| $\chi_T$ | $-\frac{1}{\rho}\left(\frac{\partial\rho}{\partial\xi}\right)_{p,T}$ | isothermal compositional expansion coefficient |
| $\omega$ | - | frequency of oscillations of fluctuations |
| $\Omega$ | - | background rotation rate |
| $Nu$ | (2.59), (3.156), (4.157) | Nusselt number |
| $Nu_Q$ | (2.84), (3.167) | Nusselt number for fixed heat flux at boundaries |
| $Ra$ | (2.62), (3.157) | Rayleigh number |
| $Ra_{comp}$ | (4.159) | compositional Rayleigh number |
| $Ra_R$ | (2.140) | radiative Rayleigh number |
| $Re$ | $\frac{\mathcal{U}L}{\nu}$ | Reynolds number |
| $Pr$ | $\frac{\nu}{\kappa}$ | Prandtl number |
| $E$ | $\frac{\nu}{2\Omega L^2}$ | Ekman number |

# Chapter 1
# The Equations of Hydrodynamics

The derivation of the fundamental equations for Newtonian fluids has been provided in many books, e.g. in Chandrasekhar (1961) and many others. Therefore here we only briefly recall the main points of the derivation, with the aim to keep the book self-consistent and set grounds for later chapters. A reader interested in the historical origins of hydrodynamics and the long process of gradual increase in the rigorousness of the description of dynamical flows since Daniel Bernoulli's *Hydrodynamica* published in 1738 is directed to Darrigol (2005).

## 1.1 General Conservation Law in a Continuous Medium

Let us take an extensive quantity $A(t)$ and introduce its density per unit volume, $a(\mathbf{x}, t)$, so that

$$A\left(t\right)=\int_{V}a\left(\mathbf{x},t\right)\mathrm{d}^{3}x. \tag{1.1}$$

The variation of $A(t)$ in a volume $V$ can be attributed only to two phenomena, that is either to sources (or sinks, which will be thought of as negative sources) of this quantity within the volume, denoted by $\sigma_A(\mathbf{x}, t)$ or the flux $\mathbf{j}_A(\mathbf{x}, t)$ of the quantity $A(t)$, which at least partly, is due to the flow of the medium. Therefore the total variation of $A(t)$ in a fixed volume $V$ can be expressed as follows

$$\frac{\mathrm{d}A}{\mathrm{d}t}=\int_{V}\frac{\partial a\left(\mathbf{x},t\right)}{\partial t}\mathrm{d}^{3}x=\int_{V}\sigma_{A}\left(\mathbf{x},t\right)\mathrm{d}^{3}x-\oint_{\partial V}\mathbf{j}_{A}(\mathbf{x},t)\cdot\hat{\mathbf{n}}\mathrm{d}\Sigma, \tag{1.2}$$



K. A. Mizerski, *Foundations of Convection with Density Stratification*, GeoPlanet: Earth and Planetary Sciences, https://doi.org/10.1007/978-3-030-63054-6_1

where $\hat{\mathbf{n}}$ is the unit normal directed outside of the surface $\partial V$ enclosing the volume $V$. Therefore a quantity is conserved in a volume $V$ if there are no sources of this quantity within the volume and the same amount of flux enters and leaves the volume. Since $V$ is arbitrary, by the use of the Gauss-Ostrogradsky divergence theorem,

$$\oint_{\partial V} \mathbf{j}_A(\mathbf{x}, t) \cdot \hat{\mathbf{n}} \mathrm{d}\Sigma = \int_V \nabla \cdot \mathbf{j}_A(\mathbf{x}, t) \mathrm{d}^3\mathbf{x}, \tag{1.3}$$

we obtain locally

$$\frac{\partial a(\mathbf{x}, t)}{\partial t} + \nabla \cdot \mathbf{j}_A(\mathbf{x}, t) = \sigma_A(\mathbf{x}, t), \tag{1.4}$$

which is a general local evolution law of a certain extensive quantity $A(t)$ in a fluid.

## 1.2 The Continuity Equation - Mass Conservation Law

At this point, to derive the law of mass conservation it is enough to substitute the mass $m(t)$ for $A(t)$ and thus the mass density $\rho(\mathbf{x}, t)$ for $a(\mathbf{x}, t)$ from the previous section, which yields

$$\frac{\partial \rho(\mathbf{x}, t)}{\partial t} + \nabla \cdot \mathbf{j}_m(\mathbf{x}, t) = \sigma_m(\mathbf{x}, t), \tag{1.5}$$

and since the flux of mass in a fluid results solely from the flow $\mathbf{u}(\mathbf{x}, t)$, it takes the form $\mathbf{j}_m(\mathbf{x}, t) = \rho(\mathbf{x}, t)\,\mathbf{u}(\mathbf{x}, t)$. Hence finally the continuity equation reads

$$\frac{\partial \rho(\mathbf{x}, t)}{\partial t} + \nabla \cdot [\rho(\mathbf{x}, t)\,\mathbf{u}(\mathbf{x}, t)] = \sigma_m(\mathbf{x}, t), \tag{1.6}$$

which expresses local mass conservation if the sources of matter $\sigma_m(\mathbf{x}, t)$ vanish at least locally.

## 1.3 The Navier-Stokes Equation - Momentum Balance

The general evolution law (1.4) applied to momentum per unit volume, $\rho\mathbf{u}$, which simply expresses the Newton's second law of dynamics, takes the form

$$\frac{\partial \rho\mathbf{u}}{\partial t} + \nabla \cdot [\rho\mathbf{u}\mathbf{u} + \mathbf{\Pi}] = \rho\mathbf{F}, \tag{1.7}$$

where , $\rho\mathbf{u}\mathbf{u} + \mathbf{\Pi}$ is the flux of momentum (tensorial, since the momentum is vectorial) with $\rho\mathbf{u}\mathbf{u}$ being the advective flux and $\mathbf{\Pi}$ the pressure tensor (pressure and frictional flux); $\mathbf{F}$ is the body force density per unit mass and the notation with explicit dependence on $(\mathbf{x}, t)$ is dropped form now on for clarity. The pressure tensor $\mathbf{\Pi}$ is equal to the negative stress tensor $\boldsymbol{\tau}$, whose components describe forces per unit surface that form between fluid elements in a flow. A force per unit surface exerted on a given fluid element, say element (1), by its neighbour, say element (2), is given by $\boldsymbol{\tau} \cdot \hat{\mathbf{n}}$, where $\hat{\mathbf{n}}$ is the unit normal to the surface separating the two fluid elements, directed from (1) to (2). For standard, isotropic Newtonian fluids the stress tensor is expressed in terms of fluid pressure $p$ and velocity gradients in the following way (cf. e.g. Batchelor 1967, Chandrasekhar 1961),[1]

$$\tau_{ij} = -\Pi_{ij} = -p\delta_{ij} + 2\mu\left(G^s_{ij} - \frac{1}{3}\nabla \cdot \mathbf{u}\delta_{ij}\right) + \mu_b \nabla \cdot \mathbf{u}\delta_{ij}, \tag{1.8}$$

where $\mu$ is the coefficient of dynamic shear viscosity, $\mu_b$ is the bulk viscosity associated with expansion (compression) processes, the subscripts $ij$ denote the cartesian components of tensors, $\delta_{ij}$ denotes a unitary tensor. Moreover,

$$G_{ij} = \frac{\partial u_i}{\partial x_j}, \qquad G^s_{ij} = \frac{1}{2}\left(\frac{\partial u_i}{\partial x_j} + \frac{\partial u_j}{\partial x_i}\right), \tag{1.9}$$

is the velocity gradient tensor and its symmetric part; note, that often kinematic coefficients of shear and bulk viscosities are used, defined as $\nu = \mu/\rho$ and $\nu_b = \mu_b/\rho$. The dissipative part of the stress tensor given in (1.8), that is $2\mu\left(G^s_{ij} - \frac{1}{3}\nabla \cdot \mathbf{u}\delta_{ij}\right) + \mu_b \nabla \cdot \mathbf{u}\delta_{ij}$, describes the viscous stresses in the fluid. Under the assumption, that there are no mass sources in the entire fluid volume, $\sigma_m(\mathbf{x}, t) = 0$, the fundamental momentum balance (1.7) takes the form

$$\rho\left[\frac{\partial \mathbf{u}}{\partial t} + (\mathbf{u} \cdot \nabla)\,\mathbf{u}\right] = \nabla \cdot \boldsymbol{\tau} + \rho \mathbf{F}. \tag{1.10}$$

[1] The fundamental assumption for Newtonian fluids is that the dissipative part of the stress tensor, say $\boldsymbol{\tau}_\mu$, associated with frictional effects due to the flow, is linearly related to the tensor of deformation rate, that is the flow velocity gradient tensor, since it is the presence of velocity gradients which is necessary and sufficient for frictional forces to appear. Once this assumption is made, the final form of the constitutive relation (1.8) for an isotropic fluid is simply an outcome of symmetry of the stress tensor, which follows from the angular momentum balance (cf. the next Sect. 1.3.1) and very basic properties of tensorial objects known from the group theory. The latter is simply the Curie's principle 1984 (cf. Chalmers 1970; de Groot and Mazur 1984), namely that in an isotropic system only those tensorial objects, which at rotations of a system of reference transform according to the same irreducible (therefore distinct) representations of the rotation groups can be linearly related. Since the trace and the symmetric traceless part of a tensor transform differently, one arrives at the constitutive relations $\mathrm{Tr}\boldsymbol{\tau}_\mu = -\mu_b \mathrm{Tr}\mathbf{G} = -\mu_b \nabla \cdot \mathbf{u}$ and $\boldsymbol{\tau}^s_\mu - (1/3)(\mathrm{Tr}\boldsymbol{\tau}_\mu)\mathbf{I} = -2\mu(\mathbf{G}^s - (1/3)(\nabla \cdot \mathbf{u})\mathbf{I})$, which are equivalent to (1.8). In the given relations the superfix $s$ denotes a symmetric part of a tensor, $\mathbf{G}$ is the velocity gradient, as in (1.9), $\mathbf{I}$ is the unitary matrix and a negative sign was introduced in front of the right hand sides, since the coefficients $\mu > 0$ and $\mu_b > 0$ describe frictional, therefore dissipative effects.

After introduction of the stress tensor form for a Newtonian fluid (1.8), we arrive at the well-known Navier-Stokes equation with non-uniform viscosity

$$\rho\left[\frac{\partial \mathbf{u}}{\partial t}+(\mathbf{u}\cdot\nabla)\,\mathbf{u}\right]=-\nabla p+\rho\mathbf{F}+\mu\nabla^2\mathbf{u}+\left(\frac{\mu}{3}+\mu_b\right)\nabla\left(\nabla\cdot\mathbf{u}\right)$$
$$+2\nabla\mu\cdot\mathbf{G}^s+\nabla\left(\mu_b-\frac{2}{3}\mu\right)\nabla\cdot\mathbf{u}. \tag{1.11}$$

### 1.3.1 The Angular Momentum Balance

It is straightforward to verify, that the stress tensor $\boldsymbol{\tau}$ for Newtonian fluids given in (1.8) is symmetric. As mentioned earlier, its symmetry results from the balance of angular momentum in a fluid. From the general evolution law for a continuous medium (1.4) we obtain for the angular momentum

$$\frac{\partial}{\partial t}\left(\mathbf{x}\times\rho\mathbf{u}\right)+\nabla\cdot\mathbf{j}_L=\rho\mathbf{x}\times\mathbf{F}, \tag{1.12}$$

where the flux of angular momentum $\mathbf{j}_L$ satisfies for any volume $V$ within the fluid

$$\int_V \mathbf{j}_L \mathrm{d}^3\mathbf{x}=\int_V \nabla\cdot\left[\rho\left(\mathbf{x}\times\mathbf{u}\right)\mathbf{u}\right]\mathrm{d}^3\mathbf{x}+\oint_{\partial V}\mathbf{x}\times\left(\boldsymbol{\tau}\cdot\hat{\mathbf{n}}\right)\mathrm{d}\Sigma. \tag{1.13}$$

In the above the first term on the right hand side is the advective flux and the second is the angular momentum generated by pressure and friction. We can apply the divergence theorem to the latter, which yields

$$\oint_{\partial V}\mathbf{x}\times\left(\boldsymbol{\tau}\cdot\hat{\mathbf{n}}\right)\mathrm{d}\Sigma=\int_V \mathbf{x}\times\nabla\cdot\boldsymbol{\tau}\mathrm{d}^3\mathbf{x}+\int_V \mathbf{a}_\tau \mathrm{d}^3\mathbf{x}, \tag{1.14}$$

where $(\mathrm{a}_\tau)_i=\epsilon_{ijk}\tau^a_{jk}$, that is $\mathbf{a}_\tau=2[\tau^a_{23},\ -\tau^a_{13},\ \tau^a_{12}]$ and the superscript $a$ denotes the antisymmetric part of a tensor. This means that the angular momentum evolution law (1.12) can be written as follows

$$\frac{\partial}{\partial t}\left(\mathbf{x}\times\rho\mathbf{u}\right)+\nabla\cdot\left[\rho\left(\mathbf{x}\times\mathbf{u}\right)\mathbf{u}\right]=\mathbf{x}\times\nabla\cdot\boldsymbol{\tau}+\rho\mathbf{x}\times\mathbf{F}+\mathbf{a}_\tau. \tag{1.15}$$

The left hand side of the above Eq. (1.15) can be manipulated to give

$$\frac{\partial}{\partial t}\left(\mathbf{x}\times\rho\mathbf{u}\right)+\nabla\cdot\left[\rho\left(\mathbf{x}\times\mathbf{u}\right)\mathbf{u}\right]=\mathbf{x}\times\left[\frac{\partial\rho\mathbf{u}}{\partial t}+\nabla\cdot\left(\rho\mathbf{u}\mathbf{u}\right)\right], \tag{1.16}$$

so that the angular momentum balance can be expressed in the following way

$$\mathbf{x} \times \left[ \frac{\partial \rho \mathbf{u}}{\partial t} + \nabla \cdot (\rho \mathbf{u}\mathbf{u} - \boldsymbol{\tau}) - \rho \mathbf{F} \right] = \mathbf{a}_{\tau}. \tag{1.17}$$

On the other hand the general momentum balance (1.7) must be satisfied, which yields

$$\mathbf{a}_{\tau} = 0 \Rightarrow \boldsymbol{\tau}^{a} = 0, \tag{1.18}$$

and hence the antisymmetric part of the stress tensor must vanish or, in other words, the stress tensor is necessarily symmetric. Simply for the sake of completeness we may provide a final, general expression for the angular momentum flux $(j_L)_{im} = \epsilon_{ijk} x_j (\rho u_k u_m + \tau_{km})$.

## 1.4 The Energy Equation

The total energy density per unit mass in a fluid volume $V$ is

$$e = \frac{1}{2}\mathbf{u}^2 + \psi + \varepsilon, \tag{1.19}$$

where $\mathbf{u}^2/2$ is the kinetic energy density, $\psi$ the potential energy resulting from presence of conservative body forces, $\mathbf{F} = -\nabla\psi$, which we will assume stationary, and $\varepsilon$ denotes the internal energy of the fluid. The general local evolution law (1.4) for the total energy reads

$$\frac{\partial \rho e}{\partial t} + \nabla \cdot (\rho e \mathbf{u} + \mathbf{j}_{\mathrm{mol}}) = Q, \tag{1.20}$$

where $\rho e \mathbf{u}$ is the flux due to energy advection by the flow, $\mathbf{j}_{\mathrm{mol}}$ is the flux of energy from molecular mechanical and thermal effects and the energy sources, here denoted by $Q$ $(= \sigma_E)$, describe heating processes (absorbed heat per unit volume per unit time) such as e.g. the radioactive heating, thermal radiation, etc. To establish the formula for the molecular energy flux $\mathbf{j}_{\mathrm{mol}}$ we must realize the effects that lead to variation of the total energy. The total change of the energy $\rho e$ in the volume $V$, in the absence of energy sources $Q$, results solely from two factors, that is the heat transfer between the volume and the rest of the fluid and the total work done on the volume by the stresses described by the stress tensor $\boldsymbol{\tau}$,

$$\oint_{\partial V} \mathbf{u} \cdot \boldsymbol{\tau} \cdot \hat{\mathbf{n}} \mathrm{d}\Sigma = \int_V \nabla \cdot (\boldsymbol{\tau} \cdot \mathbf{u}) \, \mathrm{d}V. \tag{1.21}$$

The body forces are assumed conservative hence their work

$$\int_V \rho \mathbf{u} \cdot \mathbf{F} \mathrm{d}V = -\int_V \rho \mathbf{u} \cdot \nabla \psi \mathrm{d}V, \tag{1.22}$$

simply expresses the total change of the potential energy, which is due to advection only, since the body forces are also assumed stationary. Therefore the molecular flux $\mathbf{j}_{\mathrm{mol}}$ can be decomposed into two contributions. The first one comes from thermal effects and is described by the Fourier's law,

$$\mathbf{j}_T = -k\nabla T, \tag{1.23}$$

stating, that the thermal heat flux is proportional to its cause, that is the temperature gradient[2] and the coefficient of proportionality $k$ is called the thermal conduction coefficient. The second contribution to the molecular energy flux is of mechanical nature and for now will be denoted by $\mathbf{j}_{\mathrm{mech}}$. This implies the following form of the total energy flux

$$\mathbf{j}_{\rho e} = \rho e\mathbf{u} + \mathbf{j}_{\mathrm{mech}} - k\nabla T, \tag{1.24}$$

which allows to rewrite the energy evolution Eq. (1.20) in the form

$$e\left[\frac{\partial \rho}{\partial t} + \nabla\cdot(\rho\mathbf{u})\right] + \rho\left(\frac{\partial e}{\partial t} + \mathbf{u}\cdot\nabla e\right) = -\nabla\cdot\mathbf{j}_{\mathrm{mech}} + \nabla\cdot(k\nabla T) + Q, \tag{1.25}$$

and in the absence of mass sources within the volume the term in the square brackets on the left hand side of (1.25) vanishes. However, as said, in the absence of energy sources $Q$, the energy change in a volume $V$ can result only from the heat exchanged with the surroundings and from the total work done by the stresses on the fluid volume, therefore must be equal to

$$\nabla\cdot(k\nabla T) + \nabla\cdot(\boldsymbol{\tau}\cdot\mathbf{u}), \tag{1.26}$$

which implies

$$\mathbf{j}_{\mathrm{mech}} = -\boldsymbol{\tau}\cdot\mathbf{u}. \tag{1.27}$$

The total, advective time derivative of the kinetic energy

$$\rho\left(\frac{\partial}{\partial t} + \mathbf{u}\cdot\nabla\right)\frac{1}{2}\mathbf{u}^2 = \rho\mathbf{u}\cdot\left(\frac{\partial}{\partial t} + \mathbf{u}\cdot\nabla\right)\mathbf{u}, \tag{1.28}$$

with the use of the momentum balance (1.10) can be rearranged into

$$\rho\mathbf{u}\cdot\left(\frac{\partial}{\partial t} + \mathbf{u}\cdot\nabla\right)\mathbf{u} = \mathbf{u}\cdot\nabla\cdot\boldsymbol{\tau} + \rho\mathbf{u}\cdot\mathbf{F} = \nabla\cdot(\boldsymbol{\tau}\cdot\mathbf{u}) - \boldsymbol{\tau}:\mathbf{G} + \rho\mathbf{u}\cdot\mathbf{F}, \tag{1.29}$$

where $\mathbf{G}$ is the velocity gradient tensor defined in (1.9) and double dot denotes contraction over both indices, $\boldsymbol{\tau}:\mathbf{G} = \tau_{ij}G_{ij}$. Note, that because the stress tensor is

[2] Which is yet another manifestation of the Curie's principle 1984 (cf. de Groot and Mazur 1984), stating that fluxes are linear in "thermodynamic forces", which is a term used to describe the physical causative factors of the fluxes.

symmetric, we can also substitute only the symmetric part of the velocity gradient tensor into $\boldsymbol{\tau} : \mathbf{G} = \boldsymbol{\tau} : \mathbf{G}^s$. On the other hand, as already remarked, the advection of the potential energy per unit mass, stationary by assumption, is easily expressed by

$$\rho\left(\frac{\partial}{\partial t} + \mathbf{u}\cdot\nabla\right)\psi = -\rho\mathbf{u}\cdot\mathbf{F}, \tag{1.30}$$

therefore the total energy balance yields

$$\rho\left(\frac{\partial \varepsilon}{\partial t} + \mathbf{u}\cdot\nabla\varepsilon\right) = \nabla\cdot(k\nabla T) + \boldsymbol{\tau} : \mathbf{G}^s + Q. \tag{1.31}$$

The above Eq. (1.31) expresses the first law of thermodynamics, that is the change in the internal energy of a fluid in a volume $V$ is due to the heat exchanged between the volume and its surroundings, $\nabla\cdot(k\nabla T)$, and the work done by the fluid flow on the volume unbalanced by the kinetic energy change, $\boldsymbol{\tau} : \mathbf{G}^s$. We can now write down the formula for the total energy flux

$$\mathbf{j}_{\rho e} = \rho e\mathbf{u} - \boldsymbol{\tau}\cdot\mathbf{u} - k\nabla T, \tag{1.32}$$

with $Q$ being the volume sources of the total energy, whereas the flux of the internal energy and its volume sources can be defined as follows

$$\mathbf{j}_{\rho\varepsilon} = \rho\varepsilon\mathbf{u} - k\nabla T, \qquad \sigma_{\rho\varepsilon} = \boldsymbol{\tau} : \mathbf{G}^s + Q. \tag{1.33}$$

By virtue of the formula for the stress tensor in Newtonian fluids (1.8)

$$\boldsymbol{\tau} : \mathbf{G}^s = -p\nabla\cdot\mathbf{u} + 2\mu\mathbf{G}^s : \mathbf{G}^s + \left(\mu_b - \frac{2}{3}\mu\right)(\nabla\cdot\mathbf{u})^2, \tag{1.34}$$

which includes the viscous heating, that is $2\mu\mathbf{G}^s : \mathbf{G}^s + (\mu_b - 2\mu/3)(\nabla\cdot\mathbf{u})^2$. The equation for internal energy can be transformed either into an equation for the fluid's entropy $s$ or temperature $T$. The first and second laws of thermodynamics yield[3]

$$d\varepsilon = Tds - pd\left(\frac{1}{\rho}\right), \tag{1.35}$$

[3]It is well known from the second law of thermodynamics, that the differential of the internal energy for a single component system takes the form

$$d\mathcal{E} = Td\mathcal{S} - pdV,$$

where $\mathcal{E}$, $\mathcal{S}$ and $V$ are the "canonical" thermodynamic variables, that is the actual internal energy, the entropy and the volume of the system (as opposed to the mass densities $\varepsilon$, $s$ and $\rho$); division by the total mass $M = m_m N$, where $m_m$ denotes the molecular mass of the fluid particles and $N$ the number of particles, allows to transform the above into the differential for the mass density of the internal energy, which takes the form (1.35).

which implies for the total, advective derivatives, that

$$\frac{\partial \varepsilon}{\partial t} + \mathbf{u} \cdot \nabla \varepsilon = T \left( \frac{\partial s}{\partial t} + \mathbf{u} \cdot \nabla s \right) + \frac{p}{\rho^2} \left( \frac{\partial \rho}{\partial t} + \mathbf{u} \cdot \nabla \rho \right) = T \left( \frac{\partial s}{\partial t} + \mathbf{u} \cdot \nabla s \right) - \frac{p}{\rho} \nabla \cdot \mathbf{u}, \tag{1.36}$$

where the continuity Eq. (1.6) without mass sources was used. Therefore the entropy equation reads

$$\rho T \left( \frac{\partial s}{\partial t} + \mathbf{u} \cdot \nabla s \right) = \nabla \cdot (k \nabla T) + 2\mu \mathbf{G}^s : \mathbf{G}^s + \left( \mu_b - \frac{2}{3}\mu \right) (\nabla \cdot \mathbf{u})^2 + Q. \tag{1.37}$$

Next, we parametrize the fluid's entropy with temperature $T$ and pressure $p$, that is take $s = s(T, p)$ and write

$$\mathrm{d}s = \left( \frac{\partial s}{\partial T} \right)_p \mathrm{d}T + \left( \frac{\partial s}{\partial p} \right)_T \mathrm{d}p = \frac{c_p}{T} \mathrm{d}T - \frac{\alpha}{\rho} \mathrm{d}p. \tag{1.38}$$

We adopt the standard convention of denoting by $(\partial A/\partial B)_C$ a derivative of quantity $A$ with respect to quantity $B$ at constant quantity $C$. In the above Eq. (1.38) $c_p = T(\partial s/\partial T)_p$ is the heat capacity at constant pressure and

$$\alpha = -\frac{1}{\rho} \left( \frac{\partial \rho}{\partial T} \right)_p \tag{1.39}$$

is the thermal expansion coefficient, which by the Maxwell identity $\alpha = -\rho(\partial s/\partial p)_T$. Standard fluids expand with increasing temperature and therefore we will be considering only fluids with $\alpha > 0$. With the use of (1.38) applied to the advective derivatives in (1.37), we can obtain the temperature equation in the form

$$\begin{aligned} \rho c_p \left( \frac{\partial T}{\partial t} + \mathbf{u} \cdot \nabla T \right) - \alpha T \left( \frac{\partial p}{\partial t} + \mathbf{u} \cdot \nabla p \right) =& \nabla \cdot (k \nabla T) + 2\mu \mathbf{G}^s : \mathbf{G}^s \\ &+ \left( \mu_b - \frac{2}{3}\mu \right) (\nabla \cdot \mathbf{u})^2 + Q. \end{aligned} \tag{1.40}$$

On the other hand one can also parametrize the entropy with temperature $T$ and density $\rho$, which yields

$$\mathrm{d}s = \left( \frac{\partial s}{\partial T} \right)_\rho \mathrm{d}T + \left( \frac{\partial s}{\partial \rho} \right)_T \mathrm{d}\rho. \tag{1.41}$$

The first term on right hand side of the latter equation can be easily expressed by the heat capacity at constant volume, which is defined by $c_v = T(\partial s/\partial T)_\rho$. To find an expression for the entropy derivative with respect to density at constant temperature we can first use the Maxwell identity $\rho^2(\partial s/\partial\rho)_T = -(\partial p/\partial T)_\rho$ and then the implicit function theorem to get $(\partial p/\partial T)_\rho = -(\partial \rho/\partial T)_p/(\partial\rho/\partial p)_T$, so that

$$\mathrm{d}s = \frac{c_v}{T}\mathrm{d}T - \frac{\alpha}{\rho^2\beta}\mathrm{d}\rho, \tag{1.42}$$

and we have defined the isothermal compressibility coefficient

$$\beta = \frac{1}{\rho}\left(\frac{\partial\rho}{\partial p}\right)_T. \tag{1.43}$$

Finally, by the use of (1.42) and the mass conservation law (1.6) without mass sources, $\sigma_m = 0$, we obtain yet another form of the temperature equation

$$\rho c_v\left(\frac{\partial T}{\partial t} + \mathbf{u}\cdot\nabla T\right) + \frac{\alpha T}{\beta}\nabla\cdot\mathbf{u} = \nabla\cdot(k\nabla T) + 2\mu\mathbf{G}^s : \mathbf{G}^s + \left(\mu_b - \frac{2}{3}\mu\right)(\nabla\cdot\mathbf{u})^2 + Q, \tag{1.44}$$

which does not involve any other time derivatives except for time derivative of the temperature.

### 1.4.1 Production of the Total Entropy

According to the second law of thermodynamics, in the absence of the volume heat sources/sinks, $Q = 0$, the total entropy production in the system must be greater than or equal to zero, when the adiabatic insulation

$$-k\nabla T\cdot\hat{\mathbf{n}}\big|_{\partial V} = 0, \tag{1.45}$$

is assumed. In the above $\hat{\mathbf{n}}$ is the unit normal to the boundary of the entire fluid region $\partial V$. Introducing the following notation for the traceless and symmetric part of the velocity gradient tensor

$$\underset{\sim}{\mathbf{G}}^s = \mathbf{G}^s - \frac{1}{3}(\nabla\cdot\mathbf{u})\,\mathbf{I}, \tag{1.46}$$

the term describing the viscous heating in the energy balance (1.37) can be easily rearranged into the following form

$$2\mu\mathbf{G}^s : \mathbf{G}^s + \left(\mu_b - \frac{2}{3}\mu\right)(\nabla\cdot\mathbf{u})^2 = 2\mu\underset{\sim}{\mathbf{G}}^s : \underset{\sim}{\mathbf{G}}^s + \mu_b(\nabla\cdot\mathbf{u})^2, \tag{1.47}$$

which is clearly positive definite, since a straight forward calculation yields

$$\mathbf{G}^s : \mathbf{G}^s = \left(\mathcal{G}_{11}^s\right)^2 + \left(\mathcal{G}_{22}^s\right)^2 + \left(\mathcal{G}_{33}^s\right)^2 + 2\left[\left(\mathcal{G}_{12}^s\right)^2 + \left(\mathcal{G}_{13}^s\right)^2 + \left(\mathcal{G}_{23}^s\right)^2\right]. \tag{1.48}$$

On dividing the energy equation (1.37) by $T$ with a little bit of algebra one obtains the entropy equation

$$\frac{\partial}{\partial t}(\rho s) + \nabla \cdot (\rho \mathbf{u} s) = \nabla \cdot \left(\frac{k}{T}\nabla T\right) + \frac{k}{T^2}(\nabla T)^2 + 2\frac{\mu}{T}\mathbf{G}^s : \mathbf{G}^s + \frac{\mu_b}{T}(\nabla \cdot \mathbf{u})^2, \tag{1.49}$$

where the continuity Eq. (1.6) in the absence of the sources of matter, $\sigma_m = 0$,[4] was used. Assuming periodic boundary conditions in the horizontal directions with some periods $L_x$, $L_y$, and by the use of the adiabatic (1.45) and impermeability $u_z(z = 0, L) = 0$ conditions on the top and bottom boundaries, integration of the two "divergence" terms in the entropy equation (1.49) over the entire horizontally periodic volume $V = (0, L_x) \times (0, L_y) \times (0, L)$ leads to

$$\int_V \nabla \cdot (\rho \mathbf{u} s)\, \mathrm{d}^3 x = \int_{\partial V} \rho s \mathbf{u} \cdot \hat{\mathbf{n}} \mathrm{d}\Sigma = 0, \tag{1.50}$$

$$\int_V \nabla \cdot \left[\frac{k}{T}\nabla T\right] \mathrm{d}^3 x = \int_{\partial V} \frac{k}{T}\nabla T \cdot \hat{\mathbf{n}} \mathrm{d}\Sigma = 0. \tag{1.51}$$

These terms are therefore easily eliminated from the global entropy balance, which takes the form

$$\frac{\partial}{\partial t}\int_V \rho s \mathrm{d}^3 x = \int_V \frac{k}{T^2}(\nabla T)^2 \mathrm{d}^3 x + 2\int_V \frac{\mu}{T}\mathbf{G}^s : \mathbf{G}^s \mathrm{d}^3 x + \int_V \frac{\mu_b}{T}(\nabla \cdot \mathbf{u})^2 \mathrm{d}^3 x \geq 0. \tag{1.52}$$

The latter inequality is, of course, true only when the coefficients $\mu$, $\mu_b$ and $k$ associated with irreversible processes are positive. The presented calculation demonstrates in fact, that the second law of thermodynamics demands positivity of those physical parameters of the fluid.

This way we have demonstrated, that when adiabatic insulation of the fluid is assumed and $\mu > 0$, $\mu_b > 0$ and $k > 0$, the total entropy production in the fluid is positive or null. This verifies the agreement of the derived energy balance for fluids with the second law of thermodynamics.

[4]The assumption of vanishing sources of matter is necessary here, since we assume adiabatic insulation and it is thermodynamically impossible to introduce a non-zero flux of matter without simultaneous heat transfer (cf. Guminski 1974 and Sect. 4.1 of this book).

## 1.5 Fundamental Ideas in Theoretical Description of the Phenomenon of Convection

### *1.5.1 Adiabatic Gradient*

To illustrate the importance of the adiabatic state in convective flows we will introduce now a simple and basic picture of a blob rise, which we will later come back to, in order to explain fundamental aspects of convection. In a static state, i.e. when no motions are present and the thermodynamics fields $\rho$, $T$, $p$ and $s$ are stationary but height-dependent one can study the stability of a fluid layer, heated from below, by considering two infinitesimally spaced horizontal (perpendicular to gravity) fluid layers situated at heights $z_0$ and $z_0 + \mathrm{d}z$. If a fluid volume of a unitary mass $V = 1/\rho$ is taken from the lower level $z_0$ and placed slightly higher at $z_0 + \mathrm{d}z$ it will experience a thermodynamic transformation, since its temperature and pressure will start adjusting to the environment at the higher level (cf. Fig. 1.1). If the fluid volume after the transformation becomes denser than the surroundings, the gravity acts to put it back at the original level $z_0$ and then the situation is stable; in the opposite case the buoyancy is non-zero and the system looses stability. Let us take the pressure $p(z)$ and the entropy $s(z)$ as the system parameters, then the fluid volume, initially

$$V\left(p(z_0), s(z_0)\right),$$

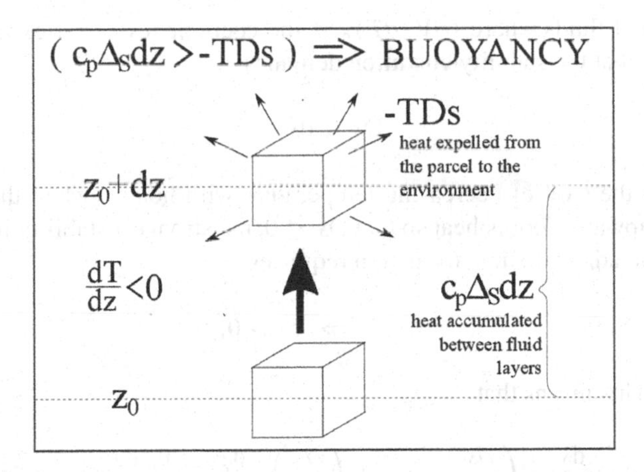


**Fig. 1.1** A schematic representation of physical conditions required for appearance of buoyancy in a fluid with negative vertical gradient of temperature. When a fluid parcel is shifted upwards between two infinitesimally distant horizontal layers, it releases heat to the environment and thus its temperature decreases, causing the density to increase. When the total heat accumulated between the layers $z_0$ and $z_0 + \mathrm{d}z$ starts to exceed the total heat released from the fluid element to its surroundings, this lost heat cannot account for an increase in density of the parcel significant enough for stability, thus the parcel's density remains smaller than that of the surroundings and buoyancy is created

after the shift and the thermodynamic transformation which adjusts the pressure to the value $p(z_0 + \mathrm{d}z)$ of the surrounding fluid at the higher level changes to

$$V\left(p(z_0 + \mathrm{d}z), s(z_0) + \mathrm{D}s\right),$$

where $s(z_0) + \mathrm{D}s$ denotes the entropy after the thermodynamic process thus $\mathrm{D}s$ is the entropy change in the process. In order for the static state to be stable we must require

$$V\left(p(z_0 + \mathrm{d}z), s(z_0) + \mathrm{D}s\right) < V\left(p(z_0 + \mathrm{d}z), s(z_0 + \mathrm{d}z)\right), \tag{1.53}$$

that the new volume of a mass unit is smaller than that of the surroundings at the new level $z_0 + \mathrm{d}z$. Since the pressure is the same on both sides of the above inequality, we may expand both sides in the entropy alone about the values at the level $z_0$ to get

$$V\left(p(z_0 + \mathrm{d}z), s(z_0)\right) + \left(\frac{\partial V}{\partial s}\right)_p \mathrm{D}s < V\left(p(z_0 + \mathrm{d}z), s(z_0)\right) + \left(\frac{\partial V}{\partial s}\right)_p \mathrm{d}s, \tag{1.54}$$

where $\mathrm{d}s = s(z_0 + \mathrm{d}z_0) - s(z_0)$. Since

$$\left(\frac{\partial V}{\partial s}\right)_p = \frac{\left(\frac{\partial V}{\partial T}\right)_p}{\left(\frac{\partial s}{\partial T}\right)_p} = \frac{T}{c_p}\left(\frac{\partial V}{\partial T}\right)_p > 0 \tag{1.55}$$

for all standard fluids,where $(\partial V/\partial T)_p$ is the coefficient of heat expansion, it is obvious now, that the stability condition demands

$$\mathrm{D}s < \mathrm{d}s, \tag{1.56}$$

and hence in the case of decreasing temperature with height, when the fluid volume moved upwards looses heat so that $\mathrm{D}s \leq 0$, the strongest stability restriction is imposed by an adiabatic transformation requiring

$$\mathrm{D}s = 0 \Rightarrow \mathrm{d}s > 0, \tag{1.57}$$

for stability. This means that

$$0 < \frac{\mathrm{d}s}{\mathrm{d}z} = \left(\frac{\partial s}{\partial T}\right)_p \frac{\mathrm{d}T}{\mathrm{d}z} + \left(\frac{\partial s}{\partial p}\right)_T \frac{\mathrm{d}p}{\mathrm{d}z} = \frac{c_p}{T}\frac{\mathrm{d}T}{\mathrm{d}z} + \alpha g, \tag{1.58}$$

where the Maxwell identity $(\partial s/\partial p)_T = -(\partial V/\partial T)_p$ and the hydrostatic balance $\mathrm{d}p/\mathrm{d}z = -g/V$ were used. As a result we obtain the following sufficient (but *not necessary*) condition for stability

$$-\frac{\mathrm{d}T}{\mathrm{d}z} < \frac{g\alpha T}{c_p}, \tag{1.59}$$

that is convection does not develop when the temperature gradient is below the adiabatic one and buoyancy forces may only start to appear when the temperature gradient exceeds that of the adiabatic profile.

In real fluids the dissipative effects are present, such as viscous friction and molecular heat conduction and therefore the thermodynamic transformation of a raised fluid element is not adiabatic and the loss of heat imposes $\mathrm{D}s < 0$. This relaxes the stability condition and the convection threshold is shifted to higher temperature gradients. Therefore a precise condition that must be satisfied for the system to be convectively stable is expressed as follows

$$-T\mathrm{D}s > -c_p\left(\frac{\mathrm{d}T}{\mathrm{d}z} + \frac{\alpha T g}{c_p}\right)\mathrm{d}z, \tag{1.60}$$

where $\mathrm{D}s < 0$ and $-\mathrm{d}T/\mathrm{d}z - \alpha T g/c_p$ becomes positive once the temperature gradient exceeds in magnitude the adiabatic one and this expression can then be called the superadiabatic temperature gradient.

*The physical meaning of the latter inequality in terms of the convective instability threshold is that the system becomes unstable as soon as the total heat per unit mass accumulated between two infinitesimally distant fluid layers starts to exceed the total heat per unit mass released to the environment by a rising fluid element on the infinitesimal distance between the layers; the fluid element, therefore, does not loose enough heat (hence is hotter than its surroundings) and the resulting energy excess is transformed into the work of the buoyancy.*

In other words immediately above the convection threshold the heat flux through the system becomes large enough, so that the molecular mechanisms are no longer capable of sustaining it and the system looses stability, since perturbed fluid elements do not loose enough heat as they rise, hence are hotter than surroundings and therefore buoyant. Note, however, that this is not the only possibility for appearance of thermally induced motion, since a time oscillatory flow may appear under some circumstances even when the condition (1.60) is still satisfied; we postpone the discussion of such cases until the end of this section.

The general stability condition (1.60) can be investigated further to yield a general expression for the threshold value of the superadiabatic temperature gradient below which the system is convectively stable and above which the instability develops. Denoting

$$\Delta_S = -\frac{\mathrm{d}T}{\mathrm{d}z} - \frac{g\alpha T}{c_p}, \tag{1.61}$$

the marginal (critical) state at convection threshold, by the use of (1.60) is described by

$$\frac{\min\left(-T\frac{\mathrm{D}s}{dt}\right)}{u_z} = c_p \Delta_{S\,crit}, \tag{1.62}$$

where the minimum of $-T\mathrm{D}s/dt$ on the left hand side is taken over all possible convective states in the vicinity of convection threshold at any point in space and $\Delta_{S\,crit}$ denotes the critical value of the superadiabatic temperature gradient $\Delta_S$ exactly at threshold.[5] In the latter equation $\mathrm{D}/dt$ denotes the total advective derivative with respect to time

$$\frac{\mathrm{D}}{dt} = \frac{\partial}{\partial t} + \mathbf{u}\cdot\nabla, \tag{1.63}$$

which describes the total change of a quantity (such as the entropy, temperature etc.) in a fluid element during its evolution in the flow. We therefore make use of the energy equation expressed in terms of the entropy (1.37) to rewrite the marginal relation (1.62) in the form

$$\Delta_{S\,crit} = \frac{\min\left[-\frac{1}{\rho}\left(\nabla\cdot(k\nabla T) + Q_{visc} + Q\right)\right]}{c_p u_z}. \tag{1.64}$$

The entire viscous heating term

$$Q_{visc} = 2\mu\mathbf{G}^s : \mathbf{G}^s + \left(\mu_b - \frac{2}{3}\mu\right)(\nabla\cdot\mathbf{u})^2, \tag{1.65}$$

is quadratically nonlinear in the velocity field $\mathbf{u}$, thus at the threshold it is very weak; consequently all the quadratic terms in the velocity can be neglected in the marginal state at leading order. This yields the following expression for the critical value of the superadiabatic temperature gradient at the convection threshold

$$\Delta_{S\,crit} = \frac{\min\left[-\frac{1}{\rho}\left(\nabla\cdot(k\nabla T) + Q\right)\right]}{c_p u_z}. \tag{1.66}$$

It must be emphasized at this stage, that the above general criterion for convective stability (1.60) and the resulting relation (1.66) are relevant only to the cases, when instability sets in without time oscillations, that is when the marginal state at convection threshold is stationary. In other words, the above reasoning and results can be directly applied to the cases, when the so-called *principle of the exchange of stabilities* formulated in Chandrasekhar (1961) pp. 24–26 is valid, that is when the growth rate of small perturbations in the vicinity of convection threshold is purely real and passes through zero from negative to positive values exactly at the threshold as the superadiabatic temperature gradient increases. But a situation, when oscilla-

[5] In cases, when the hydrostatic temperature gradient at threshold is height-dependent this determination of critical temperature jump across the fluid layer involves taking a vertical average of (1.62).

tions appear before the superadiabatic gradient (1.66) is reached is, in general, not excluded. In such a case the flow takes form of a travelling wave, which may exist even for superadiabatic gradient values significantly lower than that defined by the heat balance (1.62). It is enough then if only the advected heat per unit time and per unit mass $T\mathbf{u} \cdot \nabla s$ released by a rising fluid parcel becomes equal to $c_p \Delta_{S\,crit}$ at threshold, since the flow is then generated at the instances when the oscillatory component of the released heat $T\partial_t s$ vanishes (at other times the heat released by the parcel exceeds the one accumulated between fluid layers, thus the parcel is not buoyant). According to (1.63) we have

$$\mathbf{u} \cdot \nabla s = \frac{\mathrm{D}s}{\mathrm{d}t} - \frac{\partial s}{\partial t}, \tag{1.67}$$

therefore the final, most general expression for the critical superadiabatic gradient at convection threshold takes the form

$$\Delta_{S\,crit} = \frac{\min\left[-T\left(\frac{\mathrm{D}s}{\mathrm{d}t} - \frac{\partial s}{\partial t}\right)\right]}{u_z} \approx \frac{\min\left[T\frac{\partial s}{\partial t} - \frac{1}{\rho}\left(\nabla \cdot (k\nabla T) + Q\right)\right]}{c_p u_z}. \tag{1.68}$$

Note also, that near the marginal state, when the velocity field and the temperature perturbation are weak $T\mathbf{u} \cdot \nabla s \approx -c_p u_z \Delta_S$, since the term quadratic in the perturbations to the hydrostatic state is negligible. The issue of convection threshold will be further enlightened in later chapters, namely Chaps. 2.2 and 3.5 concerned with linear analysis of the convective instability.

### 1.5.2 Filtering Sound Waves

It is useful and often possible to filter sound waves from the wave spectrum of the system of dynamical equations, since these are very fast waves and the typical time scales of convective phenomena in many stellar and planetary interiors, including the Earth's core, likewise in many laboratory experiments are much longer than those associated with sound propagation. From the technical point of view it is very useful, because filtering sound waves is equivalent to the assumption, that the sound velocity is infinite and hence the pressure spreads infinitely fast. This allows to obtain a stationary Poisson-like problem for the pressure. In numerical simulations the sound waves are difficult to resolve because of their high frequencies, therefore an approximated set of equations with the sound waves eliminated from the wave spectrum allows for much more efficient simulations.

It will be now demonstrated, how from the technical point of view the sound waves can be removed from the spectrum of the dynamical equations, i.e. precisely which term is responsible for their presence or absence. Let us assume for simplicity, that there is no radiative heating, $Q = 0$, and the heat conduction coefficient likewise the dynamical viscosities are uniform, $k = \text{const}$, $\mu = \text{const}$, $\mu_b = \text{const}$. In such a case the full system of equations describing the dynamics of a fluid consists of the momentum balance (the Navier-Stokes equation), the continuity and energy equations and the equation of state, which can be cast in the following form

$$\rho\left[\frac{\partial \mathbf{u}}{\partial t} + (\mathbf{u}\cdot\nabla)\,\mathbf{u}\right] = -\nabla p + \rho\mathbf{g} + \mu\nabla^2\mathbf{u} + \left(\frac{\mu}{3} + \mu_b\right)\nabla(\nabla\cdot\mathbf{u}), \tag{1.69a}$$

$$\frac{\partial \rho}{\partial t} + \mathbf{u}\cdot\nabla\rho + \rho\nabla\cdot\mathbf{u} = 0, \tag{1.69b}$$

$$\rho c_v\left(\frac{\partial T}{\partial t} + \mathbf{u}\cdot\nabla T\right) - \frac{T}{\rho}\left(\frac{\partial p}{\partial T}\right)_\rho\left(\frac{\partial \rho}{\partial t} + \mathbf{u}\cdot\nabla\rho\right) = k\nabla^2 T + 2\mu\mathbf{G}^s : \mathbf{G}^s + \left(\mu_b - \frac{2}{3}\mu\right)(\nabla\cdot\mathbf{u})^2, \tag{1.69c}$$

$$\rho = \rho(p, T), \tag{1.69d}$$

Let us assume a static, spatially uniform equilibrium

$$\rho_0 = \text{const}, \quad T_0 = \text{const}, \quad p_0 = \text{const}, \quad \mathbf{u}_0 = 0, \tag{1.70}$$

and introduce small perturbations upon the equilibrium

$$\mathbf{u}(\mathbf{x}, t), \tag{1.71a}$$

$$\rho(\mathbf{x}, t) = \rho_0 + \rho'(\mathbf{x}, t), \; T(\mathbf{x}, t) = T_0 + T'(\mathbf{x}, t), \; p(\mathbf{x}, t) = p_0 + p'(\mathbf{x}, t). \tag{1.71b}$$

Linearisation of the equations with respect to the small perturbations leads to

$$\rho_0\frac{\partial \mathbf{u}}{\partial t} + \left(\frac{\partial p}{\partial \rho}\right)_T \nabla\rho' + \left(\frac{\partial p}{\partial T}\right)_\rho \nabla T' - \mu\nabla^2\mathbf{u} - \left(\frac{\mu}{3} + \mu_b\right)\nabla(\nabla\cdot\mathbf{u}) = 0, \tag{1.72a}$$

$$\frac{\partial \rho'}{\partial t} + \rho_0\nabla\cdot\mathbf{u} = 0, \tag{1.72b}$$

$$\rho_0 c_v\frac{\partial T'}{\partial t} - \frac{T}{\rho}\left(\frac{\partial p}{\partial T}\right)_\rho \frac{\partial \rho'}{\partial t} - k\nabla^2 T' = 0, \tag{1.72c}$$

It is now possible to decompose the perturbations into Fourier modes and consider single wave vector $\mathcal{K}$,

$$\rho'(\mathbf{x},t) = \hat{\rho}e^{i(\mathcal{K}\cdot\mathbf{x}-\omega t)}, \quad T'(\mathbf{x},t) = \hat{T}e^{i(\mathcal{K}\cdot\mathbf{x}-\omega t)}, \quad \mathbf{u}(\mathbf{x},t) = \hat{\mathbf{u}}e^{i(\mathcal{K}\cdot\mathbf{x}-\omega t)}. \tag{1.73}$$

Introduction of the Fourier forms of the perturbations into the linearised equations (1.72a)–(1.72c) allows to obtain the characteristic equation, since non-zero solutions can only exist if the determinant

$$\begin{vmatrix} -\mathcal{W}_1 & 0 & 0 & \mathcal{K}_x\mathcal{W}_2 & \left(\frac{\partial p}{\partial T}\right)_\rho \mathcal{K}_x \\ 0 & -\mathcal{W}_1 & 0 & \mathcal{K}_y\mathcal{W}_2 & \left(\frac{\partial p}{\partial T}\right)_\rho \mathcal{K}_y \\ 0 & 0 & -\mathcal{W}_1 & \mathcal{K}_z\mathcal{W}_2 & \left(\frac{\partial p}{\partial T}\right)_\rho \mathcal{K}_z \\ \rho_0\mathcal{K}_x & \rho_0\mathcal{K}_y & \rho_0\mathcal{K}_z & -\omega & 0 \\ 0 & 0 & 0 & \frac{\omega T_0}{\rho_0}\left(\frac{\partial p}{\partial T}\right)_\rho & -\left(\rho_0 c_v\omega + ik\mathcal{K}^2\right) \end{vmatrix} = 0 \tag{1.74}$$

vanishes, where

$$\mathcal{W}_1 = \rho_0\omega + i\mu\mathcal{K}^2, \tag{1.75a}$$

$$\mathcal{W}_2 = \left(\frac{\partial p}{\partial \rho}\right)_T - i\left(\frac{\mu}{3} + \mu_b\right)\frac{\omega}{\rho_0}. \tag{1.75b}$$

This yields

$$\left(\rho_0\omega + i\mu\mathcal{K}^2\right)^2 \left\{\omega T_0\mathcal{K}^2\left(\frac{\partial p}{\partial T}\right)_\rho^2 + \left(\rho_0 c_v\omega + ik\mathcal{K}^2\right)\left[\rho_0\mathcal{K}^2\left(\left(\frac{\partial p}{\partial \rho}\right)_T - i\frac{\omega}{\rho_0}\left(\frac{\mu}{3} + \mu_b\right)\right) - \omega\left(\rho_0\omega + i\mu\mathcal{K}^2\right)\right]\right\} = 0 \tag{1.76}$$

The modes $\omega = -i\mu\mathcal{K}^2/\rho_0$ are purely diffusive and decaying and do not involve sound waves, therefore we will concentrate only on the part of the above dispersion relation given in the braces. The presence of sound waves in the spectrum can be most easily shown in the following way. Since the speed of sound is

$$\begin{aligned} C^2 = \left(\frac{\partial p}{\partial \rho}\right)_s &= \left(\frac{\partial p}{\partial \rho}\right)_T + \left(\frac{\partial p}{\partial T}\right)_\rho\left(\frac{\partial T}{\partial \rho}\right)_s \\ &= \left(\frac{\partial p}{\partial \rho}\right)_T + \left(\frac{\partial p}{\partial T}\right)_\rho^2 \frac{T}{\rho^2 c_v}, \end{aligned} \tag{1.77}$$

which is satisfied by virtue of the implicit function theorem which implies $(\partial T/\partial \rho)_s = -T(\partial s/\partial \rho)_T/c_v$ and the Maxwell relation $\rho^2(\partial s/\partial \rho)_T = -(\partial p/\partial T)_\rho$, the dispersion relation in the limit of vanishing diffusion, $\mu \to 0$, $\mu_b \to 0$ and $k \to 0$, can be simplified to

$$\rho_0^2 c_v \omega^3 - \rho_0^2 c_v C^2 \mathcal{K}^2 \omega = 0. \tag{1.78}$$

Thus

$$\omega = \pm C\mathcal{K}, \tag{1.79}$$

so that the phase velocity of waves is equal to the speed of sound.[6] The sound waves can be eliminated from the spectrum through elimination of the term $\partial\rho'/\partial t$ from the continuity equation (1.72b), which could be achieved e.g. by assuming

$$\frac{\rho'}{\rho_0} \sim \frac{T'}{T_0} \ll 1, \quad \omega L \sim \|\mathbf{u}\| \sim \left(\frac{T'}{T_0}\right)^{1/2}. \tag{1.80}$$

Such an assumption allows to neglect the term $\partial\rho'/\partial t$ with respect to $\rho_0\nabla\cdot\mathbf{u}$ with all the other terms in the set of perturbation equations (1.72a)–(1.72c) retained. The dispersion relation (1.76) is then modified to

$$\frac{\omega T_0}{\rho_0}\left(\frac{\partial p}{\partial T}\right)_\rho^2 + \left(\rho_0 c_v \omega + \mathrm{i}k\mathcal{K}^2\right)\left[\left(\frac{\partial p}{\partial \rho}\right)_T - \mathrm{i}\frac{\omega}{\rho_0}\left(\frac{\mu}{3} + \mu_b\right)\right] = 0, \tag{1.81}$$

and it can be easily seen, that in the non-diffusive limit $\mu \to 0$, $\mu_b \to 0$ and $k \to 0$ we get

$$\rho_0 c_v C^2 \omega = 0, \tag{1.82}$$

and hence the sound waves are eliminated from the spectrum. Finally, it is of interest to note that neglection of the density time derivative in the energy equation (1.69c) only, through assumption of $(\partial p/\partial T)_\rho = 0$ does not lead to elimination of sound waves from the spectrum, since then $(\partial p/\partial \rho)_T = C^2$ and sound wave dispersion relation $\omega = \pm C\mathcal{K}$ is the only non-zero and non-diffusive in nature root of the dispersion relation (1.76).

The two most frequently used approximations for description of convection dynamics utilize the idea presented in this section, in particular the scalings (1.80) to filter the sound waves. These are the well-known Oberbeck-Boussinesq and anelastic approximations, which will be discussed in detail in the following chapters.

---

[6]Note, that in the general dissipative case, $\mu \neq 0$, $\mu_b \neq 0$ and $k \neq 0$ the sound waves are damped by the all three dissipative processes. In particular in the long-wavelength limit $\mathcal{K} \ll \rho_0 C/\mu$ the dispersion relation (1.76) can be solved by means of asymptotic expansions in the wave number and the sound modes are characterized by

$$\omega = \pm C\mathcal{K} - \mathrm{i}\frac{1}{2\rho_0}\left[\frac{4}{3}\mu + \mu_b + k\left(\frac{1}{c_v} - \frac{1}{c_p}\right)\right]\mathcal{K}^2.$$

The dissipative damping of sound waves is also often formulated in terms of spatial absorption, that is diminishing of the waves intensity as it travels a certain distance in the fluid; in other words such a formulation involves complex wave vector and real frequency, but the absorption coefficient is essentially the same and involves all three dissipation coefficients (cf. Landau and Lifshitz 1987, Chap. 79 on "*Absorption of sound*", Eq. (79.6)).

## Review Exercises

**Exercise 1** Which fundamental physical laws are responsible for symmetry of the stress tensor, $\tau_{ij} = \tau_{ji}$?

**Exercise 2** In the Cattaneo (1948) model the heat flux possesses a correction with respect to the Fourier's law (1.23) and takes the form $\mathbf{j}_T = -k\nabla T + \sigma\nabla\partial_t T$. This leads to the following hyperbolic temperature equation (cf. Straughan 2011)

$$\frac{\partial^2 T}{\partial t^2} + \frac{1}{\tau}\frac{\partial T}{\partial t} - \frac{k}{\rho c_p \tau}\nabla^2 T = 0,$$

with $\tau = \sigma/k > 0$. Demonstrate that under periodic boundary conditions in all three spatial directions (periodic box) the solutions can take a form of damped waves, which are termed *damped heat waves*.

**Exercise 3** For the problem of Exercise 2 calculate the group velocity of the damped heat waves $\mathbf{v}_g$. Then calculate the time average over the time $\tau$ of the heat flux $\mathbf{j}_T$ associated with a single standing heat wave, assuming $T(t=0) = T_0 \cos(\mathcal{K} \cdot \mathbf{x})$ and $\partial_t T(t=0) = -T(t=0)/2\tau$.
*Hint*: Assuming the phasor of the temperature oscillations in the form $\exp[i(\mathcal{K} \cdot \mathbf{x} - \omega t)]$, the group velocity is defined as $\mathbf{v}_g = \nabla_{\mathcal{K}}(\mathfrak{Re}\,\omega)$, i.e. the gradient of the real part of the complex frequency $\mathfrak{Re}\,\omega$ with respect to the wave vector $\mathcal{K}$.

**Exercise 4** Derive an expression for the vertical temperature gradient when the entropy density per unit mass $s$ remains uniform.
*Hint*: cf. Eq. (1.58).

**Exercise 5** Consider a long wavelength limit $\mathcal{K} \ll \rho_0 C/\mu$ and demonstrate, that sound waves are damped by viscous effects.
*Hint*: cf. Eq. (1.76) and footnote 6.

## References

Batchelor, G.K. 1967. *An introduction to fluid mechanics*. Cambridge: Cambridge University Press.

Cattaneo, C. 1948. Sulla conduzione del calore. *Atti del Seminario Matematico e Fisico Modena* 3: 83–101.

Chalmers, A.F. 1970. Curie's Principle. *British Journal for the Philosophy of Science* 21 (2): 133–148.

Chandrasekhar, S. 1961. In *Hydrodynamic and hydromagnetic stability*, ed. W. Marshall, and D.H. Wilkinson., International series of monographs on physics Clarendon Press: Oxford University Press (since 1981 printed by Dover Publications).

Curie, P. 1894. Sur la symétrie dans les phénomènes physiques, symétrie d'un champ électrique et d'un champ magnétique. *Journal Physique* 3:393–415. Reprinted in "Oeuvres de Pierre Curie" (1908), pp. 118–141, Gauthier-Villars, Paris.

Darrigol, O. 2005. *Worlds of flow. A history of hydrodynamics from the Bernoullis to Prandtl*. New York: Oxford University Press.
de Groot, S.R., and P. Mazur. 1984. *Non-equilibrium thermodynamics*. New York: Dover Publications.
Guminski, K. 1974. *Termodynamika*. Warsaw: Polish Scientific Publishers PWN.
Landau, L.D., and E.M. Lifshitz. 1987. *Fluid Mechanics, Course of theoretical physics*, vol. 6. Oxford: Elsevier.
Straughan, B. 2011. *Heat waves*. New York: Springer.

# Chapter 2
# The Oberbeck-Boussinesq Convection

The simplest and at the same time the most classic approach to mathematical description of convection is the Oberbeck-Boussinesq approximation, whose applicability covers systems with low density variations that result only from variations of temperature.[1] At the turn of the nineteenth and twentieth centuries a German physicist Anton Oberbeck and a French mathematician Joseph Valentin Boussinesq worked independently on a rigorous mathematical description of buoyancy driven flows. The equations which are now known as the Oberbeck-Boussinesq approximation were first obtained by Oberbeck (1879), who derived them through formal expansion in power series in the thermal expansion coefficient $\alpha$ (the equations appear as the leading order approximation) and utilized them to describe convection in spherical geometry. Two years later (Lorenz 1881) published a study of heat transfer by a free convection in a cartesian geometry using the Oberbeck equations. Quite independently Boussinesq (1903, p. 174 of that book) obtained essentially the same equations by making a series of assumptions, which led to buoyancy force expressed solely in terms of temperature and otherwise constant density in all the equations, in particular the solenoidal constraint for the velocity field. As commonly done, we will often refer in short to the flow of fluid described by the Oberbeck-Boussinesq equations simply as Boussinesq convection.

It is of interest to give a glimpse on some of the most important literature concerning significant milestones in understanding of Boussinesq convection and systematization of knowledge, so that interested readers can broaden their horizons. The earliest experiments involving thermally driven flow in a horizontal layer of fluid heated from below date back to Thomson (1882). A more detailed and com-

[1] This will be made precise in the following section.



K. A. Mizerski, *Foundations of Convection with Density Stratification*, GeoPlanet: Earth and Planetary Sciences, https://doi.org/10.1007/978-3-030-63054-6_2

prehensive experimental study was done later by Bénard (1900).[2] In sixteen years following that seminal experimental work, Lord Rayleigh (1916) was the first one to provide analytic results concerning convective instability threshold; Pellew and Southwell (1940) developed his theory to study the influence of boundary conditions on the conditions of stability breakdown. After the two main pioneers the problem of thermal convection in a horizontal layer heated from below is now commonly referred to as the Rayleigh-Bénard problem. A survey of findings from those early stages concerning the phenomenon of convection can be found in Ostrach (1957).

A noteworthy derivation of the Oberbeck-Boussinesq equations came from Spiegel and Veronis (1960), with some later corrections regarding the magnitude of viscous heating in Veronis (1962). This derivation was based on the assumption of large scale heights with respect to the layer's depth, i.e. $L \ll |\rho/\mathrm{d}_z\rho|$, $L \ll |T/\mathrm{d}_z T|$, $L \ll |p/\mathrm{d}_z p|$, and formal expansions in the small magnitude of density stratification. Next step in mathematical formalization and obtaining full mathematical rigour in the derivation of the Boussinesq system of equations was due to Mihaljan (1962), who explicitly included the assumption of smallness of diffusivities and derived the Boussinesq equations through two-parameter expansions in $\alpha\Delta T$ and $\kappa^2/c_v L^2 \Delta T$; he also included an analysis of the energetics within the Oberbeck-Boussinesq approximation. Later on, Cordon and Velarde (1975) and Velarde and Cordon (1976) utilized similar expansion parameters, with the particular emphasis on the viscous dissipation and large-gap effects.

A systematization of knowledge about Boussinesq convection has continued throughout a substantial collection of books on the topic. A quick review of some of the most important contributions includes:

(1) Chandrasekhar (1961) - a comprehensive analysis of the linearised Boussinesq equations in the very weak amplitude regime close to convection onset; the effects of background rotation and magnetic field have been thoroughly considered and the case of convection onset in spheres and spherical shells was analysed.
(2) Gershuni and Zhukhovitskii (1972, eng. trans. 1976) - a general book on convective instability, with a strong focus on the influence of various physical effects on the conditions and flow structure at the threshold of convection. This includes e.g. the effects of geometry, consideration of binary mixtures, internal heat sources, effects of longitudinal temperature gradient, the study of convection in porous medium saturated with fluid (convective filtration) or the thermocapillary effect. Here, for the sake of a quick reference of readers and self-consistency and completeness of this book, in Sect. 2.2 we provide a compendium of some of the most important, known results for the linear regime at convection onset in different physical configurations.
(3) Joseph (1976) - a fully rigorous mathematical description of convection under the Oberbeck-Boussinesq approximation including the bifurcation theory and the global stability analysis through a general energy theory of stability; the

[2] However, Bénard studied in fact a problem significantly influenced by a temperature-dependent surface tension, which is now known as the Bénard-Maragoni convection.

effects of geometry, the magnetic field and a study of turbulent convection in porous materials are included.

(4) Straughan (2004) - mathematically fully rigorous application of the energy method to Boussinesq convection. Various physical circumstances are considered, such as e.g. convection in porous media, internal heating, surface tension, the micropolar model of suspensions, electric and magnetic fields and temperature dependent viscosity. The pattern selection problem for Boussinesq convection is also considered.

(5) Getling (1980) - a seminal work containing a very comprehensive study and review of the weakly nonlinear convection and pattern selection near the onset

The primary goal of this chapter is to provide a consistent framework and reference for later chapters and gather knowledge on the development of Boussinesq convection from linear stages, through the weakly nonlinear ones up to a fully nonlinear regime. Let us recall, that the early (Malkus 1954) experiments, confirmed by many later studies clearly show that development of convection with increasing driving occurs through a sequence of consecutive instabilities, which build upon successive, gradually more complicated flows until a fully turbulent state is reached. The aforementioned (Getling 1998) book is a wonderful reference for description of the problem of pattern selection by the convective flow near onset, i.e. the changes in stable flow structure when the driving is continuously enhanced, but weak. A number of significant contributions in this field come from a German scientist Friedrich Busse. We only mention a few out of a large collection of his works, that is the influential studies Busse (1970) and Busse and Cuong (1977) on the onset of rotating convection in spherical geometry and Busse (1969, 1978), Busse and Riahi (1980) and Clever and Busse (1994) on nonlinear effects and heat transfer estimates. We conclude the short review by recalling the distinguished work of Grossmann and Lohse (2000), later updated in Stevens et al. (2013), who gathered the experimental, numerical and theoretical results concerning fully developed, nonlinear Boussinesq convection, systematized them and constructed a consistent dynamical picture together with a theory of heat transfer (including estimates of the magnitude of the flow) for Boussinesq convection with strong driving.

In this chapter we attempt to present a general picture of Boussinesq convection and explain the approaches undertaken in the description at different stages of convection development. We start from the rigorous derivation of the Boussinesq system of equations and a study of general energetic properties of Boussinesq convection, with general definitions of the Rayleigh and Nusselt numbers. Next the linear regime near the onset is described under various physical conditions and we proceed to explain the weakly nonlinear approach and major approaches to the pattern selection problem near onset. Finally we review the (Grossmann and Lohse 2000) theory of fully developed convection. A comprehensive but short summary is offered in the last section of this chapter.

## 2.1 Derivation of the Oberbeck-Boussinesq Equations

With the aim of derivation of one of the most standard approximations of the hydrodynamic equations constructed to describe convective flows, that is the so-called Boussinesq approximation for convection, we first restate the system of Navier-Stokes, mass continuity and energy equations, supplied by the equation of state

$$\rho\left[\frac{\partial \mathbf{u}}{\partial t}+(\mathbf{u}\cdot\nabla)\,\mathbf{u}\right]=-\nabla p+\rho\mathbf{g}+\mu\nabla^2\mathbf{u}+\left(\frac{\mu}{3}+\mu_b\right)\nabla\left(\nabla\cdot\mathbf{u}\right)$$
$$+2\nabla\mu\cdot\mathbf{G}^s+\nabla\left(\mu_b-\frac{2}{3}\mu\right)\nabla\cdot\mathbf{u}, \tag{2.1a}$$

$$\frac{\partial\rho}{\partial t}+\mathbf{u}\cdot\nabla\rho+\rho\nabla\cdot\mathbf{u}=0, \tag{2.1b}$$

$$\rho c_v\left(\frac{\partial T}{\partial t}+\mathbf{u}\cdot\nabla T\right)+\frac{\alpha T}{\beta}\nabla\cdot\mathbf{u}=\nabla\cdot(k\nabla T)+2\mu\mathbf{G}^s:\mathbf{G}^s+\left(\mu_b-\frac{2}{3}\mu\right)(\nabla\cdot\mathbf{u})^2+Q, \tag{2.1c}$$

$$\rho=\rho(p,T), \tag{2.1d}$$

with $G^s_{ij}=(\partial_j u_i+\partial_i u_j)/2$ being the symmetric part of the velocity field gradient tensor, $\mathbf{G}^s:\mathbf{G}^s=G^s_{ij}G^s_{ij}$ and $Q$ denoting heat sources, e.g. due to radiation. All the coefficients such as $\mu$, $\mu_b$, $c_v$ and $k$ have been assumed nonuniform for generality, with the exception that the horizontal variations of the thermal conductivity coefficient $k$ (i.e. the variations in the plane perpendicular to gravity acceleration) are neglected and thus $k$ is assumed to be a function of height only,

$$k=k(z). \tag{2.2}$$

The approach of Spiegel and Veronis (1960) will be generally followed with some changes, in particular a general equation of state will be considered and a more detailed explanation of all the assumptions made will be provided. Let us assume the $z$-axis of the coordinate system in the vertical direction, so that the gravitational acceleration points in the negative $z$-direction, i.e. $\mathbf{g}=-g\hat{\mathbf{e}}_z$. The thermodynamic variables, such as density, temperature, pressure and entropy in the considered case of time-independent boundary conditions can be split into a hydrostatic part, which is $z$-dependent only and a small but fully spatially and time-dependent fluctuation. The former will be denoted by a single upper tilde, i.e.

$$\tilde{\rho},\ \tilde{T},\ \tilde{p},\ \tilde{s}\quad \text{denote hydrostatic parts,}$$

and the fluctuations will be denoted by primes,

$$\rho',\ T',\ p',\ s',\quad \text{denote fluctuations.}$$

The hydrostatic part can be further split into a mean, $\bar{T}$, and a correction, which involves the variations in the hydrostatic state; the latter one will be denoted by an upper double tilde in the following, $\tilde{\tilde{T}}$, and in the case of temperature it allows to satisfy the imposed inhomogeneous boundary conditions, which drive the convective flow. Consequently homogeneous boundary conditions are imposed on the temperature fluctuation $T'$. Hence the thermodynamic variables are represented as follows

$$\rho(\mathbf{x}, t) = \bar{\rho} + \tilde{\tilde{\rho}}(z) + \rho'(\mathbf{x}, t), \tag{2.3a}$$

$$T(\mathbf{x}, t) = \bar{T} + \tilde{\tilde{T}}(z) + T'(\mathbf{x}, t), \tag{2.3b}$$

$$p(\mathbf{x}, t) = \bar{p} + \tilde{\tilde{p}}(z) + p'(\mathbf{x}, t), \tag{2.3c}$$

where quantities marked by the upper bar, that is $\bar{\rho}$, $\bar{T}$ and $\bar{p}$ are space averages (upper bar will denote the full spatial average in the following) of $\rho(\mathbf{x}, t)$, $T(\mathbf{x}, t)$ and $p(\mathbf{x}, t)$ respectively; the quantities with an upper double tilde correspond to hydrostatic state variation thus in the absence of motion ($z$-dependent only, since $k = k(z)$ is a sole function of height by assumption) and primed variables denote the fluctuations resulting from motion. The central point of the Boussinesq approximation is the *first simplifying assumption* that the scale heights

$$D_\rho = \left|\frac{1}{\bar{\rho}}\frac{\mathrm{d}\tilde{\tilde{\rho}}}{\mathrm{d}z}\right|^{-1}, \quad D_T = \left|\frac{1}{\bar{T}}\frac{\mathrm{d}\tilde{\tilde{T}}}{\mathrm{d}z}\right|^{-1}, \quad D_p = \left|\frac{1}{\bar{p}}\frac{\mathrm{d}\tilde{\tilde{p}}}{\mathrm{d}z}\right|^{-1}, \tag{2.4}$$

are all much larger than the thickness $L$ of the layer of fluid, which undergoes convection

$$L \ll \min\left\{D_\rho, D_T, D_p\right\}, \tag{2.5}$$

everywhere in the fluid domain. Using this assumption we define a small parameter

$$\epsilon = \frac{\Delta\tilde{\tilde{\rho}}}{\bar{\rho}} \ll 1, \tag{2.6}$$

which is small by virtue of integration of relation $L/D_\rho \ll 1$ from the level of minimal to the level of maximal density within the fluid layer and the density jump between these layers is denoted by $\Delta\tilde{\tilde{\rho}}$.

A *second crucial restriction* that we have to impose by assumption is that the fluctuations of thermodynamic variables induced by the convective motions are of the same order of magnitude or smaller than the static variation, i.e.

$$\left|\frac{\rho'}{\Delta\tilde{\tilde{\rho}}}\right| \lesssim \mathcal{O}(1), \quad \left|\frac{T'}{\Delta\tilde{\tilde{T}}}\right| \lesssim \mathcal{O}(1), \quad \left|\frac{p'}{\Delta\tilde{\tilde{p}}}\right| \lesssim \mathcal{O}(1), \tag{2.7}$$

which, in turn leads to

$$\left|\frac{\rho'}{\bar{\rho}}\right| \lesssim \mathcal{O}(\epsilon), \quad \left|\frac{T'}{\bar{T}}\right| \lesssim \mathcal{O}(\epsilon), \quad \left|\frac{p'}{\bar{p}}\right| \lesssim \mathcal{O}(\epsilon). \tag{2.8}$$

This assumption means that (2.3a)–(2.3c) could be understood as formal power series in $\epsilon$, most possibly asymptotic in nature. Therefore mathematically the first few terms of the expansion, hence the leading order of the fluctuating parts as well, can be expected to provide a satisfactory approximation of the full solution, since $\epsilon$ can always be chosen small enough. However, in modeling real systems, in which the parameter $\epsilon = \Delta\tilde{\tilde{\rho}}/\bar{\rho}$ is set, care must be taken to verify a posteriori consistency with assumption (2.8), although there does not seem to be any numerical evidence for fluctuations ever to significantly exceed the static variation.

The static, motionless state is described with the use of the dynamical equations (2.1a) and (2.1c) by

$$\frac{\partial \tilde{\tilde{p}}}{\partial x} = \frac{\partial \tilde{\tilde{p}}}{\partial y} = 0, \quad \frac{\partial \tilde{\tilde{p}}}{\partial z} = -\bar{\rho}g - \tilde{\tilde{\rho}}g, \tag{2.9a}$$

$$\nabla \cdot \left(k\nabla\tilde{\tilde{T}}\right) = -\tilde{Q}. \tag{2.9b}$$

Furthermore, we expand the equation of state (2.1d) in Taylor series to obtain

$$\rho = \bar{\rho}\left[1 - \bar{\alpha}\left(T - \bar{T}\right) + \bar{\beta}\left(p - \bar{p}\right) + \mathcal{O}\left(\epsilon^2\right)\right], \tag{2.10}$$

where

$$\bar{\alpha} = -\frac{1}{\bar{\rho}}\left(\frac{\partial\rho}{\partial T}\right)\bigg|_{(T,\,p)=(\bar{T},\,\bar{p})}, \quad \bar{\beta} = \frac{1}{\bar{\rho}}\left(\frac{\partial\rho}{\partial p}\right)\bigg|_{(T,\,p)=(\bar{T},\,\bar{p})}, \tag{2.11}$$

which by the use of (2.3a)–(2.3c) implies for both the static density and the density fluctuation

$$\frac{\tilde{\tilde{\rho}}}{\bar{\rho}} = \bar{\beta}\tilde{\tilde{p}} - \bar{\alpha}\tilde{\tilde{T}} + \mathcal{O}\left(\epsilon^2\right), \tag{2.12a}$$

$$\frac{\rho'}{\bar{\rho}} = \bar{\beta}p' - \bar{\alpha}T' + \mathcal{O}\left(\epsilon^2\right), \tag{2.12b}$$

since the static variation and fluctuations due to the flow have to be balanced separately. In the above $\bar{\alpha}$ is the thermal expansion coefficient and $\bar{\beta}$ the isothermal compressibility coefficient related to the isothermal speed of sound $C_T$

$$\bar{\beta} = \frac{1}{\bar{\rho}C_T^2}. \tag{2.13}$$

Furthermore, making use of (2.3a) and (2.6) in the mass conservation equation (2.1b) we obtain the first significant result of the Boussinesq approximation for convective flows, namely that the flow divergence is negligibly small in terms of any natural frequency in the system, such as e.g. $k/\bar{\rho}c_pL^2$, $\mu/\bar{\rho}L^2$, $\mu_b/\bar{\rho}L^2$ or $\sqrt{g/L}$,

$$\begin{aligned}
\nabla\cdot\mathbf{u} &= -\left(\frac{\partial}{\partial t}+\mathbf{u}\cdot\nabla\right)\frac{\tilde{\bar{\rho}}+\rho'}{\bar{\rho}}+\mathcal{O}\left(\epsilon^2\frac{\mathscr{U}}{L}\right)\\
&= -\epsilon\left(\frac{\partial}{\partial t}+\mathbf{u}\cdot\nabla\right)\frac{\tilde{\bar{\rho}}+\rho'}{\Delta\tilde{\bar{\rho}}}+\mathcal{O}\left(\epsilon^2\frac{\mathscr{U}}{L}\right)\\
&= 0+\mathcal{O}\left(\epsilon\frac{\mathscr{U}}{L}\right),
\end{aligned} \tag{2.14}$$

where $\mathscr{U}$ is the velocity scale defined later in (2.17).

### 2.1.1 Momentum Balance

On inserting the static state equation (2.9a) into the Navier-Stokes equation (2.1a) and utilizing the negligibility of the flow divergence (2.14) one obtains

$$\rho\left[\frac{\partial\mathbf{u}}{\partial t}+(\mathbf{u}\cdot\nabla)\,\mathbf{u}\right]=-\nabla p'-\rho' g\hat{\mathbf{e}}_z+\mu\nabla^2\mathbf{u}+2\nabla\mu\cdot\mathbf{G}^s+\mathcal{O}\left(\mu\epsilon\frac{\mathscr{U}}{L^2}\right). \tag{2.15}$$

Next, dividing the latter equation by $\bar{\rho}$ and using the definition of the small parameter $\epsilon=\Delta\tilde{\bar{\rho}}/\bar{\rho}$, as in (2.6), and keeping only the leading order terms the momentum balance simplifies to

$$\frac{\partial\mathbf{u}}{\partial t}+(\mathbf{u}\cdot\nabla)\,\mathbf{u}=-\frac{1}{\bar{\rho}}\nabla p'-\epsilon\frac{\rho'}{\Delta\tilde{\bar{\rho}}}g\hat{\mathbf{e}}_z+\nu\nabla^2\mathbf{u}+2\nabla\nu\cdot\mathbf{G}^s, \tag{2.16}$$

where $\nu=\mu/\bar{\rho}$ is the kinematic viscosity. The buoyancy force $\epsilon g\rho'/\Delta\tilde{\bar{\rho}}\hat{\mathbf{e}}_z$ drives the motions and thus the flow acceleration must be of the order $\epsilon g$, i.e. necessarily much smaller than the acceleration of gravity. This implies the following for the convective velocity and time scales

$$\mathscr{U}\sim\epsilon^{1/2}\sqrt{gL},\qquad \mathscr{T}\sim\epsilon^{-1/2}\sqrt{\frac{L}{g}}, \tag{2.17}$$

and immediately for the viscosity

$$\mu_b/\bar{\rho}\lesssim\nu\sim\epsilon^{1/2}\sqrt{gL}L \tag{2.18}$$

(the bulk viscosity does not exceed shear viscosity in the order of magnitude). Furthermore, the buoyancy term together with vertical pressure gradient in (2.16) can be rearranged with the aid of the relation between density, temperature and pressure fluctuations (2.12b) and the definition of $\epsilon$ in (2.6) to give

$$\begin{aligned} -\frac{1}{\bar{\rho}}\frac{\partial p'}{\partial z} - \frac{\rho'}{\bar{\rho}}g &= -\frac{1}{\bar{\rho}}\frac{\partial p'}{\partial z} - \bar{\beta}p'g + \bar{\alpha}T'g \\ &= -\frac{1}{\bar{\rho}}\left(\frac{\partial p'}{\partial z} + \frac{p'}{H}\right) + \bar{\alpha}T'g, \end{aligned} \tag{2.19}$$

where

$$H = \frac{1}{g\bar{\rho}\bar{\beta}} = \frac{C_T^2}{g} \tag{2.20}$$

has a dimension of length. In the case of a perfect gas equation of state, i.e.[3] $p = \rho RT$ the compressibility coefficient $\bar{\beta} = 1/\bar{p}$ and the quantity $H$ is a thickness defined by a hydrostatic pressure balance in a system with uniform mean density $\bar{\rho}$. On the other hand the definition of the pressure scale height supplied by the hydrostatic balance (2.9a) implies

$$D_p = \left[\frac{g\bar{\rho}}{\bar{p}}\left(1 + \frac{\tilde{\tilde{\rho}}}{\bar{\rho}}\right)\right]^{-1} = \frac{\bar{p}}{g\bar{\rho}}\frac{1}{1 + \epsilon\frac{\tilde{\tilde{\rho}}}{\Delta\rho}}, \tag{2.21}$$

therefore

$$H = D_p + \mathcal{O}(\epsilon) \gg L. \tag{2.22}$$

The above relation (2.22) is, in fact, more general and does not apply solely to fluids whose thermodynamic properties are described by the perfect gas equation of state. All that is necessary is that in terms of the small parameter $\epsilon$ being a measure of weak density stratification, $\partial\rho/\partial p \sim \rho/p$ or equivalently $C_T^2 \sim p/\rho$; this then implies that $H$ and $D_p$ are of the same order of magnitude. It is therefore necessary from the point of view of derivation of the Boussinesq equations to make a *third assumption* concerning the fluid thermodynamic properties, namely that

$$\frac{\sqrt{gL}}{C_T} = \mathcal{O}\left(\epsilon^{1/2}\right), \quad \text{or equivalently} \quad \frac{1}{g\bar{\rho}\bar{\beta}L} = \mathcal{O}\left(\epsilon^{-1}\right), \tag{2.23}$$

obviously satisfied by a perfect gas by virtue of the assumption (2.6); of course this means that the Mach number,

$$Ma = \frac{\mathcal{U}}{C} = \mathcal{O}(\epsilon) \tag{2.24}$$

[3] $R = k_B/m_m$ is the specific gas constant; $m_m$ is the molecular mass, $k_B$ the Boltzmann constant.

is small and scales linearly with $\epsilon$. It has to be emphasized, however, that the third assumption is by no means as fundamental as the previous two expressed in (2.6) and (2.8), but in fact reflects a rather typical experimental situation. As a consequence the term $p'/H$ in Eq. (2.19) is necessarily much smaller than the vertical variation rate of the pressure fluctuation

$$\frac{p'}{H} \ll \frac{\partial p'}{\partial z}, \tag{2.25}$$

and thus finally the leading order momentum balance (2.16) takes the form

$$\frac{\partial \mathbf{u}}{\partial t} + (\mathbf{u} \cdot \nabla)\, \mathbf{u} = -\frac{1}{\bar{\rho}} \nabla p' + g\bar{\alpha} T' \hat{\mathbf{e}}_z + \nu \nabla^2 \mathbf{u} + 2\nabla \nu \cdot \mathbf{G}^s. \tag{2.26}$$

Relation (2.22) allowed to simplify the buoyancy force through neglection of the small pressure fluctuation and retaining only the temperature fluctuation. This in turn allows to simplify also the relation (2.12b) between the fluctuations of thermodynamic variables to

$$\frac{\rho'}{\bar{\rho}} = -\bar{\alpha} T' + \mathcal{O}\left(\epsilon^2\right). \tag{2.27}$$

A straightforward consequence of the considerations leading to derivation of the final form of the Navier-Stokes equation under the Boussinesq approximation is

$$\frac{\tilde{\tilde{\rho}}}{\bar{\rho}} = \mathcal{O}(\epsilon), \qquad \frac{\rho'}{\bar{\rho}} = \mathcal{O}(\epsilon), \tag{2.28a}$$

$$\frac{\tilde{\tilde{T}}}{\bar{T}} = \mathcal{O}(\epsilon), \qquad \frac{T'}{\bar{T}} = \mathcal{O}(\epsilon), \tag{2.28b}$$

$$\frac{\tilde{\tilde{p}}}{\bar{p}} = \mathcal{O}(\epsilon), \qquad \frac{p'}{\bar{p}} = \mathcal{O}\left(\epsilon^2\right), \tag{2.28c}$$

$$\frac{\bar{\rho} g L}{\bar{p}} = \mathcal{O}(\epsilon). \tag{2.28d}$$

In particular the last relations (2.28d) and (2.28c) imply that the pressure unperturbed by fluid motion in Boussinesq systems is very strong.

Finally, we provide an example of a specific solution of (2.9a) and (2.9b) for the hydrostatic reference state in Boussinesq convection for the case of an ideal gas, $p = \rho RT$, with $Q = 0$, $\kappa = \text{const}$, which can now be fully understood from the point of view of orderings of terms,

$$\frac{\tilde{\tilde{T}}}{\bar{T}} = \frac{\Delta T}{\bar{T}} \left(\frac{1}{2} - \frac{z}{L}\right), \tag{2.29a}$$

$$\frac{\tilde{\tilde{\rho}}}{\bar{\rho}} = \left(\frac{\bar{\rho}gL}{\bar{p}} - \frac{\Delta T}{\bar{T}}\right)\left[\left(\frac{1}{2} - \frac{z}{L}\right) - \frac{\bar{\rho}gL}{2\bar{p}}\frac{z}{L}\left(1 - \frac{z}{L}\right)\right] + \mathcal{O}\left(\epsilon^3\right), \tag{2.29b}$$

$$\frac{\tilde{\tilde{p}}}{\bar{p}} = \frac{\bar{\rho}gL}{\bar{p}}\left(\frac{1}{2} - \frac{z}{L}\right) - \frac{\bar{\rho}gL}{2\bar{p}}\left(\frac{\bar{\rho}gL}{\bar{p}} - \frac{\Delta T}{\bar{T}}\right)\frac{z}{L}\left(1 - \frac{z}{L}\right) + \mathcal{O}\left(\epsilon^3\right), \tag{2.29c}$$

where $\Delta T > 0$ is the total temperature jump across the fluid layer, positive by definition ($\Delta T = T_B - T_T$). All the three variables $\tilde{\tilde{T}}/\bar{T}$, $\tilde{\tilde{\rho}}/\bar{\rho}$ and $\tilde{\tilde{p}}/\bar{p}$ are of course of the order $\mathcal{O}(\epsilon)$ and are all linear functions of $z$ at leading order. The density and pressure in the hydrostatic state possess higher order corrections and in particular the quadratic corrections

$$-\frac{\bar{\rho}gL}{2\bar{p}}\left(\frac{\bar{\rho}gL}{\bar{p}} - \frac{\Delta T}{\bar{T}}\right)\frac{z}{L}\left(1 - \frac{z}{L}\right) = \mathcal{O}\left(\epsilon^2\right), \tag{2.30}$$

are still important as the pressure quadratic correction contributes to the hydrostatic balance (2.9a) at the order $\mathcal{O}(\epsilon)$. We observe, that in many experimental situations $gL/R\Delta T$, which in terms of the small parameter is of the order $\epsilon^0$, is in fact likely to be significantly less than unity, since the specific gas constant $R$ for laboratory liquids is of the order $10^2$ J/kgK and the typical temperature gradients in laboratory are at the order 10 K/m. This implies, that as a result of the fluid's thermal expansion the density gradient in the hydrostatic state,

$$L\frac{\mathrm{d}}{\mathrm{d}z}\frac{\tilde{\tilde{\rho}}}{\bar{\rho}} = \frac{\Delta T}{\bar{T}} - \frac{\bar{\rho}gL}{\bar{p}} + \mathcal{O}\left(\epsilon^2\right) = \mathcal{O}(\epsilon), \tag{2.31}$$

is likely to be positive in the vicinity of convection threshold, although small. However, if the Boussinesq system remains close to adiabatic, that is $\Delta T/L - g/c_p \ll \Delta T/L$, then it is a simple matter to demonstrate, that $gL/R\Delta T \approx c_p/R > 1$ and thus the density in the hydrostatic reference state decreases slightly with height.

### *2.1.2 Energy Balance*

Utilizing the static state equation (2.9b) and the general energy balance (2.1c) the equation for the temperature fluctuation can be written in the form

$$\rho c_v\left(\frac{\partial T'}{\partial t} + \mathbf{u}\cdot\nabla T\right) + \frac{\alpha T}{\beta}\nabla\cdot\mathbf{u} = \nabla\cdot\left(k\nabla T'\right) + 2\mu\mathbf{G}^s : \mathbf{G}^s + \left(\mu_b - \frac{2}{3}\mu\right)(\nabla\cdot\mathbf{u})^2 + Q', \tag{2.32}$$

where

$$Q' = Q - \tilde{Q}. \tag{2.33}$$

Next we introduce the state equations (2.12a) and (2.27) into the flow divergence estimate (2.14) to obtain

$$\frac{\alpha T}{\beta}\nabla\cdot\mathbf{u}=\frac{\bar{\alpha}\bar{T}}{\bar{\beta}}\left(\frac{\partial}{\partial t}+\mathbf{u}\cdot\nabla\right)\left[\bar{\alpha}\left(\tilde{\tilde{T}}+T'\right)-\bar{\beta}\tilde{\tilde{p}}\right]+\mathcal{O}\left(\bar{p}\epsilon^2\frac{\mathcal{U}}{L}\right). \quad (2.34)$$

This can be further simplified by making use of the hydrostatic balance in (2.9a), which allows to write

$$-\bar{\alpha}\bar{T}\left(\frac{\partial}{\partial t}+\mathbf{u}\cdot\nabla\right)\tilde{\tilde{p}}=u_z g\bar{\rho}\bar{\alpha}\bar{T}+\mathcal{O}\left(\bar{p}\epsilon^2\frac{\mathcal{U}}{L}\right), \quad (2.35)$$

and therefore

$$\frac{\alpha T}{\beta}\nabla\cdot\mathbf{u}=\frac{\bar{\alpha}^2\bar{T}}{\bar{\beta}}\left(\frac{\partial}{\partial t}+\mathbf{u}\cdot\nabla\right)\left(\tilde{\tilde{T}}+T'\right)+u_z g\bar{\rho}\bar{\alpha}\bar{T}+\mathcal{O}\left(\bar{p}\epsilon^2\frac{\mathcal{U}}{L}\right). \quad (2.36)$$

Inserting the latter expression (2.36) into the energy balance (2.32), at the leading order yields

$$\begin{aligned}\left(\bar{\rho}\bar{c}_v+\frac{\bar{\alpha}^2\bar{T}}{\bar{\beta}}\right)\left(\frac{\partial T'}{\partial t}+\mathbf{u}\cdot\nabla T\right)+u_z g\bar{\rho}\bar{\alpha}\bar{T} =&\nabla\cdot\left(k\nabla T'\right)+2\mu\mathbf{G}^s:\mathbf{G}^s\\ &+\left(\mu_b-\frac{2}{3}\mu\right)(\nabla\cdot\mathbf{u})^2+Q'.\end{aligned} \quad (2.37)$$

Now we recall a thermodynamic identity

$$\begin{aligned}c_p-c_v&=T\left[\left(\frac{\partial s}{\partial T}\right)_p-\left(\frac{\partial s}{\partial T}\right)_\rho\right]=T\left[\left(\frac{\partial s}{\partial \rho}\right)_T\left(\frac{\partial \rho}{\partial T}\right)_p\right]\\ &=-\frac{T}{\rho^2}\left(\frac{\partial p}{\partial T}\right)_\rho\left(\frac{\partial \rho}{\partial T}\right)_p=\frac{\alpha^2 T}{\rho\beta},\end{aligned} \quad (2.38)$$

where we have used the Maxwell's identity $\rho^2\left(\partial s/\partial\rho\right)_T=-\left(\partial p/\partial T\right)_\rho$ and the implicit function theorem $\left(\partial p/\partial T\right)_\rho=\alpha/\beta$. First of all this allows to simplify the factor in front of the temperature time derivative,

$$\bar{\rho}\bar{c}_v+\frac{\bar{\alpha}^2\bar{T}}{\bar{\beta}}=\bar{\rho}\bar{c}_p. \quad (2.39)$$

Secondly, since according to (2.28d) we can estimate the order of magnitude of the right hand side in (2.38) as $\bar{\alpha}^2\bar{T}/\bar{\beta}\bar{\rho}\sim\bar{\alpha}^2\bar{T}\bar{p}/\bar{\rho}\sim\epsilon^{-1}gL\bar{\alpha}^2\bar{T}\sim\epsilon^{-1}gL/\bar{T}$, it follows that

$$\bar{c}_p \sim \bar{c}_v \sim \epsilon^{-1}\frac{gL}{\bar{T}}. \tag{2.40}$$

Hence, a comparison of the orders of magnitude of the advective term with viscous heating in Eq. (2.37)

$$\left|2\mu\mathbf{G}^s : \mathbf{G}^s + \left(\mu_b - \frac{2}{3}\mu\right)(\nabla\cdot\mathbf{u})^2\right| \sim \epsilon^{3/2}\bar{\rho}\frac{(gL)^{3/2}}{L}, \tag{2.41a}$$

$$\left|\left(\bar{\rho}\bar{c}_v + \frac{\bar{\alpha}^2\bar{T}}{\bar{\beta}}\right)\left(\frac{\partial T'}{\partial t} + \mathbf{u}\cdot\nabla T\right)\right| \sim \left|u_z g\bar{\rho}\bar{\alpha}\bar{T}\right| \sim \epsilon^{1/2}\bar{\rho}\frac{(gL)^{3/2}}{L}, \tag{2.41b}$$

leading to

$$\frac{\left|2\mu\mathbf{G}^s : \mathbf{G}^s + \left(\mu_b - \frac{2}{3}\mu\right)(\nabla\cdot\mathbf{u})^2\right|}{\left|\left(\bar{\rho}\bar{c}_v + \frac{\bar{\alpha}^2\bar{T}}{\bar{\beta}}\right)\left(\frac{\partial T'}{\partial t} + \mathbf{u}\cdot\nabla T\right)\right|} \sim \epsilon, \tag{2.42}$$

allows to conclude that the viscous heating provides a negligible contribution to the energy balance of Boussinesq systems. On the other hand, the process of molecular heat transfer has to be included in the energy balance, since it is a crucial process of temperature relaxation, which requires that

$$k \sim \epsilon^{-1/2}\frac{\bar{\rho}}{\bar{T}}(gL)^{3/2}L, \tag{2.43}$$

and the thermal diffusivity

$$\kappa = \frac{k}{\bar{\rho}\bar{c}_p} \sim \epsilon^{1/2}\sqrt{gL}L, \tag{2.44}$$

in consistency with the estimate of viscous diffusivity in (2.18). Therefore finally the leading order form of the temperature equation reads

$$\frac{\partial T'}{\partial t} + \mathbf{u}\cdot\nabla T' + u_z\left(\frac{\mathrm{d}\tilde{\bar{T}}}{\mathrm{d}z} + \frac{g\bar{\alpha}\bar{T}}{\bar{c}_p}\right) = \nabla\cdot\left(\kappa\nabla T'\right) + \frac{Q'}{\bar{\rho}\bar{c}_p}, \tag{2.45}$$

where $-g\bar{\alpha}\bar{T}/\bar{c}_p$ is the adiabatic gradient, i.e. temperature gradient corresponding to constant entropy per unit mass.[4] This completes the derivation of the Boussinesq system of equations given in (2.26), (2.14) and (2.45) and supplied by the static state equations (2.9a), (2.9b) and (2.12a). We note, that the isobaric heat capacity $c_p$ for standard laboratory liquids is of the order of $10^3$ J/kgK, thus for $L \sim 1$ m

[4] We note, that the final form of the temperature equation (2.45) could also be derived directly from the second form of the energy equation given in (1.40) and the following estimate $-\bar{\alpha}\bar{T}\left(\partial_t p' + \mathbf{u}\cdot\nabla\tilde{\bar{p}} + \mathbf{u}\cdot\nabla p'\right) = u_z g\bar{\rho}\bar{\alpha}\bar{T} + \mathcal{O}\left(\bar{p}\epsilon^2\mathcal{U}/L\right)$, obtained by virtue of the imposed scalings likewise the hydrostatic balance (2.9a).

and $\left|\mathrm{d}\tilde{\tilde{T}}/\mathrm{d}z\right| = |\Delta T/L| \sim 10\,\mathrm{K/m}$ one obtains that $gL/\Delta T c_p \sim 10^{-3}$ is very small. Taking into account that typically $\bar{\alpha}\bar{T} \lesssim 1$, the term involving the adiabatic gradient in Eq. (2.45) is typically negligible. However, it is not necessarily the case in natural convective systems, since e.g. in the Earth's core the adiabatic gradient is huge, of the order $10^{-4}$ K/m and very close to the static state gradient. This results from the fact, that vigorous convection is so efficient in transporting heat, that after billions of years the established temperature gradient which drives the convection in natural large-scale systems such as planetary and stellar interiors can not exceed too much the adiabatic gradient.

Finally, it has to be stressed, that the two assumptions stated in (2.6) and (2.8) lead to much smaller kinetic energy $\sim \epsilon\bar{\rho}gL$ than the internal (or thermal) energy in the system $\sim \epsilon^{-1}\bar{\rho}gL$ with the difference of two orders of magnitude in $\epsilon$. Even when considering only the varying part of the thermal energy, it is still of the order of $c_v\tilde{\tilde{T}} \sim \bar{\rho}gL$, which is $\epsilon^{-1}$ times greater than the kinetic energy. An already stated consequence of this is the negligibility of viscous heating in the energy balance.

For the sake of completeness the entropy equation can now be easily written down

$$\bar{\rho}\bar{T}\left(\frac{\partial s}{\partial t} + \mathbf{u}\cdot\nabla s\right) = \nabla\cdot\left(k\nabla T'\right) + Q'. \tag{2.46}$$

Expanding about the mean state, as in (2.10)

$$s = \bar{s} - \bar{\alpha}\frac{p-\bar{p}}{\bar{\rho}} + \bar{c}_p\frac{T-\bar{T}}{\bar{T}} + \mathcal{O}\left(\epsilon\frac{gL}{\bar{T}}\right), \tag{2.47}$$

where the Maxwell's identity $\rho^2(\partial s/\partial p)_T = (\partial\rho/\partial T)_p = -\rho\alpha$ was used, one obtains by virtue of (2.28c)

$$\tilde{\tilde{s}} = -\frac{\bar{\alpha}}{\bar{\rho}}\tilde{\tilde{p}} + \bar{c}_p\frac{\tilde{\tilde{T}}}{\bar{T}} + \mathcal{O}\left(\epsilon\frac{gL}{\bar{T}}\right), \qquad s' = \bar{c}_p\frac{T'}{\bar{T}} + \mathcal{O}\left(\epsilon\frac{gL}{\bar{T}}\right), \tag{2.48}$$

so that under the Boussinesq approximation the entropy fluctuation is the same as the fluctuation of temperature up to a constant factor. This allows to simplify the entropy equation, at leading order, to

$$\bar{\rho}\bar{T}\left(\frac{\partial s'}{\partial t} + \mathbf{u}\cdot\nabla s'\right) + \bar{\rho}\bar{c}_p u_z\left(\frac{\mathrm{d}\tilde{\tilde{T}}}{\mathrm{d}z} + \frac{g\bar{\alpha}\bar{T}}{\bar{c}_p}\right) = \nabla\cdot\left(k\nabla T'\right) + Q'. \tag{2.49}$$

or

$$\frac{\partial s'}{\partial t} + \mathbf{u}\cdot\nabla s' + \frac{\bar{c}_p}{\bar{T}}u_z\left(\frac{\mathrm{d}\tilde{\tilde{T}}}{\mathrm{d}z} + \frac{g\bar{\alpha}\bar{T}}{\bar{c}_p}\right) = \nabla\cdot\left(\kappa\nabla s'\right) + \frac{Q'}{\bar{\rho}\bar{T}}. \tag{2.50}$$

### 2.1.3 Energetic Properties of Boussinesq Systems

Gathering all the necessary equations the complete and closed system of approximate Boussinesq equations at leading order reads

$$\frac{\partial \mathbf{u}}{\partial t} + (\mathbf{u} \cdot \nabla) \mathbf{u} = -\frac{1}{\bar{\rho}} \nabla p' + g\bar{\alpha} T' \hat{\mathbf{e}}_z + \nu \nabla^2 \mathbf{u} + 2 \nabla \nu \cdot \mathbf{G}^s, \tag{2.51a}$$

$$\nabla \cdot \mathbf{u} = 0, \tag{2.51b}$$

$$\frac{\partial T'}{\partial t} + \mathbf{u} \cdot \nabla T' + u_z \left( \frac{\mathrm{d}\tilde{\bar{T}}}{\mathrm{d}z} + \frac{g\bar{\alpha}\bar{T}}{\bar{c}_p} \right) = \nabla \cdot (\kappa \nabla T) + \frac{Q}{\bar{\rho}\bar{c}_p}, \tag{2.51c}$$

where the static temperature equation $\nabla \cdot (k \nabla \tilde{\bar{T}}) = -\tilde{Q}$ has been incorporated back into the energy balance. Let us consider a plane layer of fluid of thickness $L$, say periodic in the horizontal directions with periods $L_x$, $L_y$ and take the $z$-axis vertical aligned with the gravity $\mathbf{g} = -g\hat{\mathbf{e}}_z$. Applying a horizontal average

$$\langle \cdot \rangle_h = \frac{1}{L_x L_y} \int_{-L_x/2}^{L_x/2} \int_{-L_y/2}^{L_y/2} (\cdot) \, \mathrm{d}x \mathrm{d}y, \tag{2.52}$$

to the energy equation one obtains

$$\frac{\partial}{\partial t} \langle \bar{\rho}\bar{c}_p T' \rangle_h = -\frac{\partial}{\partial z} \left[ \langle \bar{\rho}\bar{c}_p u_z T' \rangle_h - k \frac{\partial \langle T \rangle_h}{\partial z} \right] + \langle Q \rangle_h , \tag{2.53}$$

since the horizontal average of continuity equation (2.51b) results in

$$\langle u_z \rangle_h = 0. \tag{2.54}$$

Therefore the rate of change of the thermal energy is governed by vertical variation of the total heat flux (and radiative heat sources). A stationary state requires that

$$\langle \bar{\rho}\bar{c}_p u_z T' \rangle_h - k \frac{\partial \langle T \rangle_h}{\partial z} - \langle Q \rangle L = - k \frac{\partial \langle T \rangle_h}{\partial z} \bigg|_{z=0} , \tag{2.55}$$

which was obtained by integration of the stationary equation (2.53) along $'z'$ from 0 to $L$, where

$$\langle Q \rangle = \frac{1}{L_x L_y L} \int_{-L_x/2}^{L_x/2} \int_{-L_y/2}^{L_y/2} \int_0^L Q \mathrm{d}z \mathrm{d}y \mathrm{d}x \tag{2.56}$$

is the radiative heating averaged over the entire periodic fluid domain. In the absence of heat sources $Q = 0$, Eq. (2.55) states that the total, horizontally averaged heat flux entering the system at the bottom is the same at every horizontal plane (independent of height), which is a crucial feature of Boussinesq systems; in particular the total heat flux entering at the bottom, in a stationary state is equal to the total flux released through the top boundary,

$$-k\frac{\partial\langle T\rangle_h}{\partial z}\bigg|_{z=0} = -k\frac{\partial\langle T\rangle_h}{\partial z}\bigg|_{z=L}. \tag{2.57}$$

When the radiative heating is negligible with respect to advection of heat and conduction, i.e.

$$\langle Q\rangle L \ll \left\langle\bar{\rho}\bar{c}_p u_z T' - k\partial_z T\right\rangle_h, \tag{2.58}$$

the following definition of the Nusselt number, being a measure of effectiveness of heat transfer by convection seems most practically useful

$$Nu = \frac{\bar{\rho}\bar{c}_p\langle u_z T\rangle_h - k\partial_z\langle T\rangle_h - \bar{k}g\bar{\alpha}\bar{T}/\bar{c}_p}{\langle k\Delta_S\rangle}, \tag{2.59}$$

where

$$\Delta_S = -\frac{\mathrm{d}\tilde{\bar{T}}}{\mathrm{dz}} - \frac{g\bar{\alpha}\bar{T}}{\bar{c}_p} > 0 \tag{2.60}$$

is the reference temperature gradient excess with respect to the adiabatic gradient (superadiabatic gradient excess of the static state) and

$$\langle\cdot\rangle = \frac{1}{L_x L_y L}\int_{-L_x/2}^{L_x/2}\int_{-L_y/2}^{L_y/2}\int_0^L (\cdot)\,\mathrm{d}z\mathrm{d}y\mathrm{d}x, \tag{2.61}$$

is the full spatial average (the same as the upper bar). In such a way the Nusselt number is a ratio of the total superadiabatic heat flux in the convective state to the total superadiabatic heat flux in the hydrostatic reference state. The term $\bar{\rho}\bar{c}_p\langle u_z T\rangle_h$ represents the flux contribution from advection of temperature whereas $-k\partial_z\langle T\rangle_h$ from molecular conduction. Note, that the Nusselt number $Nu$ in (2.59) is in principle a function of $z$ and $t$ and becomes a constant in a stationary state, due to (2.55).

#### 2.1.3.1 Convection Driven by a Fixed Temperature Difference Between Top and Bottom Boundaries, in the Absence of Radiative Heat Sources, $Q = 0$

Accompanied by the following definition of the Rayleigh number[5]

[5]When $Q = 0$ and $\kappa = \text{const}$ the static temperature distribution is linear in the vertical coordinate $z$ and the definition of the Rayleigh number corresponds to the standard one $Ra = g\bar{\alpha}\Delta T_S L^3/\kappa\nu$, where $\Delta T_S = \Delta T - g\bar{\alpha}\bar{T}L/c_p$, $\Delta T = T_{bottom} - T_{top}$ is the temperature difference between bottom and top boundaries, $g\bar{\alpha}\bar{T}L/c_p$ expresses such a temperature difference in the adiabatic state

$$Ra = \frac{g\bar{\alpha}\left\langle \kappa\Delta_S \right\rangle L^4}{\bar{\kappa}^2\bar{\nu}}, \tag{2.62}$$

measuring the relative strength of the buoyancy forces (controlled by the static state temperature difference between top and bottom plates) with respect to diffusive effects, the above considerations allow to express the time variation of the mean kinetic energy (through taking the dot-product of the Navier-Stokes equation and **u**) in a simple form

$$\frac{\partial}{\partial t}\left(\frac{1}{2}\mathbf{u}^2\right) = -\nabla\cdot\left[\left(\frac{1}{2}\mathbf{u}^2 + \frac{p'}{\bar{\rho}}\right)\mathbf{u}\right] + g\bar{\alpha}T'u_z + \mathbf{u}\cdot\left[\nabla\cdot\left(2\nu\mathbf{G}^s\right)\right], \tag{2.63}$$

and so[6]

$$\frac{\partial}{\partial t}\left\langle\frac{1}{2}\mathbf{u}^2\right\rangle = \frac{\bar{\kappa}^2\bar{\nu}}{L^4}Ra\left(\frac{1}{L}\int_0^L Nu(z,t)\mathrm{d}z - 1\right) + g\bar{\alpha}\left\langle\kappa\frac{\partial T'}{\partial z}\right\rangle - 2\left\langle\nu\mathbf{G}^s : \mathbf{G}^s\right\rangle. \tag{2.64}$$

Of course in a stationary state $Nu = \text{const}$ and the first term on the right hand side becomes $\bar{\kappa}^2\bar{\nu}Ra(Nu-1)/L^4$. In obtaining (2.64) we have used the boundary conditions that the component of the velocity field normal to the boundaries, likewise either the tangent viscous stresses or the tangent velocity components at the boundaries are either null or periodic, i.e. either

$$\mathbf{u}\cdot\mathbf{n}|_{\partial V} = 0, \tag{2.65a}$$

$$\mathbf{n}\times\left(2\nu\mathbf{G}^s\cdot\mathbf{n}\right)\big|_{\partial V} = 0 \quad \text{or} \quad \mathbf{n}\times\mathbf{u}|_{\partial V} = 0, \tag{2.65b}$$

or the values at the boundaries are periodic (not necessarily zero), where **n** is the normal unit vector at the boundaries. This implies

$$\left\langle\nabla\cdot\left[\left(\frac{1}{2}\mathbf{u}^2 + \frac{p'}{\bar{\rho}}\right)\mathbf{u}\right]\right\rangle = 0, \quad \left\langle\nabla\cdot\left(2\nu\mathbf{u}\cdot\mathbf{G}^s\right)\right\rangle = 0, \tag{2.66}$$

and

$$\left\langle\mathbf{u}\cdot\left[\nabla\cdot\left(2\nu\mathbf{G}^s\right)\right]\right\rangle = \left\langle\nabla\cdot\left(2\nu\mathbf{u}\cdot\mathbf{G}^s\right)\right\rangle - 2\left\langle\nu\mathbf{G} : \mathbf{G}^s\right\rangle = -2\left\langle\nu\mathbf{G}^s : \mathbf{G}^s\right\rangle, \tag{2.67}$$

since the double contraction of the antisymmetric part of the velocity gradient tensor $G_{ij} = \partial u_i/\partial x_j$ with the symmetric part is necessarily zero, so that

$$\mathbf{G} : \mathbf{G}^s = \mathbf{G}^s : \mathbf{G}^s.$$

---

and in most experimental situations $\Delta T_S \approx \Delta T$, since the adiabatic gradient is negligible compared to the static one (as argued below Eq. (2.45)).

[6]Note, that the expression for the mean viscous energy dissipation is the same as that for the mean viscous heating in the energy equation.

It follows, that either the tangent stress-free or no-slip boundary conditions work for derivation of the kinetic energy evolution equation (2.64). Moreover, the mean work of the buoyancy force per unit mass has been expressed with the aid of the definitions of the Nusselt (2.59) and Rayleigh numbers (2.62) in the following way

$$\begin{aligned}\frac{g\bar{\alpha}}{L}\int_0^L \mathrm{d}z\left\langle T'u_z\right\rangle_h =& g\bar{\alpha}\left\langle\kappa\Delta_S\right\rangle Nu + \frac{g^2\bar{\alpha}^2\bar{T}\bar{\kappa}}{\bar{c}_p} + \frac{g\bar{\alpha}}{L}\int_0^L \mathrm{d}z\left\langle\kappa\frac{\partial}{\partial z}\left(\tilde{\tilde{T}}+T'\right)\right\rangle_h \\ =& \frac{\bar{\nu}\bar{\kappa}^2}{L^4}Ra\left(Nu-1\right) + g\bar{\alpha}\left\langle\kappa\frac{\partial T'}{\partial z}\right\rangle, \end{aligned} \tag{2.68}$$

since $\kappa \mathrm{d}_z\tilde{\tilde{T}}$ is uniform (independent of $\mathbf{x}$) by virtue of the static state balance and we have assumed stationarity, i.e. $Nu = \text{const}$. The last term in the latter equation is necessarily zero in the case when $\kappa$ is spatially uniform, since $\left\langle T'\right\rangle\big|_{z=0,L} = 0$, and then Eq. (2.64), in a stationary state reads

$$0 = \frac{\kappa^2\bar{\nu}}{L^4}Ra\left(Nu-1\right) - 2\left\langle\nu\mathbf{G}^s : \mathbf{G}^s\right\rangle. \tag{2.69}$$

However, if the thermal diffusivity $\kappa$ is depth-dependent this term is, in general non-zero and in particular in the fully nonlinear, turbulent regime analysed in Sect. 2.4 it can be estimated as

$$\frac{g\bar{\alpha}}{L}\int_0^L \kappa\partial_z\left\langle T'\right\rangle_h \mathrm{d}z \sim -\frac{1}{2}g\bar{\alpha}\Delta_S(\kappa_B+\kappa_T) = \frac{\bar{\kappa}^2\bar{\nu}Ra(\kappa_B+\kappa_T)\Delta_S}{2\left\langle\kappa\Delta_S\right\rangle L^4}, \tag{2.70}$$

where the subscripts $B$ and $T$ denote values of $\kappa$ taken at the bottom and top boundaries respectively.

Taking now the full spatial average of the temperature equation (2.51c) with excluded static state contribution and multiplied by $T'$

$$\frac{\partial}{\partial t}\left\langle\frac{1}{2}T'^2\right\rangle + \left\langle\nabla\cdot\left(\frac{1}{2}\mathbf{u}T'^2\right)\right\rangle - \left\langle u_z T'\Delta_S\right\rangle = \left\langle T'\nabla\cdot\left(\kappa\nabla T'\right)\right\rangle. \tag{2.71}$$

For a stationary state one obtains in a straight forward manner

$$\frac{\partial}{\partial t}\left\langle\frac{1}{2}T'^2\right\rangle = 0, \quad \left\langle\nabla\cdot\left(\frac{1}{2}\mathbf{u}T'^2\right)\right\rangle = 0, \tag{2.72}$$

where the latter comes from the assumed periodicity in horizontal directions and impermeable top and bottom boundaries; also by virtue of integration by parts and application of fixed temperature boundary conditions $T'(z = 0,\ L) = 0$,

$$\left\langle T'\nabla\cdot\left(\kappa\nabla T'\right)\right\rangle = -\left\langle\kappa\left(\nabla T'\right)^2\right\rangle. \tag{2.73}$$

Moreover, using the Nusselt number definition (2.59) the last term on the left hand side of (2.71), in a stationary state, can be cast as follows

$$\begin{aligned}\left\langle u_z T' \Delta_S \right\rangle &= \frac{1}{L}\int_0^L \mathrm{d}z \Delta_S \left[ \langle \kappa \Delta_S \rangle Nu + \left\langle \kappa \frac{\partial}{\partial z}\left(\bar{\bar{T}} + T'\right)\right\rangle_h + \frac{g\bar{\alpha}\bar{T}\bar{\kappa}}{\bar{c}_p}\right] \\ &= \langle \kappa \Delta_S \rangle \langle \Delta_S \rangle Nu + \langle \Delta_S \rangle \frac{g\bar{\alpha}\bar{T}\bar{\kappa}}{\bar{c}_p} + \left\langle \Delta_S \kappa \frac{\mathrm{d}\bar{\bar{T}}}{\mathrm{d}z}\right\rangle + \left\langle \Delta_S \kappa \frac{\partial T'}{\partial z}\right\rangle \\ &= \langle \kappa \Delta_S \rangle \langle \Delta_S \rangle (Nu - 1) + \left\langle \Delta_S \kappa \frac{\partial T'}{\partial z}\right\rangle \end{aligned} \tag{2.74}$$

where, again, we have used $\kappa \mathrm{d}_z \bar{\bar{T}} = \text{const}$, a consequence of the static state balance. Therefore a stationary convective state implies

$$\left\langle \kappa \left(\nabla T'\right)^2 \right\rangle = \langle \kappa \Delta_S \rangle \langle \Delta_S \rangle (Nu - 1) + \left\langle \Delta_S \kappa \frac{\partial T'}{\partial z}\right\rangle, \tag{2.75}$$

and hence

$$Nu = \frac{\left\langle \kappa \left(\nabla T'\right)^2 \right\rangle - \left\langle \Delta_S \kappa \frac{\partial T'}{\partial z}\right\rangle}{\langle \kappa \Delta_S \rangle \langle \Delta_S \rangle} + 1, \tag{2.76}$$

In cases, when the thermal diffusivity $\kappa$ can be considered uniform we get $\left\langle \Delta_S \kappa \frac{\partial T'}{\partial z}\right\rangle = \Delta_S \kappa \left\langle \frac{\partial T'}{\partial z}\right\rangle = 0$ and the above relation simplifies to

$$Nu = \frac{\kappa \left\langle \left(\nabla T'\right)^2 \right\rangle}{\kappa \Delta_S^2} + 1, \tag{2.77}$$

an expression for the Nusselt number in terms of a quadratic form in $\nabla T'$ - the heat flow stimulus.

Finally, by virtue of the Nusselt number estimate $Nu \leq \sqrt{Ra}/4 - 1$ valid for $Ra \geq 64$, derived in Doering and Constantin (1996) for Boussinesq equations with $Q = 0$, $\mu = \text{const}$, $k = \text{const}$ and neglection of the adiabatic gradient with respect to $\Delta T / L$ one obtains

$$\kappa \left\langle \left(\nabla T'\right)^2 \right\rangle \leq \left( \frac{\sqrt{Ra}}{4} - 2 \right) \kappa \Delta_S^2. \tag{2.78}$$

An alternative upper bound on the Nusselt number under the same assumptions $Q = 0$, $\mu = \text{const}$, $k = \text{const}$ and neglection of the adiabatic gradient, valid for $Ra \gg 1$ was obtained by Nobili (2015) and Choffrut et al. (2016), which we recall here

$$Nu = \frac{\kappa\left\langle\left(\nabla T'\right)^2\right\rangle}{\kappa\Delta_S^2} + 1 \lesssim \begin{cases} Ra^{1/3}\left(\ln Ra\right)^{1/3} & \text{for } Pr \geq Ra^{1/3}\left(\ln Ra\right)^{1/3} \\ Ra^{1/2}\left(\frac{\ln Ra}{Pr}\right)^{1/2} & \text{for } Pr \leq Ra^{1/3}\left(\ln Ra\right)^{1/3} \end{cases}. \tag{2.79}$$

The estimates (2.78) and (2.79), however, are also valid in the same form when the adiabatic gradient is included in the equations with definitions of the Nusselt and Rayleigh numbers as in (2.59) and (2.62). This is easily seen, since inclusion of the adiabatic gradient in the driving parameter $\Delta_S$ in the case of uniform fluid properties and absence of the radiative sources can be effectively interpreted as a decrease of the temperature difference $\Delta T \to \Delta T - g\bar{\alpha}\bar{T}L/\bar{c}_p$.

#### 2.1.3.2 Convection Driven by a Fixed Heat Flux at the Boundaries, in the Absence of Radiative Heat Sources, $Q = 0$

The case when the convection is driven by a fixed heat flux at the boundaries requires $\partial_z T'\big|_{z=0,L} = 0$, therefore in the absence of radiative heat sources the Eq. (2.55) corresponding to a stationary state, takes the form

$$\left\langle\bar{\rho}\bar{c}_p u_z T'\right\rangle_h - k\frac{\partial\left\langle T\right\rangle_h}{\partial z} = -k\frac{\partial\tilde{\tilde{T}}}{\partial z}\bigg|_{z=0} = -k\frac{\partial\tilde{\tilde{T}}}{\partial z} = \text{const}, \tag{2.80}$$

where the static state equation $\partial_z(k\partial_z\tilde{\tilde{T}}) = 0$ was used, and then the Nusselt number (2.59), in a stationary state, is simply unity. Furthermore, as a consequence

$$\left\langle\bar{\rho}\bar{c}_p u_z T'\right\rangle_h - k\frac{\partial\left\langle T'\right\rangle_h}{\partial z} = 0, \tag{2.81}$$

which implies that in a stationary state the mean advective heat flux is balanced by the mean fluctuation molecular flux. The definition of the Rayleigh number (2.62) is sustained

$$Ra = \frac{g\bar{\alpha}\left\langle\kappa\Delta_S\right\rangle L^4}{\bar{\kappa}^2\bar{\nu}}, \tag{2.82}$$

since $\langle\kappa\Delta_S\rangle$ is the heat flux excess with respect to the average adiabatic flux, which drives the flow. The stationary mean kinetic energy equation, by the use of (2.81), is now

$$0 = g\bar{\alpha}\left\langle\kappa\frac{\partial T'}{\partial z}\right\rangle - 2\left\langle\nu\mathbf{G}^s : \mathbf{G}^s\right\rangle. \tag{2.83}$$

The following new Nusselt number definition, as a measure of advective heat flux over the superadiabatic static flux turns out useful in description of convection driven by fixed heat flux on boundaries

$$Nu_Q = \frac{\langle \bar{\rho}\bar{c}_p u_z T' \rangle}{\langle k \Delta_S \rangle}, \tag{2.84}$$

which allows to express the stationary balance between the work of the buoyancy and viscous heating in a simple way

$$0 = \frac{\bar{\kappa}^2 \bar{\nu}}{L^4} Ra Nu_Q - 2 \langle \nu \mathbf{G}^s : \mathbf{G}^s \rangle. \tag{2.85}$$

The work of the buoyancy force in the case of uniform $\kappa$ is $g\bar{\alpha}\kappa\Delta \langle T' \rangle_h / L$, with $\Delta \langle T' \rangle_h$ denoting the difference of the mean temperature fluctuation between top and bottom boundaries and then the Nusselt number is simply

$$Nu_Q = \frac{\Delta \langle T' \rangle_h}{\Delta_S L}. \tag{2.86}$$

Furthermore, the Eq. (2.75) in the case of a fixed heat flux at the boundaries takes the form

$$\left\langle \kappa \left( \nabla T' \right)^2 \right\rangle = \left\langle \Delta_S \kappa \frac{\partial T'}{\partial z} \right\rangle, \tag{2.87}$$

and at uniform $\kappa$ one obtains

$$\left\langle \Delta_S \kappa \frac{\partial T'}{\partial z} \right\rangle = \kappa \Delta_S \left\langle \frac{\partial T'}{\partial z} \right\rangle = \kappa \Delta_S \frac{\Delta \langle T' \rangle_h}{L} = \kappa \Delta_S^2 Nu_Q \tag{2.88}$$

and therefore

$$Nu_Q = \frac{\left\langle \kappa \left( \nabla T' \right)^2 \right\rangle}{\kappa \Delta_S^2}. \tag{2.89}$$

Summarizing, the Boussinesq convection, either driven by a fixed temperature difference or a fixed heat flux at boundaries, is characterized by the thermal energy being much stronger than the kinetic one with the latter governed by a balance between the work of the buoyancy force and the viscous dissipation. The thermal energy remains uninfluenced by those effects at leading order. Moreover, the total, horizontally averaged heat flux flowing through the system is constant, i.e. independent of height.

### 2.1.4 Conservation of Mass and Values of the Mean Pressure at Boundaries

At the heart of the Boussinesq approximation lies the solenoidal constraint $\nabla \cdot \mathbf{u} = 0$, which implies sound-proof dynamics and therefore the pressure spreads with infinite velocity. Consequently, the pressure fluctuation is determined by a Poisson-type, elliptic equation

$$\nabla^2 p' = g\bar{\rho}\bar{\alpha}\frac{\partial T'}{\partial z} + \bar{\rho}\nabla\nu \cdot \nabla^2 \mathbf{u} - \bar{\rho}\nabla \cdot \left[\nabla \cdot (\mathbf{uu}) - 2\nabla\nu \cdot \mathbf{G}^s\right], \quad (2.90)$$

obtained by taking a divergence of the Navier-Stokes equation (2.51a). However, in order to fully resolve the dynamics of the equations of convection (2.51a)–(2.51c) one must keep the total mass of the fluid conserved. This means, that if we assume, that the total mass is contained in the reference state $\bar{\rho} + \tilde{\tilde{\rho}}$, we must impose

$$\langle \rho' \rangle = 0 \quad \text{at all times.} \quad (2.91)$$

The latter is *not* guaranteed by the system of equations (2.51a)–(2.51c) throughout the entire evolution, if only the initial condition is chosen to satisfy (2.91), and therefore constitutes and additional constraint, which must be imposed at every instant.

If we further make an additional assumption that the viscosity is allowed to be a function of '$z$' only, i.e. $\nu = \nu(z)$, and average the $z$-component of the Navier-Stokes equation (2.51a) over the entire periodic domain, substituting $T' = -\rho'/\bar{\rho}\bar{\alpha}$, we get

$$\langle p' \rangle_h (z = L) - \langle p' \rangle_h (z = 0) = -gL \langle \rho' \rangle \quad (2.92)$$

which by (2.91) implies

$$\langle p' \rangle_h (z = L) = \langle p' \rangle_h (z = 0) \quad \text{at all times.} \quad (2.93)$$

This constitutes a boundary condition, which must be imposed on the pressure field at every moment in time, used in tandem with the elliptic equation (2.90). Let us note, that whether or not the condition of null pressure fluctuation jump across the layer is imposed, the convective velocity field remains uninfluenced, since a shift in pressure fluctuation which is time-dependent only corresponds to a simple gauge transformation. Nevertheless, the condition (2.93) is important in order to fully resolve the dynamics of the temperature field.

An important consequence of the mass conservation constraint (2.91) and the Boussinesq relation between density and temperature fluctuations $\rho' = -\bar{\rho}\bar{\alpha}T'$, is that

$$\langle T' \rangle = 0 \quad \text{at all times} \quad (2.94)$$

must also be satisfied throughout the evolution of the system. Therefore averaging of the temperature equation (2.51c) over the entire fluid domain leads to

$$-k\frac{\partial \langle T \rangle_h}{\partial z}\bigg|_{z=0} + k\frac{\partial \langle T \rangle_h}{\partial z}\bigg|_{z=L} + \langle Q \rangle L = 0 \quad \text{at all times}, \tag{2.95}$$

so that in the absence of volume heating sources, $Q = 0$, the total heat flux entering the system at the bottom must equal the total heat flux leaving the system at the top at every instant in time (consistently with (2.57) obtained for a stationary state).

### 2.1.5 *Boussinesq Up-Down Symmetries*

For the purpose of clarity we restate here the Boussinesq approximated equations with uniform thermal and viscous diffusivity coefficients, $\nu = \text{const}$, $\kappa = \text{const}$, and no radiative heat sources $Q = 0$, in a more explicit form

$$\frac{\partial \mathbf{u}}{\partial t} + (\mathbf{u} \cdot \nabla_h)\,\mathbf{u} + u_z\frac{\partial \mathbf{u}}{\partial z} = -\frac{1}{\bar{\rho}}\nabla p' + g\bar{\alpha}T'\hat{\mathbf{e}}_z + \nu\left(\nabla_h^2\mathbf{u} + \frac{\partial^2\mathbf{u}}{\partial z^2}\right), \tag{2.96a}$$

$$\nabla_h \cdot \mathbf{u} + \frac{\partial u_z}{\partial z} = 0, \tag{2.96b}$$

$$\frac{\partial T'}{\partial t} + \mathbf{u} \cdot \nabla_h T' + u_z\left(\frac{\partial T'}{\partial z} - \Delta_S\right) = \kappa\left(\nabla_h^2 T' + \frac{\partial^2 T'}{\partial z^2}\right), \tag{2.96c}$$

where $\nabla_h$ denotes the horizontal component of the $\nabla$ operator.[7] This system of equations possesses an up-down symmetry with respect to the mid plane, i.e. under the transformation

$$z \to L - z, \tag{2.97}$$

the velocity field components and the temperature and pressure fields transform in the following way

$$u_z\,(L - z) = -u_z\,(z)\,, \quad \mathbf{u}_h\,(L - z) = \mathbf{u}_h\,(z)\,, \tag{2.98a}$$

$$T'\,(L - z) = -T'\,(z)\,, \quad p'\,(L - z) = p'\,(z)\,. \tag{2.98b}$$

In other words, when the top and bottom boundary conditions are of the same type (e.g. impermeable and either no-slip or stress-free and fixed temperature or fixed heat flux) and the initial conditions satisfy the mid-plane symmetries (2.98a), (2.98b), the full nonlinear Boussinesq equations imply, that the vertical velocity and temperature

---

[7]In Cartesian geometry $\nabla_h = (\partial_x, \partial_y)$.

perturbation are antisymmetric whereas the horizontal velocity and pressure perturbation are symmetric with respect to the mid-plane.[8] This will be called the *nonlinear up-down symmetry* or the *up-down symmetry of developed convection*. Interestingly, the linearised set of Boussinesq equations, under the assumption that the convective flow and pressure and temperature perturbations are weak, i.e.

$$\frac{\partial \mathbf{u}}{\partial t} = -\frac{1}{\bar{\rho}}\nabla p' + g\bar{\alpha}T'\hat{\mathbf{e}}_z + \nu\left(\nabla_h^2\mathbf{u} + \frac{\partial^2 \mathbf{u}}{\partial z^2}\right), \tag{2.99a}$$

$$\nabla_h \cdot \mathbf{u} + \frac{\partial u_z}{\partial z} = 0, \tag{2.99b}$$

$$\frac{\partial T'}{\partial t} - u_z \Delta_S = \kappa\left(\nabla_h^2 T' + \frac{\partial^2 T'}{\partial z^2}\right), \tag{2.99c}$$

possesses an additional opposite up-down symmetry, that is

$$u_z(L-z) = u_z(z), \quad \mathbf{u}_h(L-z) = -\mathbf{u}_h(z), \tag{2.100a}$$

$$T'(L-z) = T'(z), \quad p'(L-z) = -p'(z), \tag{2.100b}$$

so that the vertical velocity and temperature perturbation are symmetric whereas the horizontal velocity and pressure perturbation are antisymmetric with respect to the mid-plane. This symmetry corresponds to large-scale convective rolls, that is rolls on the entire scale $L$ of the fluid layer, which set in first in the weakly overcritical regime, i.e. slightly above convection threshold (c.f. next section on linear regime). This will be called the *linear up-down symmetry* or the *up-down symmetry near convection threshold*.

It can be seen easily, that in the case when both the top and bottom boundary conditions are the same, as the Rayleigh number keeps increasing hence the driving force for convection is magnified, the system passes from a linear regime to a nonlinear one, which can be associated with a change of the solution symmetry and therefore a radical change in the form of the flow and temperature distribution. When the boundary conditions break the up-down symmetry, that is the top and bottom boundaries are physically different, the solutions in general do not possess any of the above symmetries.

[8] Physically it is likely, that the initial conditions do not satisfy the mid-plane symmetries (2.98a), (2.98b), or perturbations are introduced which do not satisfy them. However, the total thermal flux at the top and bottom of the domain needs to be symmetric with respect to mid-plane (cf. (2.95)), thus only the antisymmetric mode of the mean temperature can create/experience heat flux on the boundaries. Nevertheless, the nonlinear interactions between modes can still seed the modes with opposite symmetry at least for some time. Summarizing, the modes satisfying the up-down symmetries (2.98a), (2.98b) can last infinitely long in the absence of non-symmetric perturbations, whereas the modes with opposite symmetries can not survive alone.

## 2.2 Linear Stability Analysis at Convection Threshold - Compendium of Results for Different Physical Conditions

The aim of this and two following sections is to describe and explain the development of the convective instability from its onset, through the weakly nonlinear stage with various types of convective patterns to the fully developed turbulent convection. We start with the linear analysis at convection threshold and derivation of the critical Rayleigh number for convection onset. Such a linear regime is known as the Rayleigh-Bénard problem, which has been thoroughly explained in the books of Chandrasekhar (1961), Gershuni and Zhukhovitskii (1976), Getling (1998) and it is now considered a part of a very standard knowledge on thermal convection. Therefore the derivations in the linear regime will be only briefly recalled here and the results for different types of boundary conditions with inclusion of the effects of background rotation and radiative heating will be summarized in a compact way. Attention will be focused on the discussion of the flow properties at threshold and explanation of the convective instability on physical grounds. It will be assumed, that the transport coefficients such as the viscosity $\nu$ and thermal conductivity $\kappa$, likewise the heat capacity $c_p$ are constant and the horizontal extent of the domain is large.

At the initial stage of convection development, that is close to convection threshold, the perturbations to the hydrostatic basic state are weak,

$$\frac{\mathbf{u}}{\epsilon^{1/2}\sqrt{gL}} \ll 1, \quad \text{and} \quad \frac{T'}{\tilde{\tilde{T}}} \ll 1, \tag{2.101}$$

and thus the dynamical equations can be linearised to yield the set of Eqs. (2.99a)–(2.99c). The linear problem can be solved in terms of decomposition of the small perturbation fields into normal, Fourier-type modes

$$\mathbf{u}(x, y, z, t) = \mathfrak{Re}\ \hat{\mathbf{u}}(z)\, \mathrm{e}^{\sigma t} \mathrm{e}^{\mathrm{i}(\mathcal{K}_x x + \mathcal{K}_y y)}, \quad T'(x, y, z, t) = \mathfrak{Re}\ \hat{T}(z)\, \mathrm{e}^{\sigma t} \mathrm{e}^{\mathrm{i}(\mathcal{K}_x x + \mathcal{K}_y y)}, \tag{2.102}$$

where $\sigma$ is the growth rate, thus the system becomes convectively unstable as soon as there appears at least one mode with $\mathfrak{Re}\sigma > 0$. The most general form of the solution at threshold consists of a superposition of Fourier modes of the type

$$\mathbf{u}(x, y, z, t) = \mathfrak{Re} \sum_{\substack{\mathcal{K} \\ \mathcal{K}=\mathcal{K}_{crit}}} \hat{\mathbf{u}}_{\mathcal{K}}(z)\, \mathrm{e}^{\sigma_{\mathcal{K}} t} \mathrm{e}^{\mathrm{i}(\mathcal{K}_x x + \mathcal{K}_y y)}, \tag{2.103}$$

where $\mathcal{K}_{crit}$ is the value of the marginal wave number, however, the planform of the solution remains unknown within the scope of linear theory (note, that when the system is homogeneous and isotropic in the horizontal directions, there is no preference for any horizontal direction and the amplitudes $\hat{\mathbf{u}}_{\mathcal{K}}(z)$ and the growth rates $\sigma_{\mathcal{K}}$ are the same for each Fourier mode). Taking a double curl of the Navier-Stokes equation (2.99a) allows to separate the problem for the vertical velocity component and temperature fluctuation from the equations for horizontal velocity components. The latter can be simply calculated a posteriori once $\hat{u}_z(z)$ and $\hat{T}(z)$ are known. We will first consider the case without the radiative heating. On introducing the forms of the perturbations (2.102) into the Eq. (2.99c) with $Q = 0$ and the equation for $u_z$ obtained from double curl of (2.99a) we get

$$\left[\sigma\left(\frac{d^2}{dz^2} - \mathcal{K}^2\right) - \nu\left(\frac{d^2}{dz^2} - \mathcal{K}^2\right)^2\right]\hat{u}_z = -g\bar{\alpha}\mathcal{K}^2\hat{T}, \tag{2.104a}$$

$$\left[\sigma - \kappa\left(\frac{d^2}{dz^2} - \mathcal{K}^2\right)\right]\hat{T} = \Delta_S\hat{u}_z. \tag{2.104b}$$

It can be rigorously shown (cf. Chandrasekhar 1961, §II.11, pp. 24–26), that the instability in this case occurs as a direct mode with $\sigma \in \mathbb{R}$ passing through zero. This is often called the *Principle of the exchange of stabilities*, which is satisfied in this case. Therefore the marginal state, exactly at threshold, is characterized by $\sigma = 0$, which by the use of (2.104a), (2.104b) allows to write down the vertical velocity amplitude equation in a compact form

$$\left(\frac{d^2}{dz^2} - \mathcal{K}^2\right)^3 \hat{u}_z = -\frac{Ra}{L^4}\mathcal{K}^2\hat{u}_z. \tag{2.105}$$

This equation is subject to boundary conditions and the natural choice is that the boundaries are isothermal, impermeable and either rigid or stress-free. Below, we will quickly summarize the results for each case separately, however, before this, let us first examine the general expression for the critical superadiabatic gradient at threshold (1.66); in the Boussinesq case, cf. Eqs. (2.46)–(2.50), with constant viscosity, thermal conductivity and heat capacity and without radiation it is easily reduced to[9]

$$\Delta_{S\,crit} = \kappa\frac{\min\left[-\nabla^2 T'\right]}{u_z}, \quad \text{at any } z. \tag{2.106}$$

By the use of (2.104a) we can further simplify this expression to

$$Ra_{crit} = \frac{L^4 \min\left[-\frac{1}{\mathcal{K}^2}\left(\frac{d^2}{dz^2} - \mathcal{K}^2\right)^3 u_z\right]}{u_z}, \quad \text{at any } z \tag{2.107}$$

where in the current case

[9]The condition (2.106) directly corresponds to the linearised energy equation (2.104b) with $\sigma = 0$, where the right hand side corresponds to the full advective derivative of the entropy per unit mass, since in the Boussinesq case $s' \approx \bar{c}_p T'/\bar{T}$ is equivalent to the temperature fluctuation up to the constant factor $\bar{c}_p/\bar{T}$, and $ds_0/dz = -c_p\Delta_S/\bar{T}$.

$$Ra_{crit} = \frac{g\bar{\alpha}\Delta_{S\,crit}L^4}{\kappa\nu}, \tag{2.108}$$

is the critical Rayleigh number for convection threshold. Thus the heat per unit mass released by a fluid parcel rising on an infinitesimal distance d$z$ in a time unit in the marginal state is

$$-T\frac{\mathrm{D}s}{\mathrm{d}t} = -\bar{c}_p\kappa\nabla^2 T' = \bar{c}_p\frac{\kappa\nu}{g\bar{\alpha}L^4}Ra_{crit}u_z, \tag{2.109}$$

which is equal to the total heat per unit mass accumulated between the infinitesimally distant (by $u_z\mathrm{d}t$) horizontal fluid layers, between which the perturbed parcel travels, per unit time of the rise in the marginal state, $\bar{c}_p\Delta_{S\,crit}u_z$; when $-T\mathrm{D}s/\mathrm{d}t$ falls below $\bar{c}_p\Delta_S u_z$, the system becomes unstable. This result will be used to compare the physical nature of the convective instability trigger between various specific cases, to which we turn now.

### 2.2.1 Two Isothermal, Stress-Free Boundaries, $Q = 0$

In the case of two stress-free boundaries at $z = 0,\ L$ the boundary conditions for the vertical velocity amplitude yield

$$\hat{u}_z = 0, \quad \frac{\mathrm{d}^{(2m)}\hat{u}_z}{\mathrm{d}z^{(2m)}} = 0, \tag{2.110}$$

for all natural numbers $m \in \mathbb{N}$, where the Eqs. (2.104a), (2.104b) with $\sigma = 0$ (at threshold) and $\hat{T}(z = 0,\ L) = 0$ were used. The solution of the problem (2.105) can be sought in the form of a superposition of modes of the type $\mathrm{Const}_1 \sin qz + \mathrm{Const}_2 \cos qz$, where $q$ corresponds to a set of constant coefficients determined by the boundary conditions. This allows to conclude, that the only possible solution in this case takes the form

$$\hat{u}_z(z) = A\sin\left(\frac{n\pi z}{L}\right), \tag{2.111}$$

where $n \in \mathbb{N}$ and $A$ is a constant amplitude, undetermined within the scope of linear theory. Substitution of the above form of solution into Eq. (2.105) gives $Ra = (n^2\pi^2 + \mathcal{K}^2L^2)^3/\mathcal{K}^2L^2$, which for a given $\mathcal{K}$ takes the minimal value at $n = 1$, thus

$$Ra = \frac{\left(\pi^2 + \mathcal{K}^2L^2\right)^3}{\mathcal{K}^2L^2}. \tag{2.112}$$

This already allows to draw a conclusion about the symmetry of the flow at convection threshold; since $n = 1$ at threshold, in the case of two stress-free boundaries the marginal solutions possess the *linear up-down symmetry* defined in (2.100a),

(2.100b), which corresponds to large-scale rolls. Minimization of the expression (2.112) over all possible values of $\mathcal{K}$ leads to the critical Rayleigh number for development of convective instability in the current case,

$$Ra_{crit} = \frac{27}{4}\pi^4 \approx 657.5, \quad \text{achieved at} \quad \mathcal{K}_{crit} = \frac{\pi}{\sqrt{2}L}. \tag{2.113}$$

The ratio of the horizontal to vertical thickness of a single convective roll, as depicted on Fig. 2.1a, is $\pi/\mathcal{K}_{crit}L = \sqrt{2}$, thus the rolls are slightly flattened. The result for $Ra_{crit}$ can be, of course, obtained from (2.107) and (2.111) in a straightforward way.

### 2.2.2 Two Isothermal Rigid Boundaries, $Q = 0$

Next we consider the case of two rigid boundaries with no-slip conditions at $z = 0,\ L$

$$\hat{u}_z = 0, \qquad \frac{d\hat{u}_z}{dz} = 0. \tag{2.114}$$

Because the boundaries are assumed isothermal the Eq. (2.104a) implies additionally

$$\left(\frac{d^2}{dz^2} - \mathcal{K}^2\right)^2 \hat{u}_z = 0, \tag{2.115}$$

at $z = 0,\ L$. On introducing the general form of the solution $\text{Const}_1 \sin qz + \text{Const}_2 \cos qz$ into the Eq. (2.105) it can be easily seen, that the boundary conditions (2.114) and (2.115) imply that there are two distinct classes of solutions, one satisfying the "linear" up-down symmetry (2.100a), (2.100b) and second one obeying the "developed" symmetry (2.98a), (2.98b). As demonstrated by Chandrasekhar (1961) the marginal state corresponds to the large-scale flow, with $\hat{u}_z(z)$ and $\hat{T}(z)$ even with respect to the mid-plane[10]; the $z$-dependent amplitude of the vertical velocity $\hat{u}_z$ at threshold takes the form

$$\begin{aligned}\hat{u}_z(z) =& A\cos\left(q_0\frac{2z-L}{2L}\right) - A\,0.0615\cosh\left(q_1\frac{2z-L}{2L}\right)\cos\left(q_2\frac{2z-L}{2L}\right)\\ &+ A\,0.1039\sinh\left(q_1\frac{2z-L}{2L}\right)\sin\left(q_2\frac{2z-L}{2L}\right),\end{aligned} \tag{2.116}$$

[10]This is clear from a general approach based on expansion of $\hat{u}_z(z)$ and $\hat{T}(z)$ in eigenfunctions of the problem (2.104a), (2.104b) with boundary conditions either of Dirichlet or Neumann type, which can be easily shown to form a complete set. It was demonstrated in Chandrasekhar (1961), Chap. 2, Sect. 13(a), pp. 27–31, that the eigen mode of the lowest order, that is with the largest possible wavelength in the vertical direction corresponds to the lowest $Ra$-the eigenvalue of the problem. When the top and bottom boundaries are symmetric, this mode must be symmetric with respect to the mid-plane.

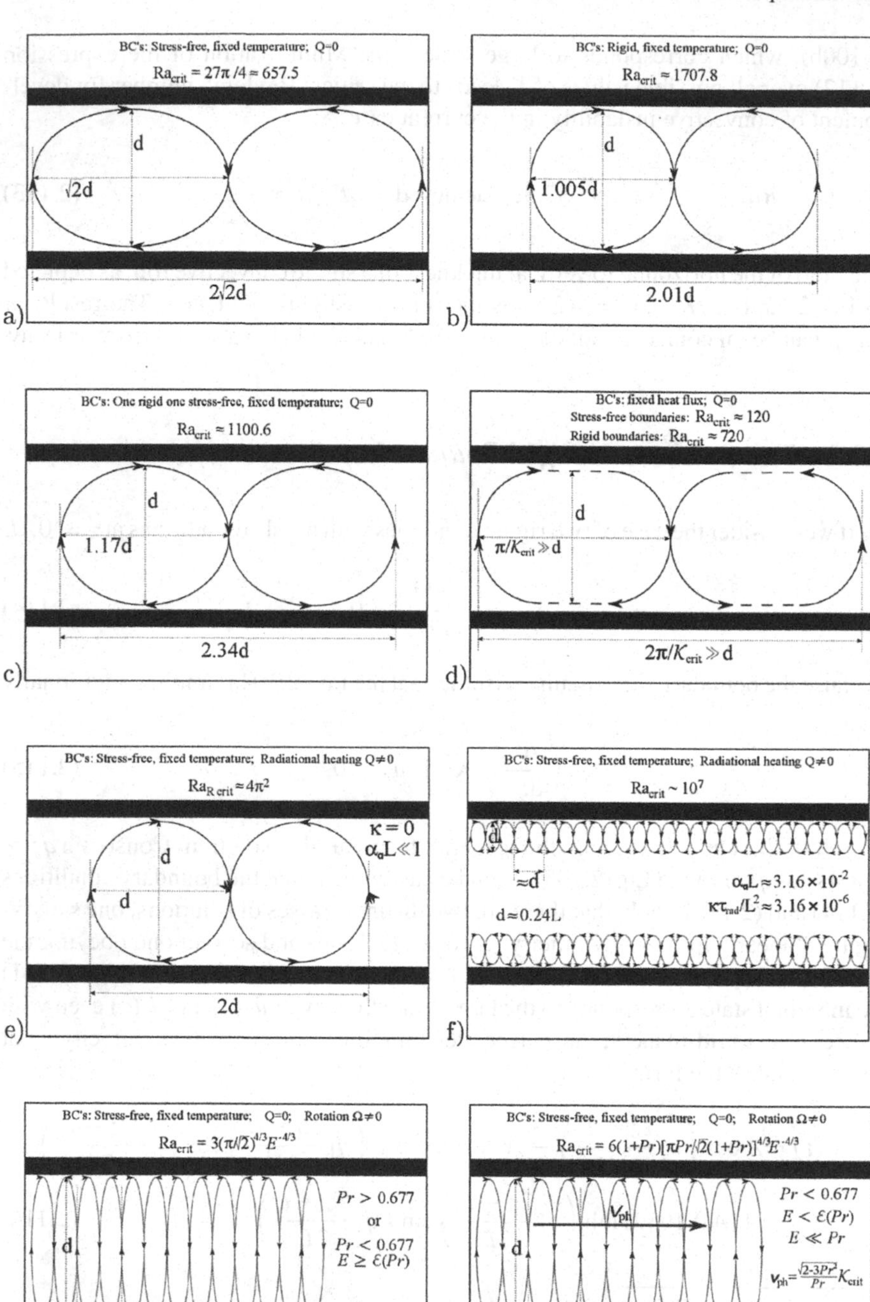
a)
BC's: Stress-free, fixed temperature; Q=0
$Ra_{crit} = 27\pi^4/4 \approx 657.5$
d
$\sqrt{2}d$
$2\sqrt{2}d$
b)
BC's: Rigid, fixed temperature; Q=0
$Ra_{crit} \approx 1707.8$
d
1.005d
2.01d
c)
BC's: One rigid one stress-free, fixed temperature; Q=0
$Ra_{crit} \approx 1100.6$
d
1.17d
2.34d
d)
BC's: fixed heat flux; Q=0
Stress-free boundaries: $Ra_{crit} \approx 120$
Rigid boundaries: $Ra_{crit} \approx 720$
d
$\pi/K_{crit} \gg d$
$2\pi/K_{crit} \gg d$
e)
BC's: Stress-free, fixed temperature; Radiational heating $Q \neq 0$
$Ra_{R\,crit} \approx 4\pi^2$
$\kappa = 0$
$\alpha_a L \ll 1$
d
d
2d
f)
BC's: Stress-free, fixed temperature; Radiational heating $Q \neq 0$
$Ra_{crit} \sim 10^7$
d
$\approx d$
$d \approx 0.24L$
$\alpha_a L \approx 3.16 \times 10^{-2}$
$\kappa\tau_{rad}/L^2 \approx 3.16 \times 10^{-6}$
g)
BC's: Stress-free, fixed temperature; Q=0; Rotation $\Omega \neq 0$
$Ra_{crit} = 3(\pi/\sqrt{2})^{4/3}E^{-4/3}$
$Pr > 0.677$
or
$Pr < 0.677$
$E \geq \mathcal{E}(Pr)$
d
$(8\sqrt{2}\pi^2 E)^{1/3}d$
h)
BC's: Stress-free, fixed temperature; Q=0; Rotation $\Omega \neq 0$
$Ra_{crit} = 6(1+Pr)[\pi Pr/\sqrt{2}(1+Pr)]^{4/3}E^{-4/3}$
$V_{ph}$
$Pr < 0.677$
$E < \mathcal{E}(Pr)$
$E \ll Pr$
$v_{ph} = \frac{\sqrt{2-3Pr^2}}{Pr}K_{crit}$
d
$[8\sqrt{2}\pi^2(1+Pr)E/Pr]^{1/3}d$

◀ **Fig. 2.1** Schematic picture of convection rolls in the marginal state for Boussinesq convection in 8 different physical situations: **a** stress-free, isothermal boundaries, $Q = 0$; **b** rigid, isothermal boundaries, $Q = 0$; **c** one rigid one stress-free boundary, both isothermal, $Q = 0$; **d** heat flux held fixed at the boundaries (either rigid or stress-free), $Q = 0$; the rolls are strongly elongated in this case; **e** and **f** demonstrate the possible effects of radiational heating $Q \neq 0$ for $\kappa = 0$ and $\kappa \neq 0$ when the boundaries are isothermal and stress-free; **g** and **h** demonstrate the effect of background rotation for isothermal and stress-free boundaries and $Q = 0$ for large and small Prandtl numbers, $Pr = \nu/\kappa$. The vertical cross-sections along the plane determined by the $z$-axis and the horizontal wave vector $\mathcal{K}$ are shown

where

$$q_0 \approx 3.9736, \quad q_1 \approx 5.1952, \quad q_2 \approx 2.1261, \tag{2.117}$$

and $A$ is an undetermined constant. The critical Rayleigh number and the wave number of the marginal mode in the case of two rigid and isothermal boundaries are

$$Ra_{crit} \approx 1707.76, \qquad \mathcal{K}_{crit} = \frac{3.12}{L}, \tag{2.118}$$

and hence the horizontal to vertical single roll thicknesses ratio is $\pi/\mathcal{K}_{crit}L \approx 1.01$, i.e. the rolls are very close to circular (cf. Fig. 2.1b). In this case the heat per unit mass released by a fluid parcel rising on an infinitesimal distance d$z$ in a time unit in the marginal state (which is equal to the maximal heat per unit mass, per unit time that can be accumulated between fluid layers before convection starts), that is $\bar{c}_p\kappa\nu Ra_{crit}u_z/g\bar{\alpha}L^4$, is higher than in the previous case. This is because the viscous stresses on the rigid walls do not allow the simple-type solutions $\hat{u}_z \sim \sin(n\pi z)$ to develop and the allowed solutions have a more complex structure (2.116), which turns out to require higher superadiabatic temperature gradients.

### 2.2.3 One Rigid and One Stress-Free Boundary, Both Isothermal, $Q = 0$

As remarked above, in the case of two rigid and isothermal boundaries there is a class of solutions with odd symmetry of $\hat{u}_z(z)$ and $\hat{T}(z)$ with respect to the mid-plane. These solutions necessarily satisfy the stress-free conditions $\hat{u}_z = 0$, $\mathrm{d}_z^2\hat{u}_z = 0$ and $(\mathrm{d}_z^2 - \mathcal{K}^2)^2\hat{u}_z = 0$ at $z = L/2$, thus the solution for the current case of one rigid and one stress-free boundary, both isothermal can be retrieved from the odd solution for the previous case, provided by Chandrasekhar (1961). The critical parameter values for convection threshold do not depend on whether the top boundary is assumed stress free and the bottom one rigid or the other way around. In the former case the solution takes the form

$$\hat{u}_z(z) = A\sin\left(q_0\frac{z-L}{2L}\right) - A\,0.0171\sinh\left(q_1\frac{z-L}{2L}\right)\cos\left(q_2\frac{z-L}{2L}\right) + A\,0.00346\cosh\left(q_1\frac{z-L}{2L}\right)\sin\left(q_2\frac{z-L}{2L}\right), \tag{2.119}$$

where

$$q_0 \approx 7.1379, \quad q_1 \approx 9.1108, \quad q_2 \approx 3.7893. \tag{2.120}$$

and $A$ is an undetermined constant. The critical Rayleigh number and the wave number of the marginal mode are in this case

$$Ra_{crit} \approx 1100.65, \qquad \mathcal{K}_{crit} = \frac{2.68}{L}, \tag{2.121}$$

and hence the horizontal to vertical single roll thicknesses ratio is $\pi/\mathcal{K}_{crit}L \approx 1.17$, i.e. the rolls are only slightly flattened (cf. Fig. 2.1c). In terms of the value of the critical Rayleigh number, it is perhaps no surprise, that this case lies in between the two previous cases.

There are, however, other types of boundary conditions and important physical effects, that have been widely considered due to relevance to particular physical settings, which have a profound effect on the stability characteristics and structure of the flow at convection threshold. E.g. instead of fixing the temperature at the top and bottom boundary the heat flux can be fixed, which is a condition well corresponding to situations found in astrophysics, where the heat flux at interfaces between layers in planetary and stellar interiors can be reasonably considered fixed for long periods of time. Another important effect in planetary mantles and stellar interiors is the radiative heating. We will first show the influence of the boundaries with fixed heat flux on the onset of convection.

### 2.2.4 Two Stress-Free, Fixed-Thermal-Flux Boundaries, $Q = 0$

The case when top and bottom boundaries are stress-free and the heat flux on both of them is fixed, that is

$$\left.\frac{\partial T'}{\partial z}\right|_{z=0,L} = 0, \tag{2.122}$$

has been thoroughly investigated and the linear analysis can be found e.g. in Jakeman (1968) and Park and Sirovich (1991). The convection threshold in this case is achieved at

$$Ra_{crit} = 120, \qquad \mathcal{K}_{crit} = 0, \tag{2.123}$$

with the trivial solution

$$\hat{u}_z = 0, \quad \text{and} \quad \hat{T} = \text{const}. \tag{2.124}$$

This means, that when the Rayleigh number only slightly exceeds the critical value, convection develops in the form of very wide (strongly flattened) large-scale roll; since the imposed boundary conditions are symmetric with respect to $z = L/2$, $\hat{u}_z(z)$ and $\hat{T}(z)$ just above the threshold possess the even symmetry with respect to the mid-plane (linear up-down symmetry), as depicted on Fig. 2.1d. The condition of fixed thermal flux physically corresponds to a situation, when the thermal conductivity of the bounding solid walls at $z = 0$, $L$ is much smaller than that of the fluid (as opposed to the isothermal walls, which can be physically realized when their thermal conductivity is much larger than the fluid's, to allow for quick temperature relaxation). The pathological structure of the critical solution can be interpreted in the context of physically realizable situations, when the thermal conductivity of the solid boundaries is always finite, which again, leads to large-scale rolls of very long wavelength at the threshold and/or in its vicinity.

Despite the triviality of the marginal solution (2.124), the situation in this case does not differ very significantly from the previous cases in terms of general qualitative description of fluid's behaviour, since the amplitude of convection is known to scale with a positive power of departure from convection threshold, $Ra - Ra_{crit}$ (see next Sect. 2.3 for weakly nonlinear estimates), thus exactly at threshold there are no motions in all the considered cases.

Furthermore, when the superadiabatic gradient is slightly above the threshold value, which in the current case implies $\mathcal{K} \ll 1$, the heat per unit mass, per unit time released by a rising fluid parcel on an infinitesimal vertical distance, which is equal to $\bar{c}_p \kappa \nu Ra_{crit} u_z / g\bar{\alpha} L^4$, is significantly smaller than in the cases with isothermal boundaries, (and at the same time also smaller than the heat per unit mass per unit time accumulated between two horizontal planes at distance $u_z \mathrm{d}t$, therefore convective flow is triggered). This means that when the boundaries are held at fixed thermal flux, to drive a convection it is necessary to accumulate much less heat between horizontal fluid layers by rising the superadiabatic gradient, than in the cases when the boundaries are held at fixed temperatures. The reason for this, is that in the case at hand the heat excess coming from perturbations cannot leave the system through boundaries thus it is easier to achieve the critical heat balance (1.62) at convection threshold as opposed to the previous cases when the heat could freely leave through isothermal boundaries.

### 2.2.5 *Two Rigid, Fixed-Thermal-Flux Boundaries, $Q = 0$*

The next case is again symmetric about the mid-plane with the heat flux fixed at both top and bottom boundaries, but this time both these boundaries are rigid (boundary conditions (2.122) and (2.114)). This problem has been addressed e.g. in Sparrow et al. (1964), Jakeman (1968), Gershuni and Zhukhovitskii (1976) and Cerisier et al. (1998), who showed, that for such a choice of boundary conditions the critical

Rayleigh number and the associated wave number for convection are

$$Ra_{crit} \approx 720, \qquad \mathcal{K}_{crit} = 0, \tag{2.125}$$

with the trivial solution

$$\hat{u}_z = 0, \quad \text{and} \quad \hat{T} = \text{const}, \tag{2.126}$$

thus, again, slightly above the threshold the temperature gradients are very small, $\hat{u}_z(z)$ and $\hat{T}(z)$ are symmetric with respect to the mid-plane (linear up-down symmetry) and the flow is organized in a pattern with large horizontal wavelength (cf. Fig. 2.1d). As expected, the rigid boundaries make the critical balance (1.62) for convection threshold harder to achieve. Still, at the same time $Ra_{crit}$ in the current case (2.125) is much smaller than in the case when both boundaries are isothermal and rigid (2.118), indicating the influence of fixed-thermal-flux boundaries, which keep the heat excess resulting from fluctuations in the system.

We now proceed to study the influence of two different physical effects, very common in nature, such as radiative heating and background rotation on the threshold of convection. We start with the effect of radiation.

### 2.2.6 The Effect of Radiative Heating, $Q \neq 0$, at Two Stress-Free, Isothermal Boundaries

We start by introducing the model of thermal radiation following Goody (1956), Goody and Yung (1989) and Goody (1995).[11] Denoting the radiative energy flux by $\mathbf{j}_{rad}$ we can express the heat per unit volume delivered to the system in a time unit by thermal radiation as

$$Q = -\nabla \cdot \mathbf{j}_{rad}, \tag{2.127}$$

with the following equation governing $\mathbf{j}_{rad}$

$$\nabla \frac{1}{\alpha_a} \nabla \cdot \mathbf{j}_{rad} - 3\alpha_a \mathbf{j}_{rad} = 4\sigma_{rad} \nabla \left(T^4\right), \tag{2.128}$$

where $\alpha_a$ is the coefficient of absorption of radiation per unit volume and $\sigma_{rad}$ is the Stefan-Boltzmann constant. The absorption coefficient, just as $\mu$, $\kappa$ and $\bar{c}_p$, will be assumed uniform in what follows. Let us first linearise the thermal source term on the right hand side of (2.128), taking into account, that under the Boussinesq approximation $T = \bar{T} + \bar{\bar{T}} + T'$ and $(\bar{\bar{T}} + T')/\bar{T} = \mathcal{O}(\epsilon) \ll 1$ according to (2.28b), which yields

[11] The details of the linear stability analysis for convective systems with radiative heating can be found in Goody (1956), Getling (1998), Goody and Yung (1989) and Goody (1995), with perhaps most comprehensive treatment by Getling (1980) and Larson (2001).

$$\nabla T^4 = 4\bar{T}^3 \nabla\left(\tilde{\tilde{T}} + T'\right) + \mathcal{O}\left(\epsilon \frac{\bar{T}^4}{L}\right), \tag{2.129}$$

and hence the Eq. (2.128) simplifies to

$$\nabla\left(\nabla \cdot \mathbf{j}_{rad}\right) - 3\alpha_a^2 \mathbf{j}_{rad} = 16\alpha_a \sigma_{rad} \bar{T}^3 \nabla\left(\tilde{\tilde{T}} + T'\right). \tag{2.130}$$

For the basic, hydrostatic state dependent on $z$ alone, $\tilde{T}(z) = \bar{T} + \tilde{\tilde{T}}(z)$ and $\tilde{\tilde{\mathbf{j}}}_{rad}(z)$ (note that $\bar{\mathbf{j}}_{rad} = 0$), the latter equation and the energy equation (2.51c) take the form

$$\kappa \frac{\mathrm{d}^2 \tilde{\tilde{T}}}{\mathrm{d}z^2} - \frac{1}{\bar{\rho}\bar{c}_p} \frac{\mathrm{d}\tilde{\tilde{j}}_{rad\,z}}{\mathrm{d}z} = 0, \tag{2.131a}$$

$$\frac{\mathrm{d}^2 \tilde{\tilde{j}}_{rad\,z}}{\mathrm{d}z^2} - 3\alpha_a^2 \tilde{\tilde{j}}_{rad\,z} = 16\alpha_a \sigma_{rad} \bar{T}^3 \frac{\mathrm{d}\tilde{\tilde{T}}}{\mathrm{d}z}, \tag{2.131b}$$

$$\tilde{\tilde{j}}_{rad\,x} = \tilde{\tilde{j}}_{rad\,y} = 0. \tag{2.131c}$$

Furthermore, let us consider the limit of a transparent fluid, when the absorption coefficient is very small, $\alpha_a \ll L^{-1}$. In such a case the term quadratic in $\alpha_a^2$ can be neglected in the Eq. (2.130). This allows for a simplification of the equation for the fluctuation of the radiative flux $\mathbf{j}'_{rad}$, by integration of the equation once with respect to $\mathbf{x}$,

$$\nabla \cdot \mathbf{j}'_{rad} \approx 16\alpha_a \sigma_{rad} \bar{T}^3 T', \tag{2.132}$$

where the constant of integration is zero, since when the temperature perturbation vanishes $T' = 0$ the perturbation to the radiative heat transfer must vanish as well. Hence the linearised thermal energy equation can now be written solely in terms of the temperature fluctuation

$$\frac{\partial T'}{\partial t} - u_z \Delta_S = \kappa \nabla^2 T' - 16\alpha_a \sigma_{rad} \bar{T}^3 T'. \tag{2.133}$$

This equation together with (2.99a), (2.99b) are subject to boundary conditions, which will now be specified. We apply the Fourier decomposition as in (2.103), thus the stress-free, isothermal boundaries imply for the fluctuations

$$\hat{u}_z(z = 0,\ L) = 0, \quad \left.\frac{\partial^2 \hat{u}_z}{\partial z^2}\right|_{z=0,\ L} = 0, \quad \hat{T}(z = 0,\ L) = 0. \tag{2.134}$$

To specify the superadiabatic gradient $\Delta_S$ the basic state solution of (2.131a)–(2.131c) is necessary. The boundary conditions on radiative flux have been derived by Goody (1956) and thus in the case when both the top and bottom boundaries are black bodies the hydrostatic state must satisfy

$$\left(\frac{\mathrm{d}\tilde{\tilde{j}}_{rad\,z}}{\mathrm{d}z} - 2\alpha_a \tilde{\tilde{j}}_{rad\,z}\right)\Bigg|_{z=0} = 0, \quad \left(\frac{\mathrm{d}\tilde{\tilde{j}}_{rad\,z}}{\mathrm{d}z} + 2\alpha_a \tilde{\tilde{j}}_{rad\,z}\right)\Bigg|_{z=L} = 0, \tag{2.135a}$$

$$\tilde{\tilde{T}}(z=0) = \tilde{\tilde{T}}_B, \quad \tilde{\tilde{T}}(z=L) = \tilde{\tilde{T}}_T, \tag{2.135b}$$

where according to our definition $\tilde{T}(z) = \bar{T} + \tilde{\tilde{T}}(z)$ we have $\tilde{\tilde{T}}_B = \Delta T/2$ and $\tilde{\tilde{T}}_T = -\Delta T/2$. The solution for the hydrostatic balance (2.131a)–(2.131c) with the boundary conditions (2.135a), (2.135b) is provided e.g. in Larson (2001) in his Eq. (31), but note the shift of the reference system origin to the mid-plane; since the general formulae are quite cumbersome, they will not be repeated here. Particularly interesting is the limit of radiatively transparent fluid (i.e. $L^{-1} \gg \alpha_a$ ) and small thermal diffusivity, defined by

$$L^{-1} \gg \alpha_a \gg (2\pi)^2 \frac{\kappa \tau_{rad}}{3L^3}, \tag{2.136}$$

where

$$\tau_{rad} = 3\bar{\rho}\bar{c}_p L / 16\sigma_{rad}\bar{T}^3 \tag{2.137}$$

is the radiative cooling time-scale. In this limit the basic temperature profile adopts a form with strong gradients near the boundaries, indicating possible formation of boundary layers in the convective flow. The system then exhibits two distinct regimes for convective threshold: regime (I) with typical large-scale convective rolls, and regime (II) with small scale convective rolls located near boundaries. The former is achieved at small and extremely small values of $\kappa\tau_{rad}/L^2$ (at a given $\tau_{rad}$ and $\alpha_a L \ll 1$), say for $\kappa_2 < \kappa \ll L^2/\tau_{rad}$ or $\kappa < \kappa_1 \ll L^2/\tau_{rad}$, where $\kappa_1$ and $\kappa_2$ are dependent on system parameters. The regime (II) is achieved for intermediate values of $\kappa\tau_{rad}/L^2$ (at a given $\tau_{rad}$ and $\alpha_a L \ll 1$), in-between the limiting values for the regime (I), i.e. for $\kappa_1 < \kappa < \kappa_2$. Both regimes satisfy the linear up-down symmetry (2.100a), (2.100b). The dependence of the radiative Rayleigh number at threshold on the wavenumber of perturbations, $Ra_R(\mathcal{K})$, can be obtained from a numerical solution of the linear eigenvalue problem at threshold, cf. Getling (1998), (Chap. 7.3.2) and Larson (2001). Under the conditions corresponding to regime (II) the curve $Ra_R(\mathcal{K})$ possesses two minima. The first minimum, around $\mathcal{K}L = \pi$, corresponds to the large-scale convective rolls extending over the entire depth of the fluid layer, whereas the second one, with larger $\mathcal{K}L$, to the small-scale rolls near the boundaries, as depicted on Fig. 2.1e and f (in fact the figure e depicts the situation for $\kappa = 0$, which qualitatively resembles that of regime (I); figure f is based on parameter values calculated in Chap. 7.3.2 of Getling (1998)). The position of the second minimum, that is the critical wave number for the regime (II) increases with the value of the parameter $\alpha_a L^3/\kappa\tau_{rad}$, so that when the latter parameter increases, the convection rolls in the regime (II) are horizontally thinning. Depending on the system parameter values either one of the minima can become a global minimum corresponding to the critical Rayleigh number for convection.

When $\kappa$ is finite the $Ra_R(\mathcal{K})$ dependence, where $Ra_R$ denotes the Rayleigh number in the presence of radiative heating, possesses only one minimum around $\mathcal{K}L = \pi$ and the system also forms large-scale rolls of similar type as in the regime (I). Furthermore, when the thermal diffusivity approaches exactly zero, $\kappa = 0$, in the limit of a transparent fluid $\alpha_a L \ll 1$ the linear problem becomes fully analytically tractable with the results at convection threshold summarized as follows

$$\hat{T}(z) \sim \hat{u}_z(z) \sim \sin\left(\pi \frac{z}{L}\right), \tag{2.138}$$

and

$$Ra_{R\,crit} = 4\pi^2, \qquad \mathcal{K}_{crit} L = \pi, \tag{2.139}$$

where the radiative Rayleigh number is defined in the following way

$$Ra_R = \frac{g\bar{\alpha}\Delta_S L^4}{3(\alpha_a L)(L^2/\tau_{rad})\nu}, \tag{2.140}$$

$$\Delta_S = \frac{3\alpha_a L}{4 + 3\alpha_a L}\frac{\Delta T}{L} - \frac{g}{\bar{c}_p}, \qquad \tau_{rad} = \frac{3\bar{\rho}\bar{c}_p L}{16\sigma_{rad}\bar{T}^3}. \tag{2.141}$$

The parameter $\Delta_S$ denotes the interior superadiabatic temperature gradient, since in the limit $\kappa = 0$, when no thermal conduction is possible the temperature of the fluid adjacent to the boundaries, say $T_b$ (bottom) and $T_t$ (top), differs from the temperatures of the boundaries denoted as usually by $T_B$ and $T_T$, so that there is a discontinuity of temperature at the boundaries $T_B > T_b$ and $T_T < T_t$. This is simply a manifestation of a radiative thermal boundary layer, which forms in a realistic situation of small, but finite diffusion, $\kappa \ll L^2/\tau_{rad}$, due to joint action of radiation and thermal diffusivity, where the temperature decays exponentially to the value at a boundary; such a boundary layer is shrank to a discontinuity at a boundary in the limit $\kappa = 0$.

Furthermore, the $\tau_{rad}$ depends explicitly on the mean temperature $\bar{T}$, which in turn implies $Ra_R \sim \bar{T}^{-3}$, thus a radiative medium can be thermally stabilized by an increase of the mean temperature of the system. This is in contrast to all the previous, non-radiative cases considered, when only the basic temperature gradient played a role in triggering the convective instability and the value of the mean temperature had no effect on stability.

Also in this case we can estimate the heat per unit mass released by a rising fluid parcel on an infinitesimal vertical distance in a time unit in the marginal state, which for $\kappa = 0$ is equal to $\bar{c}_p 3(\alpha_a L)(L^2/\tau_{rad})\nu Ra_{R\,crit} u_z / g\bar{\alpha}L^4$. It includes the effects of thermal radiation and absorption, but note that absorption by assumption is weak, $\alpha_a \ll L^{-1}$. Comparison with previous non-radiative cases can be made when the thermal diffusivity of a non-radiative system $\kappa$ is assumed comparable with the "radiative diffusivity" parameter $3(\alpha_a L)(L^2/\tau_{rad})$ of a radiative, but thermally insulating ($\kappa = 0$) system. Then the maximal heat per unit mass which can accumulate between infinitesimally distant fluid layers before convection starts,

$\bar{c}_p 3(\alpha_a L)(L^2/\tau_{rad})\nu Ra_{R\,crit} \mathrm{d}z/g\bar{\alpha}L^4$, is significantly smaller than in all the previous, non-radiative cases ($Ra_{R\,crit}$ is only about 40), which indicates that thermal radiation has a destabilizing effect in this sense (in terms of the interior temperature gradient, since in fact the temperature difference between the top and bottom plates, i.e. $\Delta T = T_B - T_T$ may be larger at threshold in the radiative case, since $\alpha_a L \ll 1$).

When both, the thermal diffusivity $\kappa \neq 0$ and the radiative effects are included the heat per unit mass released by a rising fluid parcel on an infinitesimal vertical distance in a time unit in the marginal state can be most conveniently expressed by $\bar{c}_p u_z \Delta_{S\,crit}(z)$, since in such a case the basic state temperature gradient depends on depth.

### 2.2.7 *Constraints from Rapid Rotation at Two Stress-Free, Isothermal Boundaries, $Q = 0$*

We will now briefly comment on the effect of the Coriolis force, which is a common body force in many natural convective systems such as the planetary and stellar atmospheres and cores. When the system rotates at a uniform rate about vertical axis $\mathbf{\Omega} = \Omega\hat{\mathbf{e}}_z$ (perpendicular to the planes of the parallel boundaries) the linearised Navier-Stokes equation under the Boussinesq approximation has to include the "non-inertiality" of the system of reference, namely the Coriolis force

$$\frac{\partial \mathbf{u}}{\partial t} = -\nabla\frac{p'}{\bar{\rho}} + g\bar{\alpha}T'\hat{\mathbf{e}}_z - 2\Omega\hat{\mathbf{e}}_z \times \mathbf{u} + \nu\nabla^2\mathbf{u}; \tag{2.142}$$

the centrifugal force $-\bar{\rho}\mathbf{\Omega} \times (\mathbf{\Omega} \times \mathbf{x})$, which is potential, only modifies the hydrostatic pressure distribution $\mathrm{d}_z\bar{\tilde{p}} = -(\bar{\rho} + \tilde{\bar{\rho}})g + \bar{\rho}(\mathbf{\Omega} \times \mathbf{x})^2/2$, where (2.9b) and (2.12a) still hold. Thorough analysis of the linear regime at convection threshold for rotating systems together with the derivation of the form of the dynamical equations in rotating systems can be found in §III of Chandrasekhar (1961). One can easily see, that by taking a curl of the Eq. (2.142) and letting $\nu = 0$ one obtains $\partial_z u_z = 0$, thus for impermeable boundaries, $u_z(z = 0,\ L) = 0$, rapid rotation leading to domination of the Coriolis force over the viscous friction inhibits convection. More precisely, when the so-called Ekman number

$$E = \nu/2\Omega L^2 \tag{2.143}$$

is small $E \ll 1$, the rotational effects tend to suppress the vertical convective motions.[12] When the background rotation is present the system can no longer be sufficiently described by the vertical velocity component and the temperature only, since the horizontal components of velocity can not be separated anymore. Introducing the vorticity $\boldsymbol{\zeta} = \nabla \times \mathbf{u}$ the linearised equations for the $z$-dependent amplitudes

[12]This is a manifestation of the well-known Taylor-Proundman theorem.

defined as in (2.102) take the form

$$\left[\sigma\left(\frac{\mathrm{d}^2}{\mathrm{d}z^2}-\mathcal{K}^2\right)-\nu\left(\frac{\mathrm{d}^2}{\mathrm{d}z^2}-\mathcal{K}^2\right)^2\right]\hat{u}_z+2\Omega\frac{\mathrm{d}\hat{\zeta}_z}{\mathrm{d}z}=-g\bar{\alpha}\mathcal{K}^2\hat{T}, \tag{2.144a}$$

$$\left[\sigma-\nu\left(\frac{\mathrm{d}^2}{\mathrm{d}z^2}-\mathcal{K}^2\right)\right]\hat{\zeta}_z=2\Omega\frac{\mathrm{d}\hat{u}_z}{\mathrm{d}z}, \tag{2.144b}$$

$$\left[\sigma-\kappa\left(\frac{\mathrm{d}^2}{\mathrm{d}z^2}-\mathcal{K}^2\right)\right]\hat{T}=\Delta_S\hat{u}_z. \tag{2.144c}$$

Two cases can be distinguished. In the first one the instability sets in as stationary convection with $\Im\mathfrak{m}\sigma = 0$ and $\sigma$ passing through zero (thus the principle of exchange of stabilities holds). This is always the case, when the Prandtl number $Pr = \nu/\kappa$ satisfies $Pr > \mathscr{P}r \approx 0.677$; for $Pr < \mathscr{P}r$ there always exists a finite value of the Ekman number $E = \mathscr{E}(Pr)$, such that for $E \geq \mathcal{E}(Pr)$ convection still sets in as a stationary flow. The second case is obtained for $E < \mathcal{E}(Pr)$, when the flow is oscillatory at the onset with $\Im\mathfrak{m}\sigma \neq 0$ and $\Re\mathfrak{e}\sigma$ passing through zero. In the first case ($Pr > \mathscr{P}r$) or ($Pr < \mathscr{P}r$ and $E \geq \mathcal{E}(Pr)$), when the marginal flow is stationary

$$u_z = \Re\mathfrak{e}\ A\sin\left(\pi\frac{z}{L}\right)\mathrm{e}^{\mathrm{i}(\mathcal{K}_x x+\mathcal{K}_y y)}, \tag{2.145}$$

where $A$ is an undetermined constant and in the limit of rapid rotation $E \ll 1$ one obtains

$$Ra_{crit} \approx 3\left(\frac{\pi}{\sqrt{2}E}\right)^{4/3}, \qquad \mathcal{K}_{crit}L \approx \left(\frac{\pi}{\sqrt{2}E}\right)^{1/3}. \tag{2.146}$$

In the second case ($Pr < \mathscr{P}r$ and $E < \mathcal{E}(Pr)$) the purely oscillatory marginal state[13] in the rapidly rotating limit at finite $Pr$, that is when $EPr^{-1} \ll 1$, is described by

$$u_z = \Re\mathfrak{e}\ A\sin\left(\pi\frac{z}{L}\right)\mathrm{e}^{\mathrm{i}\omega t}\mathrm{e}^{\mathrm{i}(\mathcal{K}_x x+\mathcal{K}_y y)}, \tag{2.147}$$

where $A$ is an undetermined constant, and

$$\omega = \frac{\sqrt{2-3Pr^2}}{Pr}\left(\frac{Pr}{1+Pr}\frac{\pi}{\sqrt{2}E}\right)^{2/3}, \tag{2.148}$$

[13]It should be noted, however, that in all the cases studied above exactly at threshold the amplitude of the marginal states vanishes, and the flow in the form of linear solutions is observed only slightly above the threshold value of the temperature gradient, cf. the next Sect. 2.3 on weakly nonlinear analysis.

$$Ra_{crit} = 6\,(1+Pr)\left(\frac{Pr}{1+Pr}\frac{\pi}{\sqrt{2}E}\right)^{4/3}, \qquad \mathcal{K}_{crit}L = \left(\frac{Pr}{1+Pr}\frac{\pi}{\sqrt{2}E}\right)^{1/3}. \tag{2.149}$$

In both cases the state satisfies the linear up-down symmetry (2.100a), (2.100b). Moreover, the results for both regimes (2.146) and (2.149) indicate the stabilizing role of rotation, since $Ra_{crit} \to \infty$ as $E \to 0$, thus strong rotation inhibits vertical convective heat transfer. Because in the limit $E \ll 1$ the threshold value of the temperature gradient is very high, as indicated by $Ra_{crit} \gg 1$ the heat per unit mass released by a rising fluid parcel on an infinitesimal vertical distance in a time unit in the marginal state, $\bar{c}_p \kappa \nu Ra_{R\,crit} u_z / g\bar{\alpha} L^4$, is extremely high. In other words the rapid background rotation allows for very high values of molecular heat flux before the vertically perturbed fluid parcels become buoyant. The form of the convection rolls in marginal rapidly rotating convection is schematically depicted on Fig. 2.1g and h.

### 2.2.8 Summary

We can quickly summarize the effects of different types of boundaries, radiation and background rotation as follows. The rigid walls stabilize the system by reducing its ability to develop flow near the boundaries. Thermally insulating walls exert a destabilizing effect by keeping the perturbation heat flux in the system, thus making it easier for a vertically perturbed fluid parcel to become buoyant. Furthermore, the effect of thermal radiation is, in general, complex, but for a radiatively transparent fluid, $\alpha_a \ll L^{-1}$, in the limit of weak thermal diffusion, $\kappa \ll L^2/\tau_{rad}$, it can be concluded, that radiation convectively destabilizes the system in the sense, that smaller interior superadiabatic temperature gradients are sufficient for convection threshold; this however, may correspond to larger temperature difference $\Delta T = T_B - T_T$ between the top and bottom boundaries, since the temperature jumps in thin radiative boundary layers in the hydrostatic state are finite, but the interior temperature gradient is proportional to $\alpha_a L \ll 1$. Finally the effect of rapid rotation is to inhibit the vertical flow, thus is strongly stabilizing.

A fairly general relation, that may come useful for straightforward calculation of the critical Rayleigh number for convection once the vertical structure of the flow close to threshold is known, valid for $\kappa =$ const and isothermal, impermeable and either stress-free or no-slip boundaries can be obtained from Eqs. (2.144a)–(2.144b) and the temperature equation with heat sources

$$\left[\sigma - \kappa\left(\frac{\mathrm{d}^2}{\mathrm{d}z^2} - \mathcal{K}^2\right)\right]\hat{T} = \Delta_S \hat{u}_z + \frac{\hat{Q}}{\bar{\rho}\bar{c}_p}, \tag{2.150}$$

in a similar manner to Chandrasekhar's (1961) Eq. (253), p. 125 in that book. In the above it has been assumed, that just as for all the perturbations the heat source perturbation satisfies $Q' = \hat{Q}(z)\exp\left[\mathrm{i}\left(\mathcal{K}\cdot\mathbf{x}_h + \omega t\right)\right]$. Introducing $\sigma = \mathrm{i}\omega$, $\omega \in \mathbb{R}$ at threshold and

$$F(z) = \left(\frac{\mathrm{d}^2}{\mathrm{d}z^2} - \mathcal{K}^2\right)\left(\frac{\mathrm{d}^2}{\mathrm{d}z^2} - \mathcal{K}^2 - \mathrm{i}\frac{\omega}{\nu}\right)\hat{u}_z - \frac{2\Omega}{\nu}\frac{\mathrm{d}\hat{\zeta}}{\mathrm{d}z} = \frac{g\bar{\alpha}\mathcal{K}^2}{\nu}\hat{T}, \tag{2.151}$$

the two Eqs. (2.144a) and (2.150) can be reduced to

$$\left(\frac{\mathrm{d}^2}{\mathrm{d}z^2} - \mathcal{K}^2 - \frac{\mathrm{i}\omega}{\kappa}\right)F = -\frac{g\bar{\alpha}\Delta_S}{\nu\kappa}\mathcal{K}^2\hat{u}_z - \frac{g\bar{\alpha}}{\nu\kappa}\mathcal{K}^2\frac{\hat{Q}}{\bar{\rho}\bar{c}_p}. \tag{2.152}$$

Multiplying the latter equation by $F$ and integrating with respect to the vertical variable $z$ from 0 to $L$, after integration by parts of the left hand side one obtains the Chandrasekhar's expression for the critical Rayleigh number modified here as to include the effect of heat sources (e.g. thermal radiation, radioactivity etc.),[14]

$$Ra = \frac{g\bar{\alpha}\left\langle\Delta_S(z)\right\rangle L^4}{\kappa\nu} = \frac{\int_0^L\left[\left(\frac{\mathrm{d}F}{\mathrm{d}z}\right)^2 + \left(\mathcal{K}^2 + \mathrm{i}\frac{\omega}{\kappa}\right)F^2\right]\mathrm{d}z}{\mathcal{K}^2\int_0^L\left(\frac{\Delta_S}{\langle\Delta_S\rangle}\frac{\hat{u}_z}{L^4} + \frac{\hat{Q}}{\bar{\rho}\bar{c}_p\langle\Delta_S\rangle L^4}\right)F\mathrm{d}z}. \tag{2.153}$$

With the use of the general expression on $\Delta_{S\,crit}$ (1.68) it is clear, that to obtain the actual value of the critical Rayleigh number at the convection threshold $Ra_{crit}$, the above result (2.153) has to be minimized over all possible values of the horizontal wave number $\mathcal{K}$, the inverse vertical variation scale of perturbations, say $q$, and the real frequency of oscillations $\omega(\mathcal{K}, q)$. The latter expression for the Rayleigh number corresponds exactly to the general definition (2.62) for $\kappa = \text{const}$. On the other hand, when additionally $Q = 0$ and $\boldsymbol{\Omega} = 0$ the expression for the Rayleigh number simplifies to

$$Ra = \frac{g\bar{\alpha}\Delta_S L^4}{\kappa\bar{\nu}} = \frac{L^4\int_0^L\left[\left(\frac{\mathrm{d}F}{\mathrm{d}z}\right)^2 + \mathcal{K}^2F^2\right]\mathrm{d}z}{\mathcal{K}^2\int_0^L\hat{u}_zF\mathrm{d}z} = \frac{L^4\int_0^L\left[-\hat{u}_z\left(\frac{\mathrm{d}^2}{\mathrm{d}z^2} - \mathcal{K}^2\right)^3\hat{u}_z\right]\mathrm{d}z}{\mathcal{K}^2\int_0^L\hat{u}_z^2\mathrm{d}z}. \tag{2.154}$$

At this stage, we recall here the Chandrasekhar's (1961) statement from §II, p. 34, below his equation (185) about the physical conditions for convective instability trigger in non-rotating, Boussinesq systems (for which by construction the thermal energy greatly exceeds the kinetic one):

---

[14] Note, that alternatively Eq. (2.152) could be multiplied simply by $\hat{u}_z$ and integrated over the range of $z$, which gives

$$Ra = \int_0^L\left[\hat{u}_z\left(\frac{\mathrm{i}\omega}{\kappa} + \mathcal{K}^2 - \frac{\mathrm{d}^2}{\mathrm{d}z^2}\right)F\right]\mathrm{d}z \Big/ \mathcal{K}^2\int_0^L\left(\frac{\Delta_S}{\langle\Delta_S\rangle}\frac{\hat{u}_z^2}{L^4} + \frac{\hat{Q}\hat{u}_z}{\bar{\rho}\bar{c}_p\langle\Delta_S\rangle L^4}\right)\mathrm{d}z$$

*Instability occurs at the minimum temperature gradient at which a balance can be steadily maintained between the kinetic energy dissipated by viscosity and the work done by the buoyancy force, which in general both include the effect of heating sources $Q$.*

Finally, it is of interest to provide a general relation, which can be used to calculate the growth rate of perturbations for the case when heat sources $Q$ and background rotation $\mathbf{\Omega}$ are present. By the use of Eqs. (2.144a), (2.144b) and (2.150) one obtains

$$\begin{aligned}\left(\frac{\mathrm{d}^2}{\mathrm{d}z^2}-\mathcal{K}^2\right)&\left[\sigma-\kappa\left(\frac{\mathrm{d}^2}{\mathrm{d}z^2}-\mathcal{K}^2\right)\right]\left[\sigma-\nu\left(\frac{\mathrm{d}^2}{\mathrm{d}z^2}-\mathcal{K}^2\right)\right]^2\hat{u}_z\\ &+4\Omega^2\frac{\mathrm{d}^2}{\mathrm{d}z^2}\left[\sigma-\kappa\left(\frac{\mathrm{d}^2}{\mathrm{d}z^2}-\mathcal{K}^2\right)\right]\hat{u}_z\\ &=-g\bar{\alpha}\mathcal{K}^2\left[\sigma-\nu\left(\frac{\mathrm{d}^2}{\mathrm{d}z^2}-\mathcal{K}^2\right)\right]\left(\Delta_S\hat{u}_z+\frac{\hat{Q}}{\bar{\rho}\bar{c}_p}\right). \qquad (2.155)\end{aligned}$$

Of course a relation between $\hat{Q}$ and $\hat{T}$ and thus effectively between $\hat{Q}$ and $\hat{u}_z$ must be specified by the physical properties of the system, such as e.g. in the case of thermal radiation by (2.132). However, if there are no volume heat sources, $Q=0$, and the boundaries are assumed stress-free and isothermal, the solution for the class of most unstable modes near threshold takes the form $\hat{u}_z \sim \sin\left(m\pi\frac{z}{L}\right)\mathrm{e}^{\sigma t}$ with $m=1$ and the relation for the growth rate greatly simplifies,

$$\begin{aligned}\left(\pi^2+\mathcal{K}^2L^2\right)&\left[\frac{\sigma L^2}{\kappa}+\left(\pi^2+\mathcal{K}^2L^2\right)\right]\left[\frac{\sigma L^2}{\nu}+\left(\pi^2+\mathcal{K}^2L^2\right)\right]^2\\ &+\left(\frac{\pi}{E}\right)^2\left[\frac{\sigma L^2}{\kappa}+\left(\pi^2+\mathcal{K}^2L^2\right)\right]\\ &=Ra\mathcal{K}^2L^2\left[\frac{\sigma L^2}{\nu}+\left(\pi^2+\mathcal{K}^2L^2\right)\right]. \qquad (2.156)\end{aligned}$$

Furthermore, if the rotation is neglected, $\mathbf{\Omega}=\mathbf{0}$, the dispersion relation takes the form

$$\left(\pi^2+\mathcal{K}^2L^2\right)\left[\frac{\sigma L^2}{\kappa}+\left(\pi^2+\mathcal{K}^2L^2\right)\right]\left[\frac{\sigma L^2}{\nu}+\left(\pi^2+\mathcal{K}^2L^2\right)\right]-Ra\mathcal{K}^2L^2=0, \qquad (2.157)$$

which leads to

$$\frac{\sigma L^2}{\kappa}=-\frac{1+Pr}{2}\left(\pi^2+\mathcal{K}^2L^2\right)+\sqrt{\left(\frac{1-Pr}{2}\right)^2\left(\pi^2+\mathcal{K}^2L^2\right)^2+RaPr\frac{\mathcal{K}^2L^2}{\pi^2+\mathcal{K}^2L^2}}, \qquad (2.158)$$

where only the positive root was provided. The latter expression allows to calculate the growth rate of unstable modes in the vicinity of threshold, $Ra - Ra_{crit} \ll Ra_{crit}$ and $\mathcal{K}L - \mathcal{K}_{crit}L \ll 1$ for systems with stress-free and isothermal boundaries.

As remarked at the beginning of this section the horizontal planform of the solutions remains undetermined in the linear regime. However, in the case of a horizontally infinite layer for which neither direction nor point in a horizontal plane is preferred, one can say, based purely on this symmetry and geometrical considerations, that the only possible cell patterns consist of contiguous parallel stripes (rolls), triangles, squares or hexagons filling and fitting into the entire $xy$ plane. Indeed, because the convection cells are all contiguous and fill the entire horizontal plane, the cells are either rolls (stripes) or regular polygons and in the latter case the angle at an apex of an $n$-sided regular polygon, $\pi - 2\pi/n$ must be equal to $2\pi/m$, where $m$ is an integer, since in such a configuration the apex is shared by $m$ exactly the same regular polygons; the relation $1 - 2/n = 2/m$, where $m$ and $n$ are integers can only be satisfied for pairs $(n = 3,\ m = 6)$ or $(n = 4,\ m = 4)$ or $(n = 6,\ m = 3)$. The linear analysis does not allow to determine, which one of those cell patterns is stable. We will address the issue of horizontal pattern selection by convective flow near threshold in the following section.

## 2.3 A Word on Weakly Nonlinear Estimates

There have been many works concerned with the weakly nonlinear analysis of convection near threshold involving determination of the amplitude of convection rolls and pattern selection problems. Most of them were based on similar approaches as the one developed by Schlüter et al. (1965), that is perturbative expansions in the amplitude of the convective flow, assumed weak. A most comprehensive review of the literature and results on the topic for Rayleigh-Bénard problem with isothermal and either stress-free or rigid boundaries is provided in the excellent book of Getling (1998). A thorough analysis of the pattern selection problem for the case of fixed heat flux at boundaries can be found in the seminal paper of Knobloch (1990). A brief compendium of the results on nearly marginal convection and a comprehensive picture of patterns developed by convection as the Rayleigh number departs from threshold is provided in this section. We start by considering the classical model of Rayleigh-Bénard convection without heat sources, $Q = 0$, with uniform diffusivities $\mu = \text{const.}$, $\kappa = \text{const.}$ and stress-free, isothermal boundaries to demonstrate how the amplitude of two-dimensional rolls near the threshold can be established.

It is useful to introduce the dynamical equations in non-dimensional form. With the choice of $L$, $\kappa/L$, $L^2/\kappa$, $\bar{\rho}\kappa^2/L^2$ and $L\Delta_S$ as units of length, velocity, time, pressure fluctuation and temperature fluctuation respectively

$$\mathbf{x} = L\mathbf{x}^\sharp, \quad \mathbf{u} = \frac{\kappa}{L}\mathbf{u}^\sharp, \quad t = \frac{L^2}{\kappa}t^\sharp, \quad p' = \bar{\rho}\left(\frac{\kappa}{L}\right)^2 p^\sharp, \quad T' = L\Delta_S T^\sharp, \tag{2.159}$$

the set of dynamical equations takes the following form

$$\frac{\partial \mathbf{u}^\sharp}{\partial t^\sharp} + \left(\mathbf{u}^\sharp \cdot \nabla^\sharp\right) \mathbf{u}^\sharp = -\nabla^\sharp p^\sharp + RaPrT^\sharp \hat{\mathbf{e}}_z + Pr\nabla^{\sharp 2}\mathbf{u}^\sharp, \tag{2.160a}$$

$$\nabla^\sharp \cdot \mathbf{u}^\sharp = 0, \tag{2.160b}$$

$$\frac{\partial T^\sharp}{\partial t^\sharp} + \mathbf{u}^\sharp \cdot \nabla^\sharp T^\sharp - u_z^\sharp = \nabla^{\sharp 2} T^\sharp. \tag{2.160c}$$

We start by considering the simplest case of solutions in the form of two-dimensional rolls, and since $\nabla^\sharp \cdot \mathbf{u}^\sharp = 0$ we introduce the stream function $\psi = \psi(x^\sharp, z^\sharp)$, $\mathbf{u}^\sharp = \nabla \times \left(\psi \hat{\mathbf{e}}_y\right)$, so that

$$u_x^\sharp = -\frac{\partial \psi}{\partial z^\sharp}, \quad u_y^\sharp = 0, \quad u_z^\sharp = \frac{\partial \psi}{\partial x^\sharp}. \tag{2.161}$$

By taking a curl of the momentum equation (2.160a) and taking its $y$-component one obtains

$$\frac{\partial}{\partial t^\sharp}\nabla^{\sharp 2}\psi + \mathcal{J}\left(\zeta^\sharp, \psi\right) = RaPr\frac{\partial T^\sharp}{\partial x^\sharp} + Pr\nabla^{\sharp 4}\psi, \tag{2.162a}$$

$$-\frac{\partial T^\sharp}{\partial t^\sharp} + \mathcal{J}\left(T^\sharp, \psi\right) = -\frac{\partial \psi}{\partial x^\sharp} - \nabla^{\sharp 2} T^\sharp, \tag{2.162b}$$

where

$$\zeta^\sharp = -\nabla^{\sharp 2}\psi \tag{2.163}$$

is the $y$-component of the flow vorticity and

$$\mathcal{J}\left(f, g\right) = \frac{\partial f}{\partial x^\sharp}\frac{\partial g}{\partial z^\sharp} - \frac{\partial f}{\partial z^\sharp}\frac{\partial g}{\partial x^\sharp} \tag{2.164}$$

denotes the Jacobian. As we know from the linear considerations, in the problem at hand of classical Rayleigh-Bénard convection with stress-free, isothermal boundaries, the instability sets in through a stationary mode and thus near convection threshold the time scale of flow evolution is slow. To establish the time scale of evolution of the amplitude of convection we expand the growth rate $\sigma^\sharp = \sigma L^2/\kappa$ given in (2.158) in powers of the departure from threshold defined in the following way

$$\eta = \frac{Ra - Ra_{crit}}{Ra_{crit}} \ll 1, \tag{2.165}$$

and $\delta\mathcal{K}^\sharp = \mathcal{K}L - \mathcal{K}_{crit}L$ to obtain (cf. also Getling 1998 or Fauve 2017)

$$\sigma^\sharp = \frac{3\pi^2 Pr}{2\left(1 + Pr\right)}\left(\eta - \frac{8}{3\pi^2}\delta\mathcal{K}^{\sharp 2}\right). \tag{2.166}$$

It is evident from (2.166), that the time scale of evolution of the most unstable mode with $\mathcal{K} = \mathcal{K}_{crit}$, thus $\delta\mathcal{K}^\sharp = 0$ is slow, of the order $\eta^{-1}L^2/\kappa$. Hence we introduce $\tau = \eta t^\sharp$ for the slow time-scale corresponding to the dynamics of the amplitude of convection. The growth rate of the most unstable mode, at the initial stage of evolution, when the dynamics is governed by the linear equations is

$$\sigma_0^\sharp = \frac{\eta}{\tau_0}, \quad \text{where} \quad \frac{1}{\tau_0} = \frac{3\pi^2 Pr}{2\,(1+Pr)}. \tag{2.167}$$

The main task of the weakly nonlinear analysis is to establish the equation for time evolution of the amplitude of fluctuations, which includes the effect of nonlinearities and thus allows for saturation after the initial stage of exponential amplification.

The next step is the expansion of all the dependent variables in powers of the departure from threshold, which we assume in the following form

$$\psi = \eta^r\psi_1 + \eta^{2r}\psi_2 + \eta^{3r}\psi_3 + \ldots, \quad \zeta^\sharp = \eta^r\zeta_1 + \eta^{2r}\zeta_2 + \eta^{3r}\zeta_3 + \ldots, \tag{2.168a}$$

$$T^\sharp = \eta^r T_1 + \eta^{2r} T_2 + \eta^{3r} T_3 + \ldots, \tag{2.168b}$$

where $r$ is a positive rational number. Introducing the above expansions, $\tau = \eta t^\sharp$ and (2.165) into the Eqs. (2.162a), (2.162b) and (2.163) and dividing by $\eta^r$ leads to

$$\begin{aligned}
&\eta\frac{\partial}{\partial\tau}\left(\nabla^{\sharp 2}\psi_1 + \eta^r\nabla^{\sharp 2}\psi_2\right) + \eta^r\mathcal{J}\left(\zeta_1,\, \psi_1\right) + \eta^{2r}\left[\mathcal{J}\left(\zeta_1,\, \psi_2\right) + \mathcal{J}\left(\zeta_2,\, \psi_1\right)\right] \\
&\quad = Ra_{crit}\left(1+\eta\right) Pr\frac{\partial}{\partial x^\sharp}\left(T_1 + \eta^r T_2 + \eta^{2r} T_3\right) \\
&\quad + Pr\nabla^{\sharp 4}\left(\psi_1 + \eta^r\psi_2 + \eta^{2r}\psi_3\right) + \mathcal{O}\left(\eta^{3r}\right),
\end{aligned} \tag{2.169a}$$

$$\begin{aligned}
&-\eta\frac{\partial}{\partial\tau}\left(T_1 + \eta^r T_2\right) + \eta^r\mathcal{J}\left(T_1,\, \psi_1\right) + \eta^{2r}\left[\mathcal{J}\left(T_1,\, \psi_2\right) + \mathcal{J}\left(T_2,\, \psi_1\right)\right] \\
&\quad = -\frac{\partial}{\partial x^\sharp}\left(\psi_1 + \eta^r\psi_2 + \eta^{2r}\psi_3\right) - \nabla^{\sharp 2}\left(T_1 + \eta^r T_2 + \eta^{2r} T_3\right) \\
&\quad + \mathcal{O}\left(\eta^{3r}\right),
\end{aligned} \tag{2.169b}$$

$$\zeta_n = -\nabla^{\sharp 2}\psi_n \quad \text{for all } n = 1,\, 2,\, 3,\, \ldots \tag{2.169c}$$

Balancing the leading order terms (order unity terms) gives the linear balance which allows to obtain the critical Rayleigh number and the spatial structure of $\psi_1$ and $T_1$, but not the amplitude

$$Ra_{crit}Pr\frac{\partial T_1}{\partial x^\sharp} + Pr\nabla^{\sharp 4}\psi_1 = 0, \tag{2.170a}$$

$$-\frac{\partial\psi_1}{\partial x^\sharp} - \nabla^{\sharp 2}T_1 = 0, \tag{2.170b}$$

$$\zeta_1 = -\nabla^{\sharp 2}\psi_1, \tag{2.170c}$$

The solution for the most unstable mode takes the form (cf. Sect. 2.2.1 on the relevant linear problem)

$$\psi_1 = A(\tau)\sin\left(\pi z^\sharp\right)\cos\left(\mathcal{K}^\sharp_{crit}x^\sharp\right), \tag{2.171a}$$

$$T_1 = -\frac{\mathcal{K}^\sharp_{crit}}{\pi^2 + \mathcal{K}^{\sharp 2}_{crit}} A(\tau)\sin\left(\pi z^\sharp\right)\sin\left(\mathcal{K}^\sharp_{crit}x^\sharp\right), \tag{2.171b}$$

$$\zeta_1 = \left(\pi^2 + \mathcal{K}^{\sharp 2}_{crit}\right)\psi_1. \tag{2.171c}$$

with

$$Ra_{crit} = \frac{\left(\pi^2 + \mathcal{K}^{\sharp 2}_{crit}\right)^3}{\mathcal{K}^{\sharp 2}_{crit}} = \frac{27}{4}\pi^4, \quad \mathcal{K}^\sharp_{crit} = \frac{\pi}{\sqrt{2}}, \tag{2.172}$$

and for the sake of clarity only the $\cos\left(\mathcal{K}^\sharp_{crit}x^\sharp\right)$ mode was chosen. The next order balance involves terms of the order $\eta^r$ and since at this stage we do not know yet the value of $r$, we must also include in the balance the terms of the order $\eta$; supplied by $\mathcal{J}(\zeta_1, \psi_1) = (\pi^2 + \mathcal{K}^{\sharp 2}_{crit})\mathcal{J}(\psi_1, \psi_1) = 0$ and $\mathcal{J}(T_1, \psi_1) = -\mathcal{K}^{\sharp 2}_{crit}\pi A^2 \sin(2\pi z^\sharp)/2(\pi^2 + \mathcal{K}^{\sharp 2}_{crit})$ this yields

$$Ra_{crit}Pr\frac{\partial T_2}{\partial x^\sharp} + Pr\nabla^{\sharp 4}\psi_2 = \eta^{1-r}\frac{\partial}{\partial \tau}\left(\nabla^{\sharp 2}\psi_1\right) - \eta^{1-r}Ra_{crit}Pr\frac{\partial T_1}{\partial x^\sharp}, \tag{2.173a}$$

$$-\frac{\partial \psi_2}{\partial x^\sharp} - \nabla^{\sharp 2}T_2 = -\eta^{1-r}\frac{\partial T_1}{\partial \tau} - \frac{\pi \mathcal{K}^{\sharp 2}_{crit}}{2\left(\pi^2 + \mathcal{K}^{\sharp 2}_{crit}\right)} A^2 \sin\left(2\pi z^\sharp\right), \tag{2.173b}$$

$$\zeta_2 = -\nabla^{\sharp 2}\psi_2, \tag{2.173c}$$

At this stage the standard procedure in the approach based on perturbative expansions involves application of the Fredholm Alternative Theorem, which in the case at hand means, that non-trivial solutions of the latter set of equations can only exist if the solution to the linear problem on the left hand side is orthogonal to the non-homogenity on the right hand side (cf. Korn and Korn 1961). First, however, we must introduce a certain inner product on the space of solutions and show, that the linear operator on the left hand side of (2.173a), (2.173b) is self-adjoint. We write down the Eqs. (2.173a), (2.173b) in the form

$$\mathfrak{L}\mathbf{v} = \mathfrak{N}, \tag{2.174}$$

where

$$\mathbf{v}=\begin{bmatrix}\psi_2\\ T_2\end{bmatrix},\quad \mathfrak{L}=\begin{bmatrix}Pr\nabla^{\sharp 4} & Ra_{crit}Pr\frac{\partial}{\partial x^\sharp}\\ -\frac{\partial}{\partial x^\sharp} & -\nabla^{\sharp 2}\end{bmatrix},\tag{2.175}$$

$$\mathfrak{N}=\begin{bmatrix}\eta^{1-r}\frac{\partial}{\partial\tau}\left(\nabla^{\sharp 2}\psi_1\right)-\eta^{1-r}Ra_{crit}Pr\frac{\partial T_1}{\partial x^\sharp}\\ -\eta^{1-r}\frac{\partial T_1}{\partial\tau}-\frac{\pi\mathcal{K}^{\sharp 2}_{crit}}{2\left(\pi^2+\mathcal{K}^{\sharp 2}_{crit}\right)}A^2\sin\left(2\pi z^\sharp\right)\end{bmatrix},\tag{2.176}$$

and define the inner product of $\mathbf{v}_\alpha=[\psi_\alpha,\ T_\alpha]$ and $\mathbf{v}_\beta=[\psi_\beta,\ T_\beta]$ in the following way[15]

$$\left\langle\mathbf{v}_\alpha,\ \mathbf{v}_\beta\right\rangle=\int_0^{2\pi/\mathcal{K}^\sharp_{crit}}\mathrm{d}x^\sharp\int_0^1\mathrm{d}z^\sharp\left(\psi_\alpha\psi_\beta+Ra_{crit}PrT_\alpha T_\beta\right).\tag{2.177}$$

Indeed, it is now easy to verify, that with such a definition the linear operator $\mathfrak{L}$ is self-adjoint, that is $\left\langle\mathfrak{L}\mathbf{v}_\alpha,\ \mathbf{v}_\beta\right\rangle=\left\langle\mathbf{v}_\alpha,\ \mathfrak{L}\mathbf{v}_\beta\right\rangle$ is satisfied. We can, therefore utilize the Fredholm Alternative Theorem and write

$$\begin{aligned}\eta^{1-r}&\int_0^{2\pi/\mathcal{K}^\sharp_{crit}}\mathrm{d}x^\sharp\int_0^1\mathrm{d}z^\sharp\left[\frac{\partial}{\partial\tau}\left(\nabla^{\sharp 2}\psi_1\right)-Ra_{crit}Pr\frac{\partial T_1}{\partial x^\sharp}\right]\sin\left(\pi z^\sharp\right)\cos\left(\mathcal{K}^\sharp_{crit}x^\sharp\right)\\ &+Ra_{crit}Pr\frac{\mathcal{K}^\sharp_{crit}}{\pi^2+\mathcal{K}^{\sharp 2}_{crit}}\int_0^{2\pi/\mathcal{K}^\sharp_{crit}}\mathrm{d}x^\sharp\int_0^1\mathrm{d}z^\sharp\left[\eta^{1-r}\frac{\partial T_1}{\partial\tau}\right.\\ &\left.+\frac{\pi\mathcal{K}^{\sharp 2}_{crit}}{2\left(\pi^2+\mathcal{K}^{\sharp 2}_{crit}\right)}A^2\sin\left(2\pi z^\sharp\right)\right]\sin\left(\pi z^\sharp\right)\sin\left(\mathcal{K}^\sharp_{crit}x^\sharp\right)=0.\end{aligned}\tag{2.178}$$

Since

$$\int_0^{2\pi/\mathcal{K}^\sharp_{crit}}\mathrm{d}x^\sharp\sin\left(\mathcal{K}^\sharp_{crit}x^\sharp\right)\int_0^1\mathrm{d}z^\sharp\sin\left(2\pi z^\sharp\right)\sin\left(\pi z^\sharp\right)=0,\tag{2.179}$$

we are left only with terms of the order $\eta^{1-r}$. The value of $r$ can be established as follows. If we first assume tentatively, that $r=1$, so that $\eta^{1-r}=1$, then the solvability condition involves terms from the right hand side $\mathfrak{N}$ which are all of the same order as the left hand side (originally $\eta^{2r}$) and thus from (2.178) one obtains

$$\frac{\mathrm{d}A}{\mathrm{d}\tau}=\frac{1}{\tau_0}A.\tag{2.180}$$

The latter amplitude equation does not involve any contribution from nonlinear terms in the dynamical equations and is fully linear, leading to exponential amplification of the amplitude, without the possibility for saturation. This means that the choice

[15]Note, that the linear operator $\mathfrak{L}$ is the same as in the leading order problem (2.170a), (2.170b).

$r = 1$ was wrong. Therefore the contributions from the buoyancy force and time-derivatives in the dynamical equations, both involving terms proportional to $\eta$, do not enter the second order balance (2.173a), (2.173b), but are required to enter the balance at the next and higher orders so that saturation can be achieved. Consequently the only reasonable choice is $r = 1/2$. This establishes the well known feature of nearly marginal convection, that the amplitude of the flow scales like square root of the departure from threshold, $\sqrt{(Ra - Ra_{crit})/Ra_{crit}}$.

Now, with $r = 1/2$ we solve the second-order equations (2.173a), (2.173b), which take the form

$$Ra_{crit} Pr \frac{\partial T_2}{\partial x^\sharp} + Pr \nabla^{\sharp 4} \psi_2 = 0 \tag{2.181a}$$

$$-\frac{\partial \psi_2}{\partial x^\sharp} - \nabla^{\sharp 2} T_2 = -\frac{\pi \mathcal{K}_{crit}^{\sharp 2}}{2\left(\pi^2 + \mathcal{K}_{crit}^{\sharp 2}\right)} A^2 \sin\left(2\pi z^\sharp\right), \tag{2.181b}$$

and since the right hand side is independent of $x^\sharp$ the solution takes the form[16]

$$\psi_2 = 0, \quad T_2 = -\frac{\mathcal{K}_{crit}^{\sharp 2}}{8\pi\left(\pi^2 + \mathcal{K}_{crit}^{\sharp 2}\right)} A^2 \sin\left(2\pi z^\sharp\right). \tag{2.182}$$

Next we gather the terms of the order $\eta$ in the Eqs. (2.169a)–(2.169c) to obtain

$$Ra_{crit} Pr \frac{\partial T_3}{\partial x^\sharp} + Pr \nabla^{\sharp 4} \psi_3 = \frac{\partial}{\partial \tau} \nabla^{\sharp 2} \psi_1 - Ra_{crit} Pr \frac{\partial T_1}{\partial x^\sharp}, \tag{2.183a}$$

$$-\frac{\partial \psi_3}{\partial x^\sharp} - \nabla^{\sharp 2} T_3 = -\frac{\partial T_1}{\partial \tau} + \mathcal{J}\left(T_2, \psi_1\right) \tag{2.183b}$$

$$\zeta_3 = -\nabla^{\sharp 2} \psi_3, \tag{2.183c}$$

where $\psi_2 = 0$ has been substituted. The solvability condition for the system of equations (2.183a), (2.183b) yields

$$\begin{aligned}
&\int_0^{2\pi/\mathcal{K}_{crit}^\sharp} dx^\sharp \int_0^1 dz^\sharp \left[\frac{\partial}{\partial \tau}\left(\nabla^{\sharp 2} \psi_1\right) - Ra_{crit} Pr \frac{\partial T_1}{\partial x^\sharp}\right] \sin\left(\pi z^\sharp\right) \cos\left(\mathcal{K}_{crit}^\sharp x^\sharp\right) \\
&\quad + Ra_{crit} Pr \frac{\mathcal{K}_{crit}^\sharp}{\pi^2 + \mathcal{K}_{crit}^{\sharp 2}} \int_0^{2\pi/\mathcal{K}_c^\sharp} dx^\sharp \int_0^1 dz^\sharp \left[\frac{\partial T_1}{\partial \tau}\right. \\
&\qquad \left. - \mathcal{J}\left(T_2, \psi_1\right)\right] \sin\left(\pi z^\sharp\right) \sin\left(\mathcal{K}_{crit}^\sharp x^\sharp\right) = 0.
\end{aligned} \tag{2.184}$$

[16]It is enough to take only the particular solution of the inhomogeneous problem since the solution of the homogeneous problem provides only an order $\eta$ correction to the amplitude.

which on introduction of the first-order solutions (2.171a), (2.171c) and (2.172) and

$$\mathcal{J}(T_2, \psi_1) = -\frac{\mathcal{K}_{crit}^{\sharp 3}}{4\left(\pi^2 + \mathcal{K}_{crit}^{\sharp 2}\right)} A^3 \cos\left(2\pi z^\sharp\right) \sin\left(\pi z^\sharp\right) \sin\left(\mathcal{K}_{crit}^\sharp x^\sharp\right) \tag{2.185}$$

simplifies to

$$\frac{\mathrm{d}A}{\mathrm{d}\tau} = \frac{1}{\tau_0} A - \frac{1}{24\tau_0} A^3 \tag{2.186}$$

where

$$\int_0^{2\pi/\mathcal{K}_{crit}^\sharp} \mathrm{d}x^\sharp \sin^2\left(\mathcal{K}_{crit}^\sharp x^\sharp\right) \int_0^1 \mathrm{d}z^\sharp \cos\left(2\pi z^\sharp\right) \sin^2\left(\pi z^\sharp\right) = -\frac{1}{8} \tag{2.187}$$

was used. Equation (2.186) is called the Landau equation and it governs the time evolution of the amplitude of nearly marginal solutions in the form of two-dimensional rolls. A stationary solution of this type can be easily obtained and takes the form

$$A = \pm\sqrt{24}, \tag{2.188}$$

and hence with the accuracy up to order $\mathcal{O}(\eta^{3/2})$

$$\psi = \pm\,\eta^{1/2}\sqrt{24}\sin\left(\pi z^\sharp\right)\cos\left(\frac{\pi}{\sqrt{2}}x^\sharp\right) + \mathcal{O}\left(\eta^{3/2}\right), \tag{2.189a}$$

$$T^\sharp = \mp\,\eta^{1/2}\frac{4}{\sqrt{3}\pi}\sin\left(\pi z^\sharp\right)\sin\left(\frac{\pi}{\sqrt{2}}x^\sharp\right) - \eta\frac{1}{\pi}\sin\left(2\pi z^\sharp\right) + \mathcal{O}\left(\eta^{3/2}\right). \tag{2.189b}$$

This allows to calculate the Nusselt number in such a stationary state (cf. Eq. (2.59)),

$$\begin{aligned} Nu &= \frac{\bar{\rho}\bar{c}_p \langle u_z T\rangle_h - k\partial_z \langle T\rangle_h - kg\bar{\alpha}\bar{T}/\bar{c}_p}{k\Delta_S} = 1 + \left\langle \frac{\partial\psi}{\partial x^\sharp} T^\sharp \right\rangle_h - \partial_{z^\sharp}\left\langle T^\sharp\right\rangle_h \\ &= 1 + 2\eta + \mathcal{O}\left(\eta^{3/2}\right) \approx 2\frac{Ra}{Ra_{crit}} - 1. \end{aligned} \tag{2.190}$$

Note, that the associated horizontally averaged superadiabatic temperature gradient

$$\left\langle \frac{\mathrm{d}T}{\mathrm{d}z} \right\rangle_h + \frac{g\bar{\alpha}\bar{T}}{\bar{c}_p} = -\Delta_S\left[1 + 2\eta\cos\left(\frac{2\pi z}{L}\right)\right] \tag{2.191}$$

is weakest near the mid point and strongest near the boundaries, thus the temperature profile becomes closer to adiabatic in the middle and the gradients become sharper as the boundaries are approached (see Fig. 2.2).

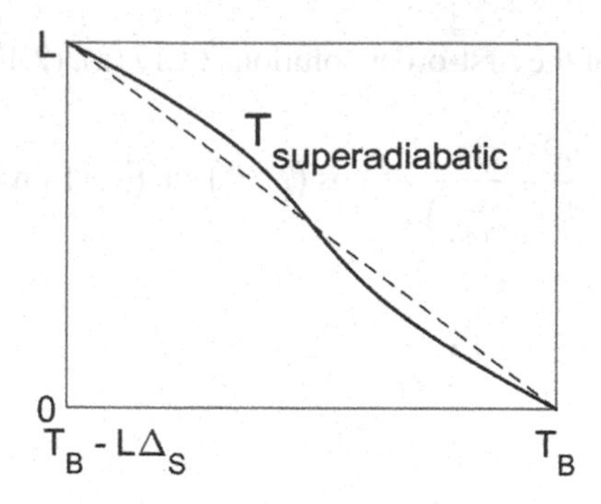


**Fig. 2.2** Vertical profile of the mean superadiabatic temperature $T_{\text{superadiabatic}} = \tilde{T} + \langle T' \rangle_h + g\bar{\alpha}\bar{T}z/\bar{c}_p$, in the weakly nonlinear regime close to convection threshold in the absence of heat sources, $Q = 0$, obtained from Eq. (2.191) for $\eta = 0.2$; the superadiabatic temperature profile in the hydrostatic reference state $\tilde{T} + g\bar{\alpha}\bar{T}z/\bar{c}_p$ is plotted with a dashed line for reference. The total mean temperature profile becomes closer to adiabatic in the middle of the fluid domain and its gradient sharpens near the boundaries

### *2.3.1 Short Introduction into the Pattern Selection Problem Near Threshold*

The selection of the horizontal planform by the convective system is typically described by stability analysis of a stationary (or more generally oscillatory) solution with a given horizontal pattern, such as e.g. the stationary two-dimensional roll solution (2.189a), (2.189b) with the amplitude determined by the Eq. (2.186) obtained via the weakly nonlinear analysis. In practice the stationary solution is perturbed $\mathbf{u}^\sharp_{st} + \delta\mathbf{u}^\sharp$, $T^\sharp_{st} + \delta T^\sharp$ and its stability with respect to infinitesimal perturbations is studied by consideration of an eigenvalue problem for the growth rate $s$ of the perturbations. By the use of the following decomposition of the velocity perturbation involving vertically irrotational and rotational parts of the horizontal component[17]

$$\delta\mathbf{u}^\sharp = \nabla^\sharp_h \Phi + \nabla^\sharp \times \left(\Psi \hat{\mathbf{e}}_z\right) + \delta u^\sharp_z \hat{\mathbf{e}}_z, \tag{2.192}$$

the eigenvalue problem can be expressed in the following way

$$\begin{aligned} s\nabla^{\sharp 2}\delta u^\sharp_z - \hat{\mathbf{e}}_z \cdot \nabla^\sharp \times \nabla^\sharp \times \left[\left(\delta\mathbf{u}^\sharp \cdot \nabla^\sharp\right)\mathbf{u}^\sharp_{st} + \left(\mathbf{u}^\sharp_{st} \cdot \nabla^\sharp\right)\delta\mathbf{u}^\sharp\right] \\ = Ra_{crit}\left(1+\eta\right) Pr \nabla^{\sharp 2}_h \delta T^\sharp + Pr \nabla^{\sharp 4}\delta u^\sharp_z, \end{aligned} \tag{2.193a}$$

[17]This can be viewed as the standard poloidal-toroidal decomposition of solenoidal fields, $\delta\mathbf{u}^\sharp = \nabla^\sharp \times \left(\Psi \hat{\mathbf{e}}_z\right) + \nabla^\sharp \times \nabla^\sharp \times \left(\Psi_{pol} \hat{\mathbf{e}}_z\right)$, where the poloidal component is separated into a horizontal part $\nabla^\sharp_h \Phi$ and a vertical part $\delta u^\sharp_z \hat{\mathbf{e}}_z$, by introduction of new variables $\Phi = \partial^\sharp_z \Psi_{pol}$ and $\delta u^\sharp_z = -\nabla^{\sharp 2}_h \Psi_{pol}$. The decomposition (2.192) is allowed (and mathematically equivalent to the poloidal-toroidal decomposition) once the divergence-free condition on the original velocity field $\nabla^\sharp \cdot \delta\mathbf{u}^\sharp = 0$ expressed in (2.193c) is imposed.

$$s\nabla_h^{\sharp 2}\Psi - \hat{\mathbf{e}}_z \cdot \nabla^\sharp \times \left[\left(\delta\mathbf{u}^\sharp \cdot \nabla^\sharp\right)\mathbf{u}_{st}^\sharp + \left(\mathbf{u}_{st}^\sharp \cdot \nabla^\sharp\right)\delta\mathbf{u}^\sharp\right] = Pr\nabla^{\sharp 2}\nabla_h^{\sharp 2}\Psi, \quad (2.193b)$$

$$\nabla_h^{\sharp 2}\Phi = -\frac{\partial \delta u_z^\sharp}{\partial z}, \quad (2.193c)$$

$$s\delta T^\sharp + \delta\mathbf{u}^\sharp \cdot \nabla^\sharp T_{st}^\sharp + \mathbf{u}_{st}^\sharp \cdot \nabla^\sharp \delta T^\sharp - \delta u_z^\sharp = \nabla^{\sharp 2}\delta T^\sharp. \quad (2.193d)$$

An approach of this type has been undertaken e.g. by Schlüter et al. (1965), Busse (1967)[18] with the results summarized in Busse (1978) and Getling (1998) for three types of planform: two-dimensional rolls, squares and hexagons, were the boundaries were assumed isothermal, impermeable and either rigid or stress-free. It has been shown, that for purely symmetric systems with respect to the mid-plane the first stable solution takes the form of two-dimensional rolls when the Rayleigh number slightly exceeds the critical value for convection threshold. These rolls become unstable when the Rayleigh number is increased even further via a variety of instability types, such as e.g. the *Zigzag instability*, which leads to sinusoidal curving of the rolls, the *Cross-roll instability* which forms a new system of rolls perpendicular to the initial ones which eventually take over or the *Eckhaus instability*. The latter instability type is the only one that does not break the two-dimensionality of the flow and leads to phase alteration, that is compression and expansion of groups of rolls along their wave vector. Which instability sets in depends on the value of the Rayleigh number and the set of wave numbers in a wave packet of the perturbation. For a comprehensive list of the types of instabilities of 2D roll structures see §**6** of Getling (1998). However, it was also reported, that when the up-down symmetry is broken by either (i) allowing for a weak temperature variation of the expansion coefficient $\alpha(T)$, of the viscosity $\nu(T)$ or of the thermal diffusivity $\kappa(T)$, (ii) imposing different boundary conditions at the top and bottom boundaries or (iii) including the effect of free surface curvature, the hexagonal pattern is preferred.

Utilizing the same decomposition as in (2.192), Manneville (1983) has derived an equation for the pattern structure function $\check{u}_{1z}$ (convection amplitude) near the convection onset for the case of stress-free, isothermal, impermeable boundaries by assuming the following expansions of variables

$$\delta u_z^\sharp = \sum_n \check{u}_{nz}\left(x^\sharp, y^\sharp, t^\sharp\right)\sin\left(n\pi z^\sharp\right), \quad (2.194a)$$

$$\delta T^\sharp = \sum_n \check{T}_{nz}\left(x^\sharp, y^\sharp, t^\sharp\right)\sin\left(n\pi z^\sharp\right), \quad (2.194b)$$

$$\Psi = \sum_n \check{\Psi}_{nz}\left(x^\sharp, y^\sharp, t^\sharp\right)\cos\left(n\pi z^\sharp\right), \quad (2.194c)$$

[18] Busse (1967) utilizes a simplifying assumption of infinite $Pr$.

and applying the Galerkin method[19] within the scope of weakly nonlinear theory. This study was concerned with the primary instabilities, thus $\mathbf{u}^\sharp_{st} = 0$ and the nonlinear terms $(\delta \mathbf{u}^\sharp \cdot \nabla^\sharp)\,\delta \mathbf{u}^\sharp$ and $\delta \mathbf{u}^\sharp \cdot \nabla^\sharp \delta T^\sharp$ likewise explicit time derivatives were included in the Eqs. (2.193a)–(2.193d) to yield[20]

$$\tau_0 \frac{\partial \check{u}_{1z}}{\partial t^\sharp} = \left[\eta - \frac{4}{3\pi^4}\left(\nabla_h^{\sharp 2} + \mathcal{K}_{crit}^{\sharp 2}\right)^2\right]\check{u}_{1z} - \frac{1}{6\pi^4}\check{u}_{1z}\left[\left(\nabla_h^\sharp \check{u}_{1z}\right)^2 + \mathcal{K}_{crit}^{\sharp 2}\check{u}_{1z}^2\right]$$
$$- \tau_0\left(\check{u}_{0x}\frac{\partial}{\partial x^\sharp} + \check{u}_{0y}\frac{\partial}{\partial y^\sharp}\right)\check{u}_{1z}, \tag{2.195}$$

$$\left(\frac{\partial}{\partial t^\sharp} - Pr\nabla_h^{\sharp 2}\right)\nabla_h^{\sharp 2}\check{\Psi}_0 = \frac{1}{\mathcal{K}_{crit}^{\sharp 2}}\left[\frac{\partial \check{u}_{1z}}{\partial y^\sharp}\frac{\partial\left(\nabla_h^{\sharp 2}\check{u}_{1z}\right)}{\partial x^\sharp} - \frac{\partial \check{u}_{1z}}{\partial x^\sharp}\frac{\partial\left(\nabla_h^{\sharp 2}\check{u}_{1z}\right)}{\partial y^\sharp}\right], \tag{2.196}$$

where $\mathcal{K}_{crit}^\sharp = 2/\sqrt{\pi}$ and

$$\check{u}_{0x} = \frac{\partial \check{\Psi}_0}{\partial y^\sharp}, \qquad \check{u}_{0y} = -\frac{\partial \check{\Psi}_0}{\partial x^\sharp} \tag{2.197}$$

so that $\check{\Psi}_0$ is the stream function of the horizontal flow $(\check{u}_{0x}, \check{u}_{0y})$, z-independent by definition (2.194a)–(2.194c). After solving the amplitude equations the flow and temperature are given by

$$u_x^\sharp = \check{u}_{0x} + \frac{2}{\pi}\frac{\partial \check{u}_{1z}}{\partial x}\cos(\pi z) - \frac{3(8+3Pr)}{64Pr\pi^4}\frac{\partial}{\partial x}\left[\left(\nabla_h \check{u}_{1z}\right)^2 + \mathcal{K}_{crit}^2\check{u}_{1z}^2\right]\cos(2\pi z), \tag{2.198a}$$

$$u_y^\sharp = \check{u}_{0y} + \frac{2}{\pi}\frac{\partial \check{u}_{1z}}{\partial y}\cos(\pi z) - \frac{3(8+3Pr)}{64Pr\pi^4}\frac{\partial}{\partial y}\left[\left(\nabla_h \check{u}_{1z}\right)^2 + \mathcal{K}_{crit}^2\check{u}_{1z}^2\right]\cos(2\pi z), \tag{2.198b}$$

$$u_z^\sharp \approx \check{u}_{1z}\sin(\pi z) + \frac{3(8+3Pr)}{128Pr\pi^5}\nabla_h^2\left[\left(\nabla_h \check{u}_{1z}\right)^2 + \mathcal{K}_{crit}^2\check{u}_{1z}^2\right]\sin(2\pi z), \tag{2.198c}$$

$$T^\sharp = \frac{2}{3\pi^2}\check{u}_{1z}(x,y)\sin(\pi z)$$
$$+ \frac{1}{6\pi^5}\left[\frac{9(8+3Pr)}{256Pr\pi^2}\nabla_h^2 - 1\right]\left[\left(\nabla_h \check{u}_{1z}\right)^2 + \mathcal{K}_{crit}^2\check{u}_{1z}^2\right]\sin(2\pi z), \tag{2.198d}$$

[19]The Galerkin method is essentially based on expansion of variables in a complete set of basis functions, introduction of the expansions into the equations and equating their inner product with each of the basis functions to zero.

[20]Note the different temperature scale assumed by Manneville (1983), which corresponds to $L\Delta_S/Ra$.

with the accuracy up to terms of the order $\mathcal{O}(\eta^{3/2})$ (see (2.165) for the definition of $\eta$). A similar equation for the pattern structure function, but for the case of both rigid boundaries held at a fixed heat flux, when the nearly marginal flow varies on very large horizontal length scales (as known from the linear theory, cf. Sect. 2.2.5) was derived by Proctor (1981). It was later generalized by Knobloch (1990) to include asymmetric top-bottom boundary conditions and temperature variation of fluid properties (such as $\nu$, $\alpha$ and $\kappa$). They assumed doubly periodic lattice and utilized perturbative expansions in square of the inverse length scale of horizontal variation, which is equivalent to expansions in square of the wave number of the planform $\mathcal{K}^{\sharp 2} = \mathcal{K}_x^{\sharp 2} + \mathcal{K}_y^{\sharp 2} \ll 1$

$$T^{\sharp}(x, y, z, t) = T_0(x, y, t) + \mathcal{K}^2 T_2(x, y, z, t) + \ldots, \tag{2.199}$$

where the leading order term for temperature is independent of $z$ by virtue of the temperature equation. For symmetric, fixed-heat flux boundary conditions and uniform fluid properties it has been shown, that $\partial_t \sim \mathcal{K}^{\sharp 4}$ and $Ra - Ra_{crit} \sim \mathcal{K}^{\sharp 2}$ and the pattern structure function obeys the following equation

$$\frac{\partial T_0}{\partial t} = -\eta \nabla_h^2 T_0 - \frac{34}{231} \nabla_h^4 T_0 + \frac{10}{7} \nabla_h \cdot \left( |\nabla_h T_0|^2 \, \nabla_h T_0 \right), \tag{2.200}$$

where $\eta = (Ra - Ra_{crit})/Ra_{crit}$ denotes the departure from threshold, as before. Proctor (1981) showed, that small departures from constant heat flux at the boundaries lead to stability of square-cell solutions over the entire range of validity of the approximation. Knobloch (1990), however, demonstrated that temperature variation of fluid properties and asymmetric boundary conditions can result in stable patterns in the forms of rolls, squares and hexagons, depending on values of system parameters.

Finally we note an interesting result concerning upper bounds on heat transport by nearly marginal convection which extends to turbulent regime, obtained for rigid and isothermal boundaries, which was reported by Busse (1969). It was conjectured, that the dynamics of convection is controlled by stability of the boundary layers and importance of the horizontal scale of convection has been emphasized. Busse (1969) generalized the results of Howard (1963) for a variational problem of finding a maximum of the convective heat transport $\langle u_z T' \rangle$ at a given Rayleigh number, by introducing a structure of successive boundary layers to adjust the horizontal length scale of variation of the convective velocities and temperature from a boundary to the interior value. The latter was assumed comparable with the interior vertical variation length scale which allowed to minimize dissipation and maximize the heat transport. This resulted in a sequence of upper bounds for the Nusselt number realized for different ranges of the Rayleigh number, well approximated by the following three formulae

$$\begin{aligned} Nu &\leq 1 + 0.1252\, Ra^{3/8} \quad \text{for} \quad Ra < 2.642 \times 10^4, \\ Nu &\leq 1 + 0.0482\, Ra^{15/32} \quad \text{for} \quad 2.642 \times 10^4 \leq Ra < 4.644 \times 10^5, \\ Nu &\leq 1 + 0.0311\, Ra^{1/2} \quad \text{for} \quad Ra \geq 4.644 \times 10^5. \end{aligned} \tag{2.201}$$

The latter upper bound on the Nusselt number obtained by Busse (1969), namely $Nu \leq 1 + 0.0311\, Ra^{1/2}$ is indeed satisfied for all $Ra \geq 4.644 \times 10^5$, even in a fully turbulent regime for extremely high values of the Rayleigh number. This can be verified through comparison of this estimate with heat flux estimates obtained from the theory of Grossmann and Lohse (2000) and Stevens et al. (2013) for fully developed turbulent convection. However, the theory of Grossman and Lohse, based on data from a large number of laboratory and numerical experiments, provides far more precise estimates than (2.201), and therefore more useful within the ranges of their validity, which depend on both, the Rayleigh number and the Prandtl number.

## 2.4 Fully Developed Convection

The fully nonlinear, developed convection at very high Rayleigh numbers is a very common in nature, but at the same time extremely complex phenomenon, for which a complete mathematical description is not achievable. The same is of course true for many turbulent flows of a different physical origin than convection, which results from the fact, that a mathematically rigorous and complete description of a fully developed turbulence, despite intensive attempts from over a century still remains beyond the grasp of the present-day fluid mechanics. Nevertheless, the physical picture of developed convection, obtained through a mixture of application of fully rigorously derived relations, some scaling arguments and experimental/numerical data was obtained in the comprehensive study of Grossmann and Lohse (2000), later updated in Stevens et al. (2013). It is the intention of this section to utilize their theory to explain the physics and major dynamical aspects of developed convection; an interested reader is referred to the seminal work of Grossmann and Lohse (2000) for all the detailed information about scaling laws for the convective heat flux (measured by the Nusselt number) and the magnitude of the convective velocities (measured by the Reynolds number) with the Rayleigh number, and their thorough derivation.

After Grossmann and Lohse (2000) let us introduce the following simplified physical setting. The system we consider is a layer of fluid of large horizontal extent described by the perfect gas equation, confined between two flat parallel boundaries at distance $L$, with spatially uniform vertical gravity $\mathbf{g} = -g\hat{\mathbf{e}}_z$ pointing downwards and no radiative heat sources, $Q = 0$. This is the configuration often used in numerical experiments. All the physical properties of the fluid, such as $k$ (heat conductivity), $\mu$ (viscosity), $\alpha$ (thermal expansion) and the heat capacities $c_p$, $c_v$ are assumed uniform and $\mu_b = 0$. Standard no-slip conditions are imposed at the boundaries, which are impermeable. The fluid is heated from below and the boundaries are held at constant temperatures. The equations governing the evolution of such a system, under the Boussinesq approximation are given in (2.51a)–(2.51c) with $Q = 0$; note, that the adiabatic gradient in the energy equation is typically negligible in comparison with the temperature gradient of the static state in experimental situations but very large in the case of many natural systems such as the Earth's and planetary cores, stellar interiors, etc. The Nusselt and Rayleigh numbers are therefore as defined in

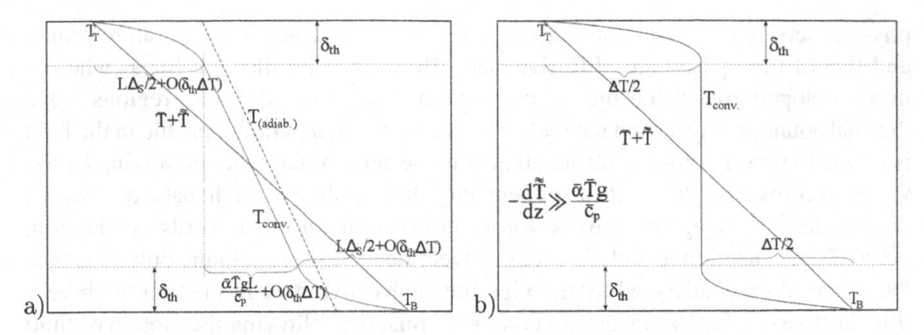


**Fig. 2.3** A schematic picture of vertical profiles of the mean temperature $T_{\text{conv.}} = \tilde{T} + \langle T' \rangle_h$, the hydrostatic reference state temperature $\tilde{T}$ and the adiabatic profile $T_{(\text{adiab.})} = -gz/c_p + \text{const}$ in fully developed convection with fixed temperature at boundaries and $Q = 0$. Figure **a** depicts the general case, whereas figure **b** was constructed to depict the experimentally most common situation, when the adiabatic gradient $g/c_p$ is negligibly small compared to the basic state temperature gradient $\Delta T/L$. The boundary layers are marked, which have the same thicknesses at top and bottom. The temperature jumps across the boundary layers are also the same at top and bottom and are approximately equal to half of the superadiabatic temperature jump across the entire fluid layer (up to corrections of order $\mathcal{O}(\delta_{th}\Delta T)$). Note, that figure **b** could also be interpreted as a profile of superadiabatic temperature $\tilde{T} + \langle T' \rangle_h + gz/c_p = T_B - \Delta_S z + \langle T' \rangle_h$, with the temperature jumps across the boundary layers $\Delta T/2$ replaced by $L\Delta_S/2 + \mathcal{O}(\delta_{th}\Delta T)$ and the top temperature $T_T$ replaced by $T_B - L\Delta_S$

(2.59) and (2.62), but since in the absence of heat sources the static temperature profile is linear in $z$, the expressions simplify to

$$Nu = \frac{\bar{\rho} c_p \langle u_z T \rangle_h - k\partial_z \langle T \rangle_h - kg/c_p}{k\Delta_S}, \tag{2.202}$$

$$Ra = \frac{g\alpha\Delta_S L^4}{\kappa\nu}, \tag{2.203}$$

where $\kappa = k/\bar{\rho}c_p = \text{const}$ and the superadiabatic gradient is now uniform,

$$\Delta_S = \frac{\Delta T}{L} - \frac{g}{\bar{c}_p} > 0, \tag{2.204}$$

and $\Delta T = T_B - T_T > 0$. Although Grossmann and Lohse (2000) do not include the adiabatic gradient in their considerations, their entire analysis and results can also be applied to the more general situation considered here, but with the static state temperature profile replaced by the superadiabatic profile $\bar{T} + \tilde{\tilde{T}}(z) + gz/c_p$, thus effectively $\Delta T$ replaced by $L\Delta_S$.

The physical picture of fully developed convection which emerges from a large number of theoretical, numerical and experimental studies can be summarised as follows. The average temperature profile in developed convection for the considered physical setting is depicted on Fig. 2.3 along with the adiabatic temperature profile and that of the hydrostatic, diffusive state. Both latter profiles are linear whereas in a developed convection the system separates into three different regions - two thermal boundary layers and the bulk. The convective temperature profile in the bulk is bound to be very close to the adiabatic one because of the efficient mixing by the vigorous convective flow, with a gradient only slightly above the adiabatic one, which nevertheless is enough to drive vigorous convection. In other words conduction, $-k\partial_z \left\langle T' \right\rangle_h$, is negligible with respect to advection, $\bar{\rho}c_p \left\langle u_z T \right\rangle_h$, in the bulk.[21] Hence the temperature is advected by the vigorous convective flow in the bulk with very little heat losses, i.e. like an almost conserved quantity following the motion of fluid parcels. This is why the bulk has been called a thermal shortcut in the literature (cf. Grossmann and Lohse 2000). The top and bottom boundary layers adjust the almost adiabatic bulk profile to the boundary conditions, that is $T_T$ and $T_B$ respectively, which results in the formation of large temperature gradients (see Fig. 2.3). In the boundary layers the flow is not as efficient as in the bulk and heat advection is balanced by the enhanced diffusion, since the horizontally averaged vertical temperature gradient can significantly exceed the adiabatic one. The temperature gradient must decrease with the distance from the boundary, therefore the thickness of the boundary layer is established by the distance from the boundary on which conduction, $-k\partial_z \left\langle T \right\rangle_h$, is strong enough to balance (or overcome) advection, $\bar{\rho}c_p \left\langle u_z T \right\rangle_h$. Bearing in mind the Boussinesq up-down symmetry this allows to estimate the total, horizontally averaged superadiabatic convective heat flux as $-kL\Delta_S/2\delta_{th}$ and thus the thickness of thermal boundary layers as

$$\frac{\delta_{th}}{L} = \frac{1}{2Nu}, \tag{2.205}$$

where the factor of a half results from the fact, that the temperature jumps across the two top and bottom boundary layers must be the same, thus equal to $L\Delta_S/2$ and both the boundary layers must have the same thicknesses.

The central idea of the Grossmann and Lohse (2000) theory for convection between rigid, isothermal boundaries introduced on the basis of a vast experimental and numerical evidence involves the existence of a mean convective flow termed the "wind of turbulence", that is a large-scale convection roll, for which the velocity scale will be denoted by $\mathscr{U}$. This scale is used to estimate the advective terms in the viscous boundary layers of the Blasius type, which in turn allows to use the standard estimate for the viscous, laminar boundary layer thickness,

$$\frac{\delta_\nu}{L} = Re^{-1/2}, \qquad Re = \frac{\mathscr{U} L}{\nu}, \tag{2.206}$$

[21] Nevertheless, the conductive and advective fluxes associated with fluctuations about the horizontal means may be in balance, as happens in the case when the total thermal dissipation is dominated by its bulk contribution considered below among other cases; the fluctuations about the horizontal means are small compared to the means in a well-mixed, turbulent bulk.

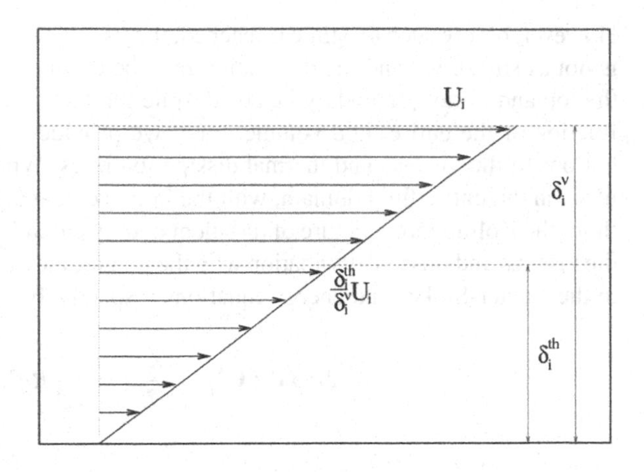


**Fig. 2.4** A schematic picture of nested boundary layers. When the thermal layer is nested in the viscous one, which occurs for $Pr > 1$, the velocity magnitude for the thermal layer needs to be rescaled with a ratio of the boundary layers thicknesses, $\delta_{th}/\delta_\nu < 1$, due to the approximately linear velocity profile in the boundary layers which adjusts the wind of turbulence $\mathscr{U}$ to the zero velocity at boundary

characterized by the Reynolds number $Re$ which measures the ratio of inertia to viscous diffusion. Similar type estimates of temperature advection and thermal diffusion can be made for the thermal boundary layers, which means that the thickness of the thermal boundary layer can be also characterized by the Péclet number, $Pe = \mathscr{U}_{th}L/\kappa$, measuring the ratio of heat advection to heat conduction, thus implying a relation between $Pe$ and $Nu$. There is a subtle difference, however, since the velocity scale $\mathscr{U}_{th}$ used to estimate heat advection is not necessarily the same as the scale $\mathscr{U}$ in the estimate of the advection of momentum, the latter being the mean large-scale convective flow (the wind of turbulence) which stirs the bulk and is responsible for creation of viscous boundary layers. The scale $\mathscr{U}_{th}$ is related to $\mathscr{U}$ and it depends on whether the viscous layer is nested inside the thermal one, $\delta_\nu < \delta_{th}$, or the thermal layer is nested inside the viscous one, $\delta_{th} < \delta_\nu$ as shown on Fig. 2.4. In the latter case, which corresponds to large Prandtl numbers $Pr = \nu/\kappa$, the wind of turbulence has to be scaled with the boundary layer thicknesses ratio $\delta_{th}/\delta_\nu < 1$, to obtain the thermal layer velocity scale in the form $\mathscr{U}_{th} = \mathscr{U}\,\delta_{th}/\delta_\nu$. On the other hand, when $\delta_\nu < \delta_{th}$ there is no need for rescaling the velocity in the thermal layer, as the layer is directly influenced by the wind, so that $\mathscr{U}_{th} = \mathscr{U}$ in this case. Moreover, as it will become evident from the following analysis, a simple estimate of the thermal layer thickness as $Pe^{-1/2}$ leading to a similar advection-diffusion balance as in the viscous layer is not always valid, which suggests a significant role of large horizontal gradients of temperature in the dynamics of thermal layers in some cases.

The main objective now is the derivation of the scaling laws for the Nusselt (2.202) and Reynolds, $Re = \mathscr{U}L/\nu$ numbers with $Ra$ (2.203) in developed convection. The form of the scaling laws crucially depends on whether the thermal and viscous dissipation take place predominantly in the boundary layers or in the bulk of the convection. At a very high Rayleigh number the strongly turbulent convection becomes dominated by a vast number of small-scale structures with strong gradients spread across the entire bulk, in which case the dissipation takes place in the entire fluid domain. Grossmann and Lohse (2000) estimate it to be the case at $Ra \gtrsim 10^{14}$.

However, before such a regime is reached, that is at $10^7 \lesssim Ra \lesssim 10^{14}$, the turbulence is not as strong yet and the dissipation may be dominated by the large gradients in the top and bottom boundary layers, despite the fact, that they occupy only a small fraction of the entire fluid volume. First, we provide estimates of the bulk contributions to the viscous and thermal dissipation rates. When strong dissipation takes place in the entire fluid domain, with the large-scale wind of turbulence stirring the fluid, the Kolmogorov picture of turbulent energy cascade can be applied to estimate the viscous and thermal dissipation with the magnitude of the relevant nonlinear term in the Navier-Stokes and energy equations respectively

$$2\nu\left\langle \mathbf{G}^s : \mathbf{G}^s\right\rangle \approx \frac{\mathscr{U}^3}{L} = \frac{\nu^3}{L^4}Re^3, \tag{2.207a}$$

$$\kappa\left\langle\left(\nabla T'\right)^2\right\rangle \approx \mathscr{U}\frac{(L\Delta_S)^2}{L} = \kappa\Delta_S^2 Pe = \kappa\Delta_S^2 Re Pr, \quad \text{when} \quad \delta_\nu < \delta_{th}, \ Pr\text{-small}, \tag{2.207b}$$

$$\kappa\left\langle\left(\nabla T'\right)^2\right\rangle \approx \frac{\delta_{th}}{\delta_\nu}\mathscr{U}\frac{(L\Delta_S)^2}{L} = \kappa\Delta_S^2 Pe = \frac{1}{2}\kappa\Delta_S^2 Nu^{-1}Re^{3/2}Pr, \quad \text{when} \quad \delta_\nu > \delta_{th}, \ Pr\text{-large}, \tag{2.207c}$$

where

$$Pr = \frac{\nu}{\kappa}, \quad Pe = \frac{\mathscr{U}_{th}L}{\kappa}, \quad \delta_\nu = LRe^{-1/2}, \tag{2.208}$$

and the thermal boundary layer thickness is given in (2.205). By the use of (2.77) we can already see, that when the total thermal dissipation can be estimated by the bulk contributions we have $Nu - 1 \approx Pe$; in turbulent convection the Nusselt number is typically much greater than unity, therefore $Nu - 1 \approx Nu$. Consequently the thickness of the thermal boundary layers can be estimated with the use of the Péclet number as $\delta_{th} \approx 1/2Pe$ for all values of the Prandtl number (that is for both cases of $\delta_\nu < \delta_{th}$ and $\delta_\nu > \delta_{th}$).[22] This suggests that in the case of bulk-dominated

[22]Note, that the temperature equation (2.51c) can in fact be rewritten in terms of the superadiabatic temperature $T_S = \tilde{T} + T' + gz/c_p = T_B - \Delta_S z + T'$, to yield

$$\frac{\partial T_S}{\partial t} + \mathbf{u}\cdot\nabla T_S = \kappa\nabla^2 T_S,$$

where we have used the current assumptions $Q = 0$, $\kappa =$ const and $\alpha T = 1$, the latter being a general property of a perfect gas. It follows, that the estimate $\left\langle(\nabla T_S)^2\right\rangle \approx \Delta_S^2 Pe$, analogous to (2.207b), (2.207c) still holds for the case of bulk-dominated thermal dissipation and since for isothermal boundaries we have $\left\langle\left(\nabla T'\right)^2\right\rangle = \left\langle(\nabla T_S)^2\right\rangle - \Delta_S^2$, inspection of (2.77) allows to obtain the estimate $Nu \approx Pe$ in an even more straightforward way. We stress again, that it is valid only for the case when thermal dissipation takes place predominantly in the bulk.

thermal dissipation the horizontal length scales of temperature variation in the thermal boundary layers are similar to the vertical scale of temperature variation (thus very small, of the order $\delta_{th}$), since advection has to balance diffusion in the boundary layers.

Next we turn to the case, when the total viscous and thermal dissipation are dominated by the contributions from boundary layers. Estimating the magnitude of velocity gradients in the boundary layers by $\mathcal{U}/\delta_\nu$ a straightforward integration of the expression for the total viscous dissipation gives

$$2\nu\left\langle \mathbf{G}^s : \mathbf{G}^s \right\rangle \approx \nu \frac{\mathcal{U}^2}{\delta_\nu^2} \frac{\delta_\nu}{L} \approx \frac{\nu^3}{L^4} Re^{5/2}, \tag{2.209}$$

where the factor of $\delta_\nu / L$ results from vertical integration and the fraction of the total volume occupied by the boundary layers gives the dominant contribution to dissipation. A similar estimate for the thermal dissipation leads to a tautology, that is

$$\kappa\left\langle (\nabla T')^2 \right\rangle \approx \kappa \frac{(L\Delta_S)^2}{\delta_{th}^2} \frac{\delta_{th}}{L} \approx \kappa \Delta_S^2 Nu, \tag{2.210}$$

which is equivalent to the rigorous relation (2.77) (up to neglection of unity with respect to $Nu \gg 1$),[23] thus no new information is provided. However, it is worth noting, that the above estimate in (2.210) can always be applied solely to the boundary layers, that is with the total average replaced by the horizontal one and vertical integration over the thermal boundary layers only, $0 < z < \delta_{th}$ and $L - \delta_{th} < z < L$. This means, that in fact the thermal dissipation in the thermal boundary layers is always of the same order of magnitude in terms of the non-dimensional numbers $Nu$ or $Ra$ as the total thermal dissipation. Hence the formerly considered regime of strong bulk dissipation in (2.207a)–(2.207c) corresponds to at most a moderate domination of the bulk contribution to the total dissipation over contributions from boundary layers; the Eqs. (2.207a)–(2.207c) are nevertheless valid, even if the boundary layer dissipation is comparable to the one in the bulk.

It is however possible for the boundary layer contribution to the total thermal dissipation to dominate over the bulk one. In such a case, to obtain the necessary relations between the numbers $Nu$, $Re$ and $Ra$, $Pr$ one needs to consider the advection-diffusion balance within the thermal boundary layers. Estimating the horizontal temperature gradients by $\Delta_S$ and the diffusive term by $\kappa L \Delta_S / \delta_{th}^2$ one obtains

$$\mathcal{U}\Delta_S = \frac{\kappa L \Delta_S}{\delta_{th}^2} \Rightarrow Nu \approx \frac{1}{2} Re^{1/2} Pr^{1/2}, \quad \text{when} \quad \delta_\nu < \delta_{th}, \ Pr\text{-small}, \tag{2.211a}$$

$$\frac{\delta_{th}}{\delta_\nu}\mathcal{U}\Delta_S = \frac{\kappa L \Delta_S}{\delta_{th}^2} \Rightarrow Nu \approx \frac{1}{2} Re^{1/2} Pr^{1/3}, \quad \text{when} \quad \delta_\nu > \delta_{th}, \ Pr\text{-large}. \tag{2.211b}$$

[23] See also the footnote 22 for $\left\langle (\nabla T_S)^2 \right\rangle \approx \Delta_S^2 Nu$, which is exactly equivalent to (2.77).

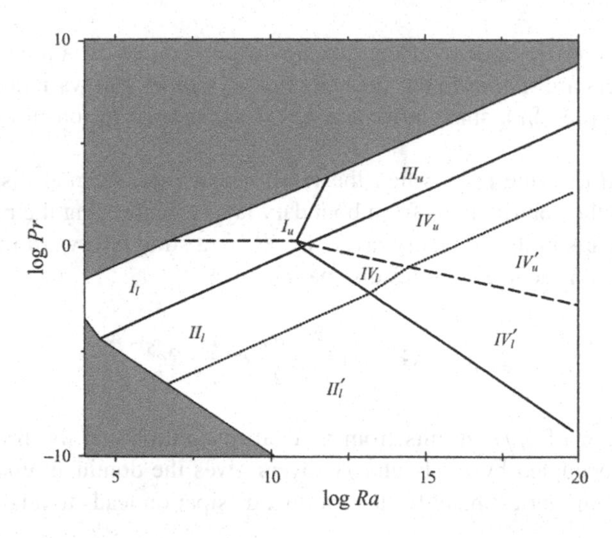


**Fig. 2.5** A diagram of distinct dynamical regimes for developed Boussinesq convection with fixed temperature on boundaries and no heat sources $Q = 0$. The different regimes are determined by different combinations of dominant contributions to viscous and thermal dissipation (cf. (2.207a)–(2.207c) for bulk estimates and (2.209) and (2.211a), (2.211b) for boundary layer estimates). The dashed line, obtained by equating $\delta_{th} = LNu^{-1}$ and $\delta_\nu = LRe^{-1/2}$ divides the parameter space into regions with low and high $Pr$ denoted by subscripts $l$ and $u$ respectively. The gray regions correspond to regimes of not fully developed turbulence (after Grossmann and Lohse 2000)

It is noteworthy, that the above estimates (2.211a), (2.211b) in conjunction with the exact Eq. (2.77) allow to estimate the thickness of the thermal boundary layers with the Péclet number as $\delta_{th} \approx Pe^{-1/2}$ for all values of the Prandtl number. This resembles the Blasius layer relation $\delta_\nu = Re^{-1/2}$, the reason for that being that the estimates of horizontal and vertical gradients ($\nabla_h \sim 1/L$, $\partial_z \sim 1/\delta_{th}$) utilized here are of same type as in the standard Blasius boundary layer theory, hence lead to the same type of final distinguished balance. We recall, that in the formerly considered regime of extremely large Rayleigh numbers we obtained $\delta_{th} \approx (2Pe)^{-1}$ (see discussion below (2.208) and the footnote 22), which suggests, that in that regime the boundary layer balance on the far left of relations (2.211a), (2.211b) has to be modified by a new estimate of the horizontal gradients, as large as $L\Delta_S/\delta_{th}$.

Having now the complete set of relations between the parameters $Nu$ and $Re$, which describe the response of the system to driving described by $Ra$ with system properties included in $Pr$, we can equate the exact expressions for viscous and thermal dissipation in (2.77) and (2.69) with their estimates for different dynamical regimes in (2.207a)–(2.207c), (2.209) and (2.211a), (2.211b). This results in four different basic regimes of turbulent convection, in which the dominant contributions to total viscous or thermal dissipation come either from the bulk or the boundary layers, which sums up to four cases symbolically denoted as $\nu_{BL}\kappa_{BL}$ (regime I), $\nu_{bulk}\kappa_{BL}$ (regime II), $\nu_{BL}\kappa_{bulk}$ (regime III) and $\nu_{bulk}\kappa_{bulk}$, (regime IV), where *BL* stands for boundary layer. Figure 2.5 taken from Grossmann and Lohse (2000) depicts the regimes on the $Pr$-$Ra$ plane. Each of the regimes is further divided into a "lower" regime achieved at small values of the Prandtl number when $\delta_\nu < \delta_{th}$ and an "upper" regime obtained at large $Pr$ when $\delta_\nu > \delta_{th}$. The scaling relations $Nu(Ra, Pr)$ and $Re(Ra, Pr)$ obtained from equating the exact expressions and estimates of dissipation are therefore distinct in all the regimes $I_l, I_u, II_l, \ldots IV_u$. E.g. in the bulk-dominated regime at small Prandtl number one obtains $Nu \sim Ra^{1/2}Pr^{1/2}$ and $Re \sim Ra^{1/2}Pr^{-1/2}$ whereas in the regime dominated by dissipation in boundary layers at small Prandtl number $Nu \sim Ra^{1/4}Pr^{1/8}$ and $Re \sim Ra^{1/2}Pr^{-3/4}$. Furthermore, the boundaries between the basic regimes $I$, $II$, $III$ and $IV$ are calculated by equating the relevant dissipation estimates, e.g. the boundaries between $I$ and $II$ and between $III$ and $IV$ are obtained by equating the viscous dissipation estimates (2.209) and (2.207a) but with the Reynolds numbers expressed in terms of the $Ra$ and $Pr$ numbers through the corresponding scalings; the boundary between $I_u$ and $III_u$ is calculated by equating (2.211b) and (2.207c), again, expressed by $Ra$ and $Pr$ with the use of the corresponding scalings, etc. The dashed line on Fig. 2.5, obtained by equating $\delta_{th} = LNu^{-1}$ and $\delta_\nu = LRe^{-1/2}$ divides the parameter space into regions with low and high $Pr$ denoted by subscripts $l$ and $u$ respectively. The primed regimes marked below the dotted line, reached at high enough values of $Ra$ are characterized by already turbulent boundary layers and therefore are all bulk-dominated. In other words the scalings in regime $II_l'$ are the same as in $IV_l$. A complete derivation and discussion of the scaling laws for all different possible dynamical regimes is provided in Grossmann and Lohse (2000), where also the prefactors for scaling relations are obtained based on comparison with laboratory and numerical experiments, later updated in Stevens et al. (2013).

## 2.5 Validity of the Approximation and Summary

Let us briefly recall all the necessary assumptions for validity of the Boussinesq approximation. *Firstly* it must be required, that the density, temperature and pressure scale heights satisfy

$$L \ll \min\left\{D_\rho,\ D_T,\ D_p\right\}, \tag{2.212}$$

everywhere in the fluid domain. Using this assumption we define the small parameter

$$\epsilon = \frac{\Delta\tilde{\tilde{\rho}}}{\bar{\rho}} \ll 1, \tag{2.213}$$

which is small by virtue of integration of relation $L/D_\rho \ll 1$ from the level of minimal to the level of maximal density within the fluid layer and the density jump between these layers is denoted by $\Delta\tilde{\tilde{\rho}}$. A *second restriction* concerns the magnitude of the fluctuations of thermodynamic variables induced by the convective motions, which effectively implies

$$\frac{\rho'}{\bar{\rho}} = \mathcal{O}(\epsilon), \quad \frac{T'}{\bar{T}} = \mathcal{O}(\epsilon), \quad \frac{p'}{\bar{p}} = \mathcal{O}\left(\epsilon^2\right), \tag{2.214}$$

$$\frac{\tilde{\tilde{\rho}}}{\bar{\rho}} = \mathcal{O}(\epsilon), \quad \frac{\tilde{\tilde{T}}}{\bar{T}} = \mathcal{O}(\epsilon), \quad \frac{\tilde{\tilde{p}}}{\bar{p}} = \mathcal{O}(\epsilon), \quad \frac{\bar{\rho} g L}{\bar{p}} = \mathcal{O}(\epsilon), \tag{2.215}$$

(an assumption, which in principle needs to be checked for consistency a posteriori) and for the magnitude of convective velocity $\mathscr{U}$, the time scales $\mathscr{T}$, the diffusivities and the heat capacity

$$\mathscr{U} \sim \epsilon^{1/2}\sqrt{gL}, \quad \mathscr{T} \sim \epsilon^{-1/2}\sqrt{\frac{L}{g}}, \tag{2.216}$$

$$\mu_b/\bar{\rho} \lesssim \nu \sim \epsilon^{1/2}\sqrt{gL}L, \quad \kappa = \frac{k}{\bar{\rho}\bar{c}_p} \sim \epsilon^{1/2}\sqrt{gL}L, \quad \bar{c}_p \sim \bar{c}_v \sim \epsilon^{-1}\frac{gL}{\bar{T}}. \tag{2.217}$$

A *third weakest restriction* concerns the fluid thermodynamic properties,

$$C_T \gg \sqrt{gL}, \quad \text{or equivalently} \quad \frac{1}{g\bar{\rho}\bar{\beta}} \gg L. \tag{2.218}$$

The latter assumption is easily satisfied for most of fluids by virtue of the first assumption (2.213), in particular obviously satisfied for a weakly stratified perfect gas. The set of above assumptions leads to the following set of momentum, energy and mass balance supplied by the equation of state, under the Boussinesq approximation at leading order in $\epsilon$

$$\frac{\partial \mathbf{u}}{\partial t} + (\mathbf{u}\cdot\nabla)\mathbf{u} = -\frac{1}{\bar{\rho}}\nabla p' + g\bar{\alpha}T'\hat{\mathbf{e}}_z + \nu\nabla^2\mathbf{u} + 2\nabla\nu\cdot\mathbf{G}^s, \tag{2.219a}$$

$$\nabla\cdot\mathbf{u} = 0, \tag{2.219b}$$

$$\frac{\partial T'}{\partial t} + \mathbf{u}\cdot\nabla T' + u_z\left(\frac{\mathrm{d}\tilde{\tilde{T}}}{\mathrm{d}z} + \frac{g\bar{\alpha}\bar{T}}{\bar{c}_p}\right) = \nabla\cdot\left(\kappa\nabla T'\right) + \frac{Q'}{\bar{\rho}\bar{c}_p}, \tag{2.219c}$$

$$\rho = \bar{\rho} + \tilde{\tilde{\rho}} + \rho' = \bar{\rho}\left[1 - \bar{\alpha}\tilde{\tilde{T}} + \bar{\beta}\tilde{\tilde{p}} - \bar{\alpha}T'\right].$$

The system (2.219a-c) is a closed set of equations for the velocity field $\mathbf{u}$, temperature fluctuation $T'$ and the pressure fluctuation $p'$, whereas the state equation can be used for calculation of the density fluctuation $\rho' = -\bar{\alpha}T'$. The equations describing the hydrostatic state (denoted by the subscript 0) are given in (2.9a), (2.9b) and (2.12a).

Finally, in what follows a short summary of main aspects of Boussinesq convection is provided. The general, useful definition of the Rayleigh number is given in the form

$$Ra = \frac{g\bar{\alpha}\langle\kappa\Delta_S\rangle L^4}{\bar{\kappa}^2\bar{\nu}}, \tag{2.220}$$

which is a measure of the magnitude of the superadiabatic gradient, thus of the driving force for convective flow. Two Boussinesq systems are dynamically equivalent when the Rayleigh numbers and the Prandtl numbers $Pr = \nu/\kappa$ are the same for both systems (likewise in the case when some additional effects are present any other nondimensional number describing the relative strength of those effects, such as e.g. the Ekman number $E = \nu/2\Omega L^2$ describing the effect of rotation); in such a case the response of the systems to driving must be the same. One possible measure of the dynamical response of a convective system to driving, in the absence of heat sources $Q = 0$ is provided by the Nusselt number, which is defined as a ratio of the total superadiabatic convective heat flux over the superadiabatic molecular heat flux in a corresponding hydrostatic state[24] in the following way

$$Nu = \frac{\bar{\rho}\bar{c}_p\langle u_z T\rangle_h - k\partial_z\langle T\rangle_h - \bar{k}g\bar{\alpha}\bar{T}/\bar{c}_p}{\langle k\Delta_S\rangle} = \frac{L}{2\delta_{th}}, \tag{2.221}$$

for isothermal boundaries and

$$Nu_Q = \frac{\left\langle\bar{\rho}\bar{c}_p u_z T'\right\rangle}{\langle k\Delta_S\rangle} = \frac{1}{2}\left(\frac{L}{\delta_{th}} - 1\right) \approx \frac{L}{2\delta_{th}}, \tag{2.222}$$

when the heat flux at boundaries is held fixed. A characteristic feature of Boussinesq systems is that when $Q = 0$ the total, horizontally averaged heat flux through every horizontal plane is the same, in other words the total heat flux is height independent. $\delta_{th}$ denotes the thickness of the thermal boundary layers which form at the top and bottom in fully developed convection with large temperature gradients to account for the large convective heat flux in the bulk of the flow.

We also recall here the results of Sect. 2.1.4. The conservation of mass implies, that in order for the density and thus temperature fluctuations to be correctly resolved, the jump of the mean pressure fluctuation across the depth of the fluid layer must vanish at all times, i.e. $\left\langle p'\right\rangle_h (z = L) - \left\langle p'\right\rangle_h (z = 0) = 0$. This constitutes a boundary condition, which must be imposed on the pressure field.

Furthermore, there are two possible types of up-down symmetry in Boussinesq convection with symmetric top-bottom boundary conditions: the *nonlinear up-down symmetry*, the only one allowed by the nonlinear equations, defined by

$$u_z(L - z) = -u_z(z), \quad \mathbf{u}_h(L - z) = \mathbf{u}_h(z), \tag{2.223}$$

$$T'(L - z) = -T'(z), \quad p'(L - z) = p'(z), \tag{2.224}$$

[24] Such a hydrostatic state is, of course, unstable, since it corresponds to the same driving $\Delta_S$ as that in the convective state.

and corresponding to developed convection, and the *linear up-down symmetry*, opposite to the previous one

$$u_z(L-z) = u_z(z), \quad \mathbf{u}_h(L-z) = -\mathbf{u}_h(z), \tag{2.225}$$

$$T'(L-z) = T'(z), \quad p'(L-z) = -p'(z), \tag{2.226}$$

allowed only by the linearised dynamical equations and corresponding to large-scale convective rolls at convection threshold. The critical Rayleigh number for convection, in the absence of heat sources $Q = 0$, for $\nu = \text{const.}$, $\kappa = \text{const.}$ and for the case when the instability sets in as stationary flow can be calculated from the following formula

$$Ra_{crit} = \frac{L^4 \min\left[-\frac{1}{\mathcal{K}^2}\left(\frac{d^2}{dz^2} - \mathcal{K}^2\right)^3 \hat{u}_z\right]}{\hat{u}_z}, \quad \text{at any } z \tag{2.227}$$

once the flow at threshold is obtained as a solution of the dynamical equations with the growth rate $\sigma = 0$. A more general formula for $Q \neq 0$, for $\nu = \nu(z)$, $\kappa = \kappa(z)$ and $\mathbf{\Omega} \neq 0$, thus including the possibility of oscillatory flow at convection threshold can be formulated to yield

$$\begin{aligned} Ra_{crit} =& \frac{\bar{\alpha} g \langle \kappa \Delta_S \rangle L^4}{\bar{\kappa}^2 \bar{\nu}} \\ =& \min \Re\mathfrak{e} \left\langle \frac{\kappa(z)}{\bar{\kappa}} \frac{\left[\mathrm{i}\omega - \kappa(z)\left(\frac{d^2}{dz^2} - \mathcal{K}^2\right)\right]\left(\frac{\bar{\alpha} g L^4 \hat{T}}{\bar{\kappa}\bar{\nu}}\right) - \frac{\bar{\alpha} g L^4 \hat{Q}}{\bar{\rho}\bar{c}_p \bar{\kappa}\bar{\nu}}}{\hat{u}_z} \right\rangle, \end{aligned} \tag{2.228}$$

where the frequency of oscillations at threshold $\omega$ as a function of the horizontal wave number $\mathcal{K}$ and the inverse vertical variation scale (vertical "wave number") of perturbations $q$ (say), $\omega = \omega(\mathcal{K}, q)$ can be determined by equating the imaginary part of the expression in the angular brackets on the right hand side of the latter equation to zero,

$$\Im\mathfrak{m} \left\{ \frac{\kappa(z)}{\bar{\kappa}} \frac{\left[\mathrm{i}\omega - \kappa(z)\left(\frac{d^2}{dz^2} - \mathcal{K}^2\right)\right]\left(\frac{\bar{\alpha} g L^4 \hat{T}}{\bar{\kappa}\bar{\nu}}\right) - \frac{\bar{\alpha} g L^4 \hat{Q}}{\bar{\rho}\bar{c}_p \bar{\kappa}\bar{\nu}}}{\hat{u}_z} \right\} = 0. \tag{2.229}$$

In the marginal state the heat per unit mass released by a rising fluid parcel on an infinitesimal vertical distance in a time unit can be expressed by $\bar{c}_p \kappa \nu Ra_{crit} u_z / g\bar{\alpha} L^4$. For growth rates of perturbations to the hydrostatic state in the vicinity of convection threshold, i.e for $Ra - Ra_{crit} \ll Ra_{crit}$, $\mathcal{K}L - \mathcal{K}_{crit}L \ll 1$, see the general relation (2.155) and Eqs. (2.158) and (2.166) for the particular case of stress-free and isothermal boundaries.

## Review Exercises

**Exercise 1** Formulate the fundamental assumptions which lead to the Boussinesq system of equations and derive the elliptic Poisson-type equation for pressure.
*Hint*: The assumptions are formulated in (2.5) and (2.8) (or 2.28a)–(2.28d), supplied by (2.18), (2.40) and (2.43).

**Exercise 2** Calculate the mean value of the entropy fluctuation in the Boussinesq convection, $\langle s' \rangle$.
*Hint*: cf. the Eq. (2.48) and Sect. 2.1.4.

**Exercise 3** Consider Boussinesq fluid of viscosity

$$\nu = \nu_0 \left(1 - \frac{z^2}{2L^2}\right),$$

convectively driven by a fixed temperature difference between bottom and top plates $\Delta T$, under uniform gravity $g$. The mean temperature $\bar{T}$, mean density $\bar{\rho}$ and the mean specific heat $\bar{c}_p$ are given; the fluid is described by the equation of state of the perfect gas. The heat flux of a hydrostatic state is also given and is denoted by $F_0$. The thermal diffusivity is uniform. Given the total viscous dissipation rate $D_\nu$ calculate the total convective heat flux in the system $F_{conv}$. Then calculate the Rayleigh number.
*Hint*: utilize the results of Sect. 2.1.3.1 to show that $F_{conv} = \langle k\Delta_S \rangle (Nu - 1) = \bar{\rho}\bar{c}_p\bar{T} D_\nu / g$ and

$$Ra = \frac{6g\Delta T^2 L^2}{5\bar{T} F_0 \nu_0}\left(1 - \frac{gL}{\bar{c}_p \Delta T}\right).$$

**Exercise 4** Calculate the growth rate of convection near threshold in a system rotating at the angular velocity $\mathbf{\Omega} = \Omega \hat{\mathbf{e}}_z$.
*Hint*: calculate roots of the cubic equation (2.156).

**Exercise 5** By the use of the weakly nonlinear theory derive the relation between the Nusselt and Rayleigh numbers near convection threshold.
*Hint*: cf. Sect. 2.3.

## References


Bénard, H. 1900. Les tourbillons cellulaires dans une nappe liquide, Reviews of Gén. *Sciences Pure and Applied* 11 (23): 1261–1271, 11 (24): 1309–1328.

Boussinesq, J. 1903. *Théorie analytique de la chaleur*, vol. 2. Paris: Gauthier-Villars.

Busse, F.H. 1967. The stability of finite amplitude cellular convection and its relation to an extremum principle. *Journal of Fluid Mechanics* 30 (4): 625–649.

Busse, F.H. 1969. On Howard's upper bound for heat transport by turbulent convection. *Journal of Fluid Mechanics* 37 (3): 457–477.

Busse, F.H. 1970. Thermal instabilities in rapidly rotating systems. *Journal of Fluid Mechanics* 44 (3): 441–460.

Busse, F.H. 1978. Non-linear properties of thermal convection. *Reports on Progress in Physics* 41: 1929–1967.

Busse, F.H., and P.G. Cuong. 1977. Convection in rapidly rotating spherical fluid shells. *Geophysical & Astrophysical Fluid Dynamics* 8 (1): 17–41.

Busse, F.H., and N. Riahi. 1980. Nonlinear convection in a layer with nearly insulating boundaries. *Journal of Fluid Mechanics* 96: 243–256.

Cerisier, P., S. Rahal, J. Cordonnier, and G. Lebon. 1998. Thermal influence of boundaries on the onset of Reyleigh-Bénard convection. *International Journal of Heat and Mass Transfer* 41: 3309–3320.

Chandrasekhar, S. 1961. In *Hydrodynamic and hydromagnetic stability*, ed. Marshall, W., and D.H. Wilkinson., International series of monographs on physics Clarendon Press: Oxford University Press (since 1981 printed by Dover Publications).

Choffrut, A., C. Nobili, and F. Otto. 2016. Upper bounds on Nusselt number at finite Prandtl number. *The Journal of Differential Equations* 260: 3860–3880.

Clever, R.M., and F.H. Busse. 1994. Steady and oscillatory bimodal convection. *Journal of Fluid Mechanics* 271: 103–118.

Cordon, R.P., and M.G. Velarde. 1975. On the (non-linear) foundations of Boussinesq approximation applicable to a thin layer of fluid. *Journal de Physique* 36 (7–8): 591–601.

Doering, C.R., and P. Constantin. 1996. Variational bounds on energy dissipation in incompressible flows. III. Convection. *Physical Review E* 53 (6): 5957–5981.

Fauve, S. 2017. Henri Bénard and pattern-forming instabilities. *Comptes Rendus Physique* 18: 531–543.

Gershuni, G.Z., and Zhukhovitskii. 1976. *Convective stability of incompressible fluids, Israel Program for Scientific Translations Jerusalem (translated from Russia: 1972)*. Moscow: Nauka.

Getling, A.V. 1980. On the scales of convection flows in a horizontal layer with radiative energy transfer. *Atmospheric and Oceanic Physics* 16: 363–365.

Getling, A.V. 1998. *Rayleigh-Bénard convection: Structures and dynamics*, vol. 11., Advanced Series in Nonlinear Dynamics Singapore: World Scientific Publishing.

Goody, R.M. 1956. The influence of radiative transfer on cellular convection. *Journal of Fluid Mechanics*. 1 (4): 424–435. (Note: Goody, R.M. 1956. Corrigendum. *Journal of Fluid Mechanics*. **1**(6): 670).

Goody, R.M. 1995. *Principles of atmospheric physics and chemistry*. New York: Oxford University Press.

Goody, R.M., and Y.L. Yung. 1989. *Atmospheric radiation: theoretical basis*. New York: Oxford University Press.

Grossmann, S., and D. Lohse. 2000. Scaling in thermal convection: a unifying theory. *Journal of Fluid Mechanics* 407: 27–56.

Howard, L.N. 1963. Heat transport by turbulent convection. *Journal of Fluid Mechanics* 17 (3): 405–432.

Jakeman, E. 1968. Convective instability in fluids of high thermal diffusivity. *Physics of Fluids* 2 (1): 10–14.

Joseph, D.D. 1976. In *Stability of fluid motions II*, vol. 28, ed. Coleman, B.D., Springer Tracts in Natural Philosophy Berlin: Springer.

Knobloch, E. 1990. Pattern selection in long-wavelength convection. *Physica D* 41 (3): 450–479.

Korn, G.A., and T.M. Korn. 1961. *Mathematical handbook for scientists and engineers*. Mineola, New York: Dover Publications.

Larson, V.E. 2001. The effects of thermal radiation on dry convective instability. *Dynamics of Atmospheres and Oceans* 34: 45–71.

Lorenz, L. 1881. Über das Leitungsvermögen der Metalle für Wärme und Elektrizität. *Annalen der Physik und Chemie* 13: 582–606.

Malkus, W.V.R. 1954. Discrete transitions in turbulent convection. *Proceedings of The Royal Society A* 225: 185–195.
Manneville, P. 1983. A two-dimensional model for three-dimensional convective patterns in wide containers. *Journal de Physique* 44 (7): 759–765.
Mihaljan, J.M. 1962. A rigorous exposition of the Boussinesq approximation applicable to a thin layer of fluid. *The Astrophysical Journal* 136 (3): 1126–1133.
Nobili, C. 2015. Rayleigh-Bénard convection: bounds on the Nusselt number, Dissertation, Universität Leipzig (available at: http://ul.qucosa.de/api/qucosa%3A14688/attachment/ATT-0/).
Oberbeck, A. 1879. Über die wärmeleitung der flüssigkeiten bei der berücksichtigung der strömungen infolge von temperaturdifferenzen. *Annual Review of Physical Chemistry* 7 (6): 271–292.
Ostrach, S. 1957. Convection phenomena in fluids heated from below. *Transactions of the ASME* 79: 299–305.
Park, H., and L. Sirovich. 1991. Hydrodynamic stability of Rayleigh-Bénard convection with constant heat flux boundary condition. *Quarterly of Applied Mathematics* XLIX (2): 313–332.
Pellew, A., and R.V. Southwell. 1940. On maintained convective motion in a fluid heated from below. *Proceedings of The Royal Society A* 176: 312–343.
Proctor, M.R.E. 1981. Planform selection by finite-amplitude thermal convection between poorly conducting slabs. *Journal of Fluid Mechanics* 113: 469–485.
Rayleigh, Lord. 1916. On convection currents in a horizontal layer of fluid when the higher temperature is on the under side. *Philosophical Magazine* 32: 529–546.
Schlüter, A., D. Lortz, and F. Busse. 1965. On the stability of steady finite amplitude convection. *Journal of Fluid Mechanics* 23 (1): 129–144.
Sparrow, E.M., R.J. Goldstein, and V.K. Jonsson. 1964. Thermal instability in a horizontal fluid layer: effect of boundary conditions and non-linear temperature profile. *Journal of Fluid Mechanics* 18 (4): 513–528.
Spiegel, E.A., and G. Veronis. 1960. On the Boussinesq approximation for a compressible fluid. *The Astrophysical Journal* 131 (2): 442–447.
Stevens, R.J.A.M., E.P. van der Poel, S. Grossmann, and D. Lohse. 2013. The unifying theory of scaling in thermal convection: the updated prefactors. *Journal of Fluid Mechanics* 730: 295–308.
Straughan, B. 2004. In *The energy method, stability, and nonlinear convection*, vol. 91, ed. Antman, S.S., J.E. Marsden, and L. Sirovich., Applied Mathematical Sciences New York: Springer.
Thomson, J. 1882. On a changing tessellated structure in certain liquids. *Proceedings of the Philosophical Society of Glasgow* 13: 464–468.
Velarde, M.G., and R.P. Cordon. 1976. On the (non-linear) foundations of Boussinesq approximation applicable to a thin layer of fluid (II). Viscous dissipation and large cell gap effects. *Journal de Physique* 37 (3): 177–182.
Veronis, G. 1962. The magnitude of the dissipation terms in the Boussinesq approximation. *The Astrophysical Journal* 135 (2): 655–656.

# Chapter 3
# Anelastic Convection

The Boussinesq approximation is applicable to thin layers of fluid, where the density variation is weak. However, in most astrophysical applications such a thin layer approximation is not satisfactory, because the typical scale heights in the system, $D_p$, $D_\rho$ and $D_T$ (cf. Eq. (2.4)) are comparable or even significantly smaller than the depth of the convective layer. Still, the phenomenon of convection is very common in natural systems and it is in fact a crucial factor in the dynamics of stellar and planetary interiors and atmospheres, since the convectively driven flow transports energy and angular momentum. Moreover, convection in electrically conducting domains within the stellar and planetary interiors (cores) is responsible for the hydromagnetic dynamo effect, in other words the generation of magnetic fields of those astrophysical bodies (see, e.g. Soward 1991, Tobias and Weiss 2007a, b). This results in a strong need for an accurate description of convection in systems with strong density stratification. The fully compressible models are very cumbersome due to inclusion of the dynamics of fast sound waves, which needs to be thoroughly resolved, hence it is desirable, that the mathematical description of planetary and stellar convection allows to filter out the sound waves, similarly as in the case of the Boussinesq approximation. In order to satisfy the needs for a sound-proof description of strongly stratified convection the anelastic approximation was formulated by Ogura and Phillips (1962) and Gough (1969). Ogura and Phillips (1962) also invented the name *anelastic* approximation, based on the fact, that what they called an 'elastic' part of the internal energy of the fluid can be neglected. From a more general point of view the distinction of the 'elastic energy' is not necessarily strictly definite, nevertheless the name anelastic approximation has been established through a wide use over many decades.

The anelastic approximation was later even more thoroughly explained and generalized to the magnetohydrodynamic case by Lantz and Fan (1999). The main idea underlying this approach is that guided by the observation that natural large-scale



K. A. Mizerski, *Foundations of Convection with Density Stratification*, GeoPlanet: Earth and Planetary Sciences, https://doi.org/10.1007/978-3-030-63054-6_3

convective systems in their long-time evolution develop states that are nearly adiabatic, and only a slight excess above the adiabatic gradient drives a very vigorous convective flow, a fundamental assumption of weak superadiabaticity of the dynamical system is put forward. This implies, that the Mach number, i.e. the ratio of convective velocity to the speed of sound is small, but the temperature and density stratification can be arbitrary. There is a large amount of scientific literature on the properties of anelastic convection. Although we do not intend here to provide a complete review of the developments on various dynamical aspects of convection under the anelastic approximation, it is without a doubt of interest to direct an interested reader to some of the most important findings. In a series of papers Gilman and Glatzmaier (1981) and Glatzmaier and Gilman (1981a, b) have studied the influence of various physical effects and conditions, such as the dissipative effects, the boundary conditions and zone depth on the dynamics of anelastic convection. They have also studied the linear convection onset in spherical shells, which was further investigated by Drew et al. (1995). It is important to point out at this stage, that recently Calkins et al. (2015) and Verhoeven and Glatzmaier (2018) have demonstrated, that in rapidly rotating low-Prandtl number systems the anelastic approximation breaks down. More precisely when the rate of background rotation is fast enough, so that the value of $\Omega L$ starts to exceed the mean speed of sound $\bar{C}$ and for $\nu/\kappa \ll 1$ the growth rate of the convective instability at threshold is greatly enhanced and the time derivative of the density fluctuation in the continuity equation ceases to be negligible. In turn the sound waves start to play a dynamical role (cf. (1.80) and the discussion below) and thus the evolution obtained from the anelastic system of equations does not capture a crucial factor of the dynamics; therefore in this limit the anelastic approximation is clearly not applicable. Nevertheless, for non-rotating systems or systems for which the background rotation is not as rapid, i.e. satisfies $\Omega L \ll \bar{C}$ the anelastic approximation was never shown to break down and the restriction $\Omega L \ll \bar{C}$ for its validity is by no means strong.

Furthermore, a notable development concerning adaptation of the anelastic formulation to the dynamics of the Earth's core, together with a comprehensive discussion of the dynamics of geophysical convection was done in a seminal paper by Braginsky and Roberts (1995). Even further developments on the topic, including comparison of the anelastic and Boussinesq approaches in the Earth's core context have been performed by Anufriev et al. (2005); a more general comparison of the two sound-proof approaches can be found in Lilly (1996).

From the point of view of dynamics of compressible atmospheres the issue of stratification in the hydrostatic reference state is an important one. Noteworthy models developed for description of dynamics of convection in compressible atmospheres include e.g. Wilhelmson and Ogura (1972), Lipps and Hemler (1982) and Durran (1989). The latter was further analysed and explained by Durran (2008), Klein (2009), Klein et al. (2010) and Klein and Pauluis (2012). A survey of different approaches to stratified convection in compressible atmospheres was provided by Bannon (1996).

In the following sections of this chapter we consider a general case of an arbitrary equation of state and provide a thorough discussion of all undertaken steps necessary to derive the anelastic equations; possible further simplifications in special cases are proposed. The linear convection close to convection threshold likewise the general energetic characteristics of anelastic convection and nonlinear heat transfer at fully developed state are described. Finally a comparison of different approaches with an adiabatic and non-adiabatic reference temperature profiles is provided and a relation between the anelastic and Boussinesq approximations is explained.

## 3.1 Derivation of the Anelastic Equations

The anelastic liquid approximation is commonly applied to the study of planetary and stellar interiors where convection takes place in large and heavy fluid regions. Those regions typically have comparable mass with the entire celestial body that they are a part of. Therefore typically gravity also significantly varies with depth. Moreover, convection disturbs the mass distribution in the fluid layers providing further corrections to the acceleration of gravity. Consequently variation of gravity should, in general, be included (allowed) in the mathematical description of anelastic convection. The full set of dynamical equations expressing the physical laws of conservation of momentum, mass and energy, supplied by the equations of state and the gravitational potential equation takes the form

$$\rho\left[\frac{\partial \mathbf{u}}{\partial t} + (\mathbf{u}\cdot\nabla)\,\mathbf{u}\right] = -\nabla p + \rho\mathbf{g} + \mu\nabla^2\mathbf{u} + \left(\frac{\mu}{3} + \mu_b\right)\nabla(\nabla\cdot\mathbf{u}) + 2\nabla\mu\cdot\mathbf{G}^s + \nabla\left(\mu_b - \frac{2}{3}\mu\right)\nabla\cdot\mathbf{u}, \tag{3.1a}$$

$$\frac{\partial \rho}{\partial t} + \nabla\cdot(\rho\mathbf{u}) = 0, \tag{3.1b}$$

$$\nabla^2\psi = 4\pi G\rho, \qquad \mathbf{g} = -\nabla\psi, \tag{3.1c}$$

$$\rho T\left(\frac{\partial s}{\partial t} + \mathbf{u}\cdot\nabla s\right) = \nabla\cdot(k\nabla T) + 2\mu\mathbf{G}^s : \mathbf{G}^s + \left(\mu_b - \frac{2}{3}\mu\right)(\nabla\cdot\mathbf{u})^2 + Q, \tag{3.1d}$$

$$\rho = \rho(p, T), \quad s = s(p, T), \tag{3.1e}$$

where $G$ is the gravitational constant and $\psi$ is the gravitational potential. Once again let us assume that the shear dynamical viscosity $\mu$, the bulk viscosity $\mu_b$, the specific heat $c_v$ and thermal conduction $k$ are nonuniform for generality, however, the latter is a function of depth only. Furthermore, we take the $z$-axis of the coordinate system so that it is perpendicular to the plates at $z = 0,\ L$. For time-independent boundary conditions we decompose the thermodynamic variables into the static (reference) state contributions, denoted by an upper tilde (now incorporating both, the mean and the vertically varying static departure from mean, previously denoted by double tilde) and fluctuations induced by convective flow denoted by prime,

$$\rho(\mathbf{x}, t) = \tilde{\rho}(z) + \rho'(\mathbf{x}, t), \tag{3.2a}$$

$$T(\mathbf{x}, t) = \tilde{T}(z) + T'(\mathbf{x}, t), \tag{3.2b}$$

$$p(\mathbf{x}, t) = \tilde{p}(z) + p'(\mathbf{x}, t), \tag{3.2c}$$

$$s(\mathbf{x}, t) = \tilde{s}(z) + s'(\mathbf{x}, t), \tag{3.2d}$$

$$\psi(\mathbf{x}, t) = \tilde{\psi}(z) + \psi'(\mathbf{x}, t). \tag{3.2e}$$

The gravitational acceleration, therefore, is also decomposed in a similar way

$$\mathbf{g}(\mathbf{x}, t) = \tilde{\mathbf{g}}(z) + \mathbf{g}'(\mathbf{x}, t) = -\tilde{g}(z)\hat{\mathbf{e}}_z + \mathbf{g}'(\mathbf{x}, t), \tag{3.3}$$

and since the density distribution in the hydrostatic state is depth-dependent only, so is the gravitational acceleration, and it is along the $z$-direction; we have chosen the direction of the $z$-axis, so that it points vertically upwards, thus $\tilde{\mathbf{g}}(z) = -\tilde{g}(z)\hat{\mathbf{e}}_z$. In consequence, from this point onwards we will call the $z$-direction vertical. Note, that this implies that the density of the bottom and top bodies occupying volumes at $z < 0$ and at $z > L$ is by assumption homogeneous in the horizontal directions, so that they only generate vertical gravity.

For the sake of simplicity, let us now make an assumption, that the top body at $z > L$ is light enough not to influence the total gravity. Such an assumption corresponds to the situation in the planetary and stellar interiors, where because of the roughly spherical geometry the exterior regions of the body do not influence gravitationally the inner ones in a significant way. This allows to say, that $\tilde{g}(z) > 0$ (if the fluid layer is not heavier than the bottom body) and therefore for the convection to take place the bottom plate must be hotter than the upper one.

Next, the Eq. (3.1c) written separately for $\tilde{\mathbf{g}}(z)$ and $\mathbf{g}'$ in the fluid region $0 < z < L$ takes the form

$$-\nabla \cdot \tilde{\mathbf{g}} = 4\pi G\tilde{\rho}, \qquad -\nabla \cdot \mathbf{g}' = 4\pi G\rho'. \tag{3.4}$$

Since $\tilde{\mathbf{g}}$ must be continuous across the top and bottom interfaces at $z = 0,\ L$, the gravitational fluctuation $\mathbf{g}'$ must vanish at the top and bottom

$$g'\big|_{z=0,\, L} = 0, \tag{3.5}$$

which is consistent with the equation $-\nabla \cdot \mathbf{g}' = 4\pi G\rho'$ and the fact, that the mass conservation equation $\partial_t(\tilde{\rho} + \rho') + \nabla \cdot [(\tilde{\rho} + \rho')\mathbf{u}] = 0$ implies $\langle \rho' \rangle = 0$. Consequently, the gravity fluctuation $\mathbf{g}'$ is of the same order of magnitude as $G\rho' L$. Integration of equation $-\nabla \cdot \tilde{\mathbf{g}} = 4\pi G\tilde{\rho}$ over a cuboid $V_f(z_{cm}; z) = \left\{(x, y, z') : |x| \leq L_x,\ |y| \leq L_y,\ z_{cm} < z' \leq z\right\}$ bounded by horizontal planes at $z = z_{cm}$, where $z_{cm}$

denotes the position of the centre of mass for the fluid layer and the bottom body, and at the height $z$, yields

$$\tilde{g}(z)L_xL_y = 4\pi GL_xL_y\left[\int_{-z_{cm}}^{0}\rho_b(z')\mathrm{d}z' + \int_0^z\tilde{\rho}(z')\mathrm{d}z'\right] = 4\pi G\left[M_{bc} + M_f(z)\right]. \tag{3.6}$$

In the above we have denoted by $M_f(z)$ the mass of the fluid occupying the volume $V_f(0; z)$ and by $M_{bc}$ the mass of the part of the bottom body contained within the region $z_{cm} < z < 0$ between the level of the centre of mass and the bottom of the fluid layer ($z_{cm}$ is expected to lie within the bottom body, so that $\tilde{\mathbf{g}}$ is directed downwards in the fluid region); $\rho_b$ denotes the density of the bottom body. It is of interest to comment on the limit, when the bottom body is much heavier than the fluid layer. When the total mass of the fluid $M_f = M_f(L)$ is negligible with respect to the mass of the bottom body $M_b$, i.e. $M_f/M_b \ll 1$, then $M_{bc} \approx M_b/2$[1] and at any $0 < z \leq L$ we get $M_f(z)/M_{bc} \ll 1$. It is then predominantly the contribution $4\pi GM_{bc}/L_xL_y$ to the gravity acceleration in (3.6) which is responsible for creation of buoyancy in the fluid and which drives the fluid motions when the bottom boundary is heated. Therefore in such a case it is legitimate to neglect the effect of the fluid on the gravity and assume, that the gravitational force, which acts on the fluid is generated solely by the presence of the bottom body, $\mathbf{g} \approx \tilde{\mathbf{g}} \approx -2\pi GM_b/L_xL_y\hat{\mathbf{e}}_z$ (the gravitational fluctuation is then of the order $g' \sim GM_f\rho'/L_xL_y\tilde{\rho}$, thus its influence is also negligible, $\tilde{\rho}g'/\rho'\tilde{g} \sim M_f/M_b$). In the flat geometry this implies, that $\mathbf{g}$ is approximately constant, but in the spherical case, corresponding to the convective zones in celestial bodies, $\mathbf{g} \sim 1/r^2$, where $r$ measures the radial distance from the centre of the body.

The general equations of the hydrostatic equilibrium take the following form

$$\frac{\mathrm{d}\tilde{p}}{\mathrm{d}z} = -\tilde{\rho}\tilde{g}, \tag{3.7a}$$

$$\frac{\mathrm{d}^2\tilde{\psi}}{\mathrm{d}z^2} = 4\pi G\tilde{\rho}, \qquad \tilde{\mathbf{g}} = -\frac{\mathrm{d}\tilde{\psi}}{\mathrm{d}z}\hat{\mathbf{e}}_z, \tag{3.7b}$$

$$\frac{\mathrm{d}}{\mathrm{d}z}\left(k\frac{\mathrm{d}\tilde{T}}{\mathrm{d}z}\right) = -\tilde{Q}, \tag{3.7c}$$

$$\tilde{\rho} = \rho(\tilde{p}, \tilde{T}), \quad \tilde{s} = s(\tilde{p}, \tilde{T}). \tag{3.7d}$$

[1]Note, that $z_{cm} = -M_bL_b/2M_{tot} + L_xL_y\int_0^L z\tilde{\rho}(z)\mathrm{d}z/M_{tot}$, where $M_{tot} = M_b + M_f$ is the total mass of the system fluid-bottom body and $L_b$ is the vertical span of the bottom body. Integration by parts leads to $\int_0^L z\tilde{\rho}(z)\mathrm{d}z = LM_f/L_xL_y - \int_0^L M_f(z)\mathrm{d}z/L_xL_y > 0$, hence in the limit $M_b \gg M_f$ the term $L_xL_y\int_0^L z\tilde{\rho}(z)\mathrm{d}z/M_{tot}$ becomes negligibly small with respect to $M_bL_b/2M_{tot}$ and we obtain $z_{cm} \approx -L_b/2$; this implies $M_{bc} \approx M_b/2$.

We elaborate on the possible forms of the static reference state later. We now introduce the two fundamental assumptions of the anelastic approximation. *Firstly*, guided by the observations that the natural large-scale convective systems such as e.g. planetary and stellar interiors or atmospheres in their long-time evolution develop states that are nearly adiabatic and only a slight excess above the adiabatic gradient drives a very vigorous convective flow we assume that

$$0 < \delta \equiv \left\langle \frac{L}{\tilde{T}} \Delta_S \right\rangle = -\left\langle \frac{L}{\tilde{T}} \left( \frac{\mathrm{d}\tilde{T}}{\mathrm{d}z} + \frac{\tilde{g}\tilde{\alpha}\tilde{T}}{\tilde{c}_p} \right) \right\rangle \ll 1, \tag{3.8}$$

i.e. thermal gradient in the fluid, which undergoes convection is only weakly superadiabatic. Note, that the general relation

$$\frac{\partial s}{\partial z} = \frac{c_p}{T} \frac{\partial T}{\partial z} - \frac{\alpha}{\rho} \frac{\partial p}{\partial z}, \tag{3.9}$$

obtained with the aid of heat capacity definition $c_p = T(\partial s/\partial T)_p$ and the Maxwell relation $\rho^2(\partial s/\partial p)_T = (\partial \rho/\partial T)_p = -\rho\alpha$ allows to write

$$\frac{\mathrm{d}\tilde{s}}{\mathrm{d}z} = \frac{\tilde{c}_p}{\tilde{T}} \left( \frac{\mathrm{d}\tilde{T}}{\mathrm{d}z} + \frac{\tilde{g}\tilde{\alpha}\tilde{T}}{\tilde{c}_p} \right), \tag{3.10}$$

thus

$$\delta = -\left\langle \frac{L}{\tilde{T}} \left( \frac{\mathrm{d}\tilde{T}}{\mathrm{d}z} + \frac{\tilde{g}\tilde{\alpha}\tilde{T}}{\tilde{c}_p} \right) \right\rangle = -\left\langle \frac{L}{\tilde{c}_p} \frac{\mathrm{d}\tilde{s}}{\mathrm{d}z} \right\rangle \ll 1. \tag{3.11}$$

*Secondly*, similarly as in the case of Boussinesq approximation we assume that the fluctuations of thermodynamic variables are much smaller than their static profiles. Since it is the departure from adiabatic state that drives the convection it is natural to assume that it is also a measure of the relative magnitude of fluctuations

$$\left| \frac{\rho'}{\tilde{\rho}} \right| \sim \left| \frac{T'}{\tilde{T}} \right| \sim \left| \frac{p'}{\tilde{p}} \right| \sim \left| \frac{s'}{\tilde{s}} \right| \sim \mathcal{O}(\delta) \ll 1, \tag{3.12}$$

an assumption justified to a large extent by numerical evidence which, however, needs to be verified for consistency *a posteriori* in each particular case.[2] Note, that

[2]Some developments concerning the magnitude of fluctuations for sound-proof equations turned out possible for the case of weak solutions. It has been shown, that either in infinite space or for boundaries absorbing the energy of acoustic waves the time-dependent solutions remain in proximity to the initial conditions in a sense of certain integral bounds; in other words the solutions do not departure too far from the initial state—cf. e.g. the book on this topic by Feireisl and Novotný (2017). These estimates, however, have not yet benefited from taking into account the damping of acoustic waves by dissipative processes, such as viscosities and thermal conduction (cf. Footnote 6 in Chap. 1).

the equilibrium entropy $\tilde{s}$ consists of two contributions, $\tilde{s} = \tilde{s}_0 + \tilde{\tilde{s}}(z)$, where the constant entropy $\tilde{s}_0 = \text{const}$ corresponds to an adiabatic state of uniform entropy. The superadiabatic vertical variation of the entropy in the hydrostatic equilibrium, denoted here by $\tilde{\tilde{s}}(z) = \mathcal{O}(\delta \tilde{s}_0)$, constitutes only a weak correction to the mean.

As a consequence the equations of state (3.1e) are now expanded about the hydrostatic equilibrium at every height

$$\rho = \tilde{\rho}\left[1 - \tilde{\alpha}\left(T - \tilde{T}\right) + \tilde{\beta}\left(p - \tilde{p}\right) + \mathcal{O}\left(\delta^2\right)\right], \tag{3.13a}$$

$$s = \tilde{s} - \tilde{\alpha}\frac{p - \tilde{p}}{\tilde{\rho}} + \tilde{c}_p\frac{T - \tilde{T}}{\tilde{T}} + \mathcal{O}\left(\tilde{c}_p\delta^2\right), \tag{3.13b}$$

(where, again the Maxwell relation $\rho^2(\partial s/\partial p)_T = (\partial \rho/\partial T)_p = -\rho\alpha$ was used) which results in

$$\frac{\rho'}{\tilde{\rho}} = -\tilde{\alpha}T' + \tilde{\beta}p' + \mathcal{O}\left(\delta^2\right), \tag{3.14a}$$

$$s' = -\tilde{\alpha}\frac{p'}{\tilde{\rho}} + \tilde{c}_p\frac{T'}{\tilde{T}} + \mathcal{O}\left(\tilde{c}_p\delta^2\right). \tag{3.14b}$$

Furthermore, a similar type of argument as the one used to establish the convective flow magnitude and time scales in the Boussinesq fluid can also be applied here. The buoyancy force in the momentum balance (3.1a) i.e. $-\tilde{g}\rho'\hat{\mathbf{e}}_z - \tilde{\rho}\nabla\psi'$ drives the fluid motion which implies the flow acceleration of the order $g\rho'/\tilde{\rho} \sim \delta g$, thus much smaller than the acceleration of gravity. This, in turn, results in the following convective velocity and time scales

$$\mathscr{U} \sim \delta^{1/2}\sqrt{\bar{g}L}, \qquad \mathscr{T} \sim \delta^{-1/2}\sqrt{\frac{L}{\bar{g}}}, \tag{3.15}$$

and hence the viscosity scales also have to be small for consistency,

$$\mu_b \lesssim \mu \sim \delta^{1/2}\bar{\rho}\sqrt{\bar{g}L}L. \tag{3.16}$$

Similarly, the thermal conductivity coefficient must satisfy

$$k \sim \delta^{1/2}\bar{\rho}\bar{c}_p\sqrt{\bar{g}L}L, \tag{3.17}$$

(note, that contrary to the Boussinesq case, in the anelastic approximation the thermal conductivity $k$ and the thermal diffusion $\kappa$ are both of the same order of magnitude in terms of the small parameter, $\delta$). It follows, that by assumption

$$\tilde{Q} \sim k\frac{\bar{T}}{L^2} \sim \delta^{1/2}\bar{c}_p\bar{\rho}\bar{T}\sqrt{\frac{\bar{g}}{L}}, \qquad \frac{Q'}{\tilde{Q}} \sim \delta, \tag{3.18}$$

has to be satisfied in order for the anelastic approximation to be valid. In other words the anelastic approximation can be applied only, when the radiogenic heating is weak enough not to drive the system too far away from the adiabatic state; i.e. $\tilde{Q}/\bar{c}_p\bar{\rho}\bar{T}\sqrt{\bar{g}/L} \sim \delta^{1/2}$ allows the fundamental assumption (3.11) to be satisfied.

An interesting consequence is that the Mach number squared, which is the squared ratio of the convective velocity scale to the mean speed of sound $\bar{C}$

$$Ma^2 = \frac{\mathscr{U}^2}{\bar{C}^2} = \frac{\mathscr{U}^2}{\left\langle\left(\frac{\partial p}{\partial \rho}\right)_s\right\rangle} = \mathcal{O}\left(\delta \bar{g} L \left\langle\tilde{\rho}\tilde{\beta}\right\rangle\right) = \mathcal{O}(\delta) \ll 1 \tag{3.19}$$

must be of the order $\delta$, hence small. In the above we have used the estimate $\Delta\tilde{p} \sim \tilde{\rho}gL$ of the hydrostatic pressure jump across the fluid layer and the thermodynamic identity

$$\begin{aligned} C^2 = \left(\frac{\partial p}{\partial \rho}\right)_s &= \left(\frac{\partial p}{\partial \rho}\right)_T + \left(\frac{\partial p}{\partial T}\right)_\rho \left(\frac{\partial T}{\partial \rho}\right)_s \\ &= \frac{1}{\rho\beta} + \left(\frac{\partial p}{\partial T}\right)_\rho^2 \frac{T}{\rho^2 c_v} \\ &= \frac{1}{\rho\beta}\left(1 + \frac{\alpha^2 T}{c_v \beta \rho}\right) = \frac{\gamma}{\rho\beta}, \end{aligned} \tag{3.20}$$

satisfied by virtue of the implicit function theorem which implies $(\partial p/\partial T)_\rho = \alpha/\beta$ and $(\partial T/\partial \rho)_s = -T(\partial s/\partial \rho)_T/c_v$ and the Maxwell relation $\rho^2(\partial s/\partial \rho)_T = -(\partial p/\partial T)_\rho$; the last equality in (3.20) is obtained on the basis of (2.38) and $\gamma = c_p/c_v$. Of course the above estimates in (3.19) correspond, in fact, to the *third assumption*, which was made in the process of derivation of the Boussinesq equations (2.23), which is typically satisfied by fluids and therefore is a weak assumption, not introducing strong restrictions. From the point of view of the state equations it means we assume, that in terms of the small parameter $\delta$ one can estimate $\partial\rho/\partial p \sim \rho/p$ etc. Note, however, that in the Boussinesq case the Mach number scales linearly with the small parameter of perturbative expansions $\epsilon$, whereas in the anelastic case the Mach number scales like square root of the small parameter $\delta$.

It is also important to note the physical consequence of the assumption (3.15) concerning the velocity and time scales of evolution. In the anelastic case the scale heights in the system are comparable with the system's vertical span $L$, i.e.

$$\frac{D_p}{L} = \left(-\frac{L}{\tilde{p}}\frac{\mathrm{d}\tilde{p}}{\mathrm{d}z}\right)^{-1} = \frac{\tilde{g}L}{R\tilde{T}} = \gamma\frac{\tilde{g}L}{\tilde{C}^2} = \mathcal{O}(1), \tag{3.21}$$

(similarly $-L\mathrm{d}_z\tilde{\rho}/\tilde{\rho} = \mathcal{O}(1)$ and $-L\mathrm{d}_z\tilde{T}/\tilde{T} = \mathcal{O}(1)$), where we have used the hydrostatic pressure balance for the reference state $\mathrm{d}_z\tilde{p} = -\tilde{\rho}g$. Therefore the value of $\sqrt{\bar{g}L}$ which has the dimension of velocity is of comparable magnitude as the mean speed of sound $\bar{C}$. It follows, that the assumed time scale $\mathscr{T} \sim \delta^{-1/2}\sqrt{L/\bar{g}}$ is of the same order of magnitude as the inertial time scale

$$\mathscr{T}_{\text{inertial}} = \frac{L}{\mathscr{U}} = Ma^{-1}\frac{L}{\bar{\bar{C}}} \sim \delta^{-1/2}\frac{L}{\bar{\bar{C}}} \sim \mathscr{T}. \tag{3.22}$$

Moreover, time scale associated with internal gravity waves

$$\mathscr{T}_{\text{grav.}} = \left(\frac{g}{\bar{\bar{T}}}\Delta_S\right)^{-1/2} \sim \delta^{-1/2}\sqrt{\frac{L}{g}} = \mathscr{T} \tag{3.23}$$

is also comparable with the chosen dynamical time scale, but all the three time scales $\mathscr{T}_{\text{inertial}}$, $\mathscr{T}_{\text{grav.}}$ and $\mathscr{T}$ are all much longer than the fast acoustic time scale $L/\bar{C}$. Therefore the following relation is satisfied

$$\mathscr{T} \sim \mathscr{T}_{\text{inertial}} \sim \mathscr{T}_{\text{grav.}} \gg \frac{L}{\bar{\bar{C}}}, \tag{3.24}$$

so that both the inertial and buoyancy effects are included in the dynamics, but the acoustic effects are filtered out.

We are now ready to write down the dynamic equations for fluctuations, i.e. with subtracted hydrostatic balance (3.7a, 3.7b)

$$\begin{aligned}(\tilde{\rho}+\rho')\left[\frac{\partial \mathbf{u}}{\partial t} + (\mathbf{u}\cdot\nabla)\,\mathbf{u}\right] =& -\nabla p' + \rho'\tilde{\mathbf{g}} - \tilde{\rho}\nabla\psi' + \mu\nabla^2\mathbf{u} + \left(\frac{\mu}{3} + \mu_b\right)\nabla\,(\nabla\cdot\mathbf{u})\\ &+ 2\nabla\mu\cdot\mathbf{G}^s + \nabla\left(\mu_b - \frac{2}{3}\mu\right)\nabla\cdot\mathbf{u},\end{aligned} \tag{3.25a}$$

$$\frac{\partial \rho'}{\partial t} + \nabla\cdot\left[(\tilde{\rho}+\rho')\,\mathbf{u}\right] = 0, \tag{3.25b}$$

$$\nabla^2\psi' = 4\pi G\rho', \tag{3.25c}$$

$$\begin{aligned}(\tilde{\rho}+\rho')\left(\tilde{T}+T'\right)\left[\frac{\partial s'}{\partial t} + \mathbf{u}\cdot\nabla\,(\tilde{s}+s')\right] =& \nabla\cdot\left(k\nabla T'\right) + 2\mu\mathbf{G}^s : \mathbf{G}^s\\ &+ \left(\mu_b - \frac{2}{3}\mu\right)(\nabla\cdot\mathbf{u})^2 + Q',\end{aligned} \tag{3.25d}$$

$$\frac{\rho'}{\tilde{\rho}} = -\tilde{\alpha}T' + \tilde{\beta}p' + \mathscr{O}\left(\delta^2\right), \quad s' = -\tilde{\alpha}\frac{p'}{\tilde{\rho}} + \tilde{c}_p\frac{T'}{\tilde{T}} + \mathscr{O}\left(\tilde{c}_p\delta^2\right), \tag{3.25e}$$

and with the aid of (3.12), (3.15), (3.16) and (3.18) take their leading order form

$$\tilde{\rho}\left[\frac{\partial \mathbf{u}}{\partial t}+(\mathbf{u}\cdot\nabla)\,\mathbf{u}\right]=-\nabla p'+\rho'\tilde{\mathbf{g}}-\tilde{\rho}\nabla\psi'+\mu\nabla^2\mathbf{u}+\left(\frac{\mu}{3}+\mu_b\right)\nabla(\nabla\cdot\mathbf{u})$$
$$+2\nabla\mu\cdot\mathbf{G}^s+\nabla\left(\mu_b-\frac{2}{3}\mu\right)\nabla\cdot\mathbf{u}, \tag{3.26a}$$

$$\nabla\cdot(\tilde{\rho}\mathbf{u})=0, \tag{3.26b}$$

$$\nabla^2\psi'=4\pi G\rho', \tag{3.26c}$$

$$\tilde{\rho}\tilde{T}\left[\frac{\partial s'}{\partial t}+\mathbf{u}\cdot\nabla\left(\tilde{s}+s'\right)\right]=\nabla\cdot\left(k\nabla T'\right)+2\mu\mathbf{G}^s:\mathbf{G}^s+\left(\mu_b-\frac{2}{3}\mu\right)(\nabla\cdot\mathbf{u})^2+Q', \tag{3.26d}$$

$$\frac{\rho'}{\tilde{\rho}}=-\tilde{\alpha}T'+\tilde{\beta}p', \qquad s'=-\tilde{\alpha}\frac{p'}{\tilde{\rho}}+\tilde{c}_p\frac{T'}{\tilde{T}}. \tag{3.26e}$$

Note, that according to Sect. 1.5.2 (cf. Eq. (1.80)) the anelastic continuity equation filters out sound waves. Making use of (3.10) the energy equation can be also written in the form

$$\tilde{\rho}\tilde{T}\left(\frac{\partial s'}{\partial t}+\mathbf{u}\cdot\nabla s'\right)-\tilde{\rho}\tilde{c}_p u_z\Delta_S=\nabla\cdot\left(k\nabla T'\right)+2\mu\mathbf{G}^s:\mathbf{G}^s$$
$$+\left(\mu_b-\frac{2}{3}\mu\right)(\nabla\cdot\mathbf{u})^2+Q'. \tag{3.27}$$

For the sake of completeness we also provide the temperature equation within the anelastic approximation (cf. (1.44))

$$\left(\tilde{\rho}+\rho'\right)c_v\left[\frac{\partial T'}{\partial t}+\mathbf{u}\cdot\nabla\left(\tilde{T}+T'\right)\right]+\frac{\alpha}{\beta}\left(\tilde{T}+T'\right)\nabla\cdot\mathbf{u}$$
$$=\nabla\cdot\left(k\nabla T'\right)+2\mu\mathbf{G}^s:\mathbf{G}^s+\left(\mu_b-\frac{2}{3}\mu\right)(\nabla\cdot\mathbf{u})^2+Q' \tag{3.28}$$

which at the leading order can be written in the following form

$$\tilde{\rho}\tilde{c}_v\left(\frac{\partial T'}{\partial t}+\mathbf{u}\cdot\nabla T'\right)+\left(\tilde{\rho}+\rho'\right)c_v u_z\frac{\mathrm{d}\tilde{T}}{\mathrm{d}z}+\frac{\alpha}{\beta}\left(\tilde{T}+T'\right)\nabla\cdot\mathbf{u}$$
$$=\nabla\cdot\left(k\nabla T'\right)+2\mu\mathbf{G}^s:\mathbf{G}^s+\left(\mu_b-\frac{2}{3}\mu\right)(\nabla\cdot\mathbf{u})^2+Q'. \tag{3.29}$$

Note, that $c_v$ in the term proportional to $\mathrm{d}\tilde{T}/\mathrm{d}z$ and $\alpha/\beta$ in the term proportional to the flow divergence are not taken at the hydrostatic equilibrium (not marked with an upper tilde), but $\alpha$ and $\beta$ and in particular also their fluctuations $\alpha'$, $\beta'$ can be obtained once the equations of state are known; the specific heat $c_v$ is an input parameter in the theory of fluid mechanics resulting from kinetic or phenomenological models, but it is known, that the specific heat often significantly depends on the temperature; once the relation $c_v(T)$ is known $\tilde{c}_v$ and $c_v'$ can be established, if necessary. We stress, that it is important to keep the terms which involve the fluctuations of the fluid's thermodynamic properties, i.e. $\tilde{\rho}c_v'u_z\mathrm{d}_z\tilde{T}$ and $\alpha'\tilde{T}\nabla\cdot\mathbf{u}/\tilde{\beta}$ and $-\tilde{\alpha}\beta'\tilde{T}\nabla\cdot\mathbf{u}/\tilde{\beta}^2$ in the energy equation, since they contribute to the leading order balance obtained at the order $\mathcal{O}(\delta\mathcal{U}\bar{\rho}\bar{g}) = \mathcal{O}\left(\delta^{3/2}\bar{\rho}\bar{g}\sqrt{\bar{g}L}\right)$ in the energy equation. Furthermore, although it is obvious that at the held degree of accuracy the terms $\frac{\alpha}{\beta}T'\nabla\cdot\mathbf{u}$, $\alpha'\tilde{T}\nabla\cdot\mathbf{u}/\tilde{\beta}$ and $-\tilde{\alpha}\beta'\tilde{T}\nabla\cdot\mathbf{u}/\tilde{\beta}^2$ can only involve the leading order expression for the flow divergence $\nabla\cdot\mathbf{u} = -\mathrm{d}_z\tilde{\rho}u_z/\tilde{\rho} + \mathcal{O}(\delta\mathcal{U}/L)$ because $\tilde{\alpha}T'\mathcal{U}/\tilde{\beta}L$, $\alpha'\tilde{T}\mathcal{U}/\tilde{\beta}L$ and $\tilde{\alpha}\beta'\tilde{T}\mathcal{U}/L\tilde{\beta}^2$ are already of the required order of magnitude $\mathcal{O}(\delta\mathcal{U}\bar{\rho}\bar{g})$, on the contrary the term $\frac{\alpha}{\beta}\tilde{T}\nabla\cdot\mathbf{u}$ must include the entire expression $\nabla\cdot\mathbf{u} = -[\nabla(\tilde{\rho}+\rho')\cdot\mathbf{u} + \partial_t\rho']/(\tilde{\rho} + \rho')$, resulting from the law of mass conservation; the corrections $\mathcal{O}(\delta\mathcal{U}/L)$ in the expression for the flow divergence are vital for this term, since $(\tilde{\alpha}\tilde{T}/\tilde{\beta})\mathcal{O}(\delta\mathcal{U}/L)$ is of the required order of magnitude. We note, however, that there are also terms such as $\tilde{\rho}\tilde{c}_v u_z\mathrm{d}_z\tilde{T}$ and

$$\frac{\tilde{\alpha}\tilde{T}}{\tilde{\beta}}\nabla\cdot\mathbf{u} = -\frac{\tilde{\alpha}\tilde{T}}{\tilde{\rho}\tilde{\beta}}u_z\frac{\mathrm{d}\tilde{\rho}}{\mathrm{d}z} + \mathcal{O}(\delta\mathcal{U}\bar{\rho}\bar{g}), \tag{3.30}$$

which alone are $\delta^{-1}$ times stronger than the rest of the terms in the energy equation, all of the order $\mathcal{O}(\delta\mathcal{U}\bar{\rho}\bar{g})$. We will demonstrate now, that the sum of the two unfitting terms appearing in the equation is in fact of the required order of magnitude $\mathcal{O}(\delta\mathcal{U}\bar{\rho}\bar{g})$. Namely with the use of the continuity equation $\partial_t\rho' + \nabla\cdot((\tilde{\rho}+\rho')\mathbf{u}) = 0$ the sum of the second and third terms on the left hand side of (3.29) can be rearranged to give

$$\begin{aligned}\left(\tilde{\rho}+\rho'\right)c_v u_z\frac{\mathrm{d}\tilde{T}}{\mathrm{d}z} + \frac{\alpha}{\beta}\left(\tilde{T}+T'\right)\nabla\cdot\mathbf{u} &= u_z\left(\tilde{\rho}c_v\frac{\mathrm{d}\tilde{T}}{\mathrm{d}z} - \frac{\alpha}{\beta}\frac{\tilde{T}}{\tilde{\rho}}\frac{\mathrm{d}\tilde{\rho}}{\mathrm{d}z}\right) + \mathcal{O}(\delta\mathcal{U}\bar{\rho}\bar{g})\\ &= \tilde{\rho}\tilde{c}_v u_z\left(\frac{\mathrm{d}\tilde{T}}{\mathrm{d}z} - \frac{\tilde{\alpha}}{\tilde{c}_v\tilde{\beta}}\frac{\tilde{T}}{\tilde{\rho}}\frac{\mathrm{d}\tilde{\rho}}{\mathrm{d}z}\right) + \mathcal{O}(\delta\mathcal{U}\bar{\rho}\bar{g}).\end{aligned} \tag{3.31}$$

Furthermore, the $z$-derivative of the basic density $\tilde{\rho}(\tilde{p},\tilde{T})$ reads

$$\frac{\mathrm{d}\tilde{\rho}}{\mathrm{d}z} = \tilde{\rho}\tilde{\beta}\frac{\mathrm{d}\tilde{p}}{\mathrm{d}z} - \tilde{\rho}\tilde{\alpha}\frac{\mathrm{d}\tilde{T}}{\mathrm{d}z}, \tag{3.32}$$

and from the definition of $\delta$ in (3.11) and the hydrostatic balance $\mathrm{d}\tilde{p}/\mathrm{d}z = -\tilde{\rho}\bar{g}$ we obtain an estimate of the vertical gradient of density in the hydrostatic state

$$\frac{\mathrm{d}\tilde{\rho}}{\mathrm{d}z} = -\tilde{\rho}^2\tilde{\beta}\tilde{g} - \tilde{\rho}\tilde{\alpha}\frac{\mathrm{d}\tilde{T}}{\mathrm{d}z} = -\left(\frac{\tilde{\rho}^2\tilde{\beta}}{\tilde{\alpha}\tilde{T}}\right)\left[\frac{\tilde{c}_v\tilde{\alpha}\tilde{T}\tilde{g}}{\tilde{c}_p} - (\tilde{c}_p - \tilde{c}_v)\frac{\tilde{T}}{L}\frac{L\Delta_S}{\tilde{T}}\right]$$
$$= -\frac{\tilde{c}_v\tilde{\beta}\tilde{\rho}^2\tilde{g}}{\tilde{c}_p} + \mathcal{O}(\bar{\rho}\delta/L) = \frac{\tilde{c}_v\tilde{\beta}\tilde{\rho}^2}{\tilde{\alpha}\tilde{T}}\frac{\mathrm{d}\tilde{T}}{\mathrm{d}z} + \mathcal{O}(\bar{\rho}\delta/L). \tag{3.33}$$

The latter can be utilized to get

$$(\tilde{\rho} + \rho')\, c_v u_z \frac{\mathrm{d}\tilde{T}}{\mathrm{d}z} + \frac{\alpha}{\beta}\left(\tilde{T} + T'\right)\nabla\cdot\mathbf{u} = \tilde{\rho}\tilde{c}_v u_z\left(\frac{\mathrm{d}\tilde{T}}{\mathrm{d}z} + \frac{\tilde{\alpha}\tilde{T}\tilde{g}}{\tilde{c}_p}\right) + \mathcal{O}(\delta\mathcal{U}\bar{\rho}\bar{g})$$
$$= -\tilde{\rho}c_v u_z\Delta_S + \mathcal{O}(\delta\mathcal{U}\bar{\rho}\bar{g}) = \mathcal{O}(\delta\mathcal{U}\bar{\rho}\bar{g}). \tag{3.34}$$

This proves consistency for the temperature equation (3.29) which constitutes a balance between terms at the order $\mathcal{O}(\delta\mathcal{U}\bar{\rho}\bar{g})$. However, we recall that when deriving (3.31) we have in fact used the full (not only leading order) continuity equation,

$$(\tilde{\rho} + \rho')\,\nabla\cdot\mathbf{u} = -\mathbf{u}\frac{\mathrm{d}\tilde{\rho}}{\mathrm{d}z} - \frac{\partial\rho'}{\partial t} - \mathbf{u}\cdot\nabla\rho', \tag{3.35}$$

since the terms

$$-\frac{\alpha\tilde{T}}{\beta}\left(\frac{\partial\rho'}{\partial t} + \mathbf{u}\cdot\nabla\rho'\right) = \mathcal{O}(\delta\mathcal{U}\bar{\rho}\bar{g}) \tag{3.36}$$

also contribute to the leading order temperature balance. This means that application of the temperature balance expressed as in (3.29) along with the leading order mass conservation law $\nabla\cdot(\tilde{\rho}\mathbf{u}) = 0$ is *not* possible, since it leads to neglection of vital terms in the temperature balance and hence the full continuity equation would have to be used. This is an undesirable situation, since in such a case the quick sound waves are not filtered out and the entire point of anelastic approximation is lost. The situation can be rectified by rearranging the temperature equation (3.29) with the use of the full mass conservation law (cf. (1.40)), which yields

$$\tilde{\rho}\tilde{c}_p\left(\frac{\partial T'}{\partial t} + \mathbf{u}\cdot\nabla T'\right) - \tilde{\alpha}\tilde{T}\left(\frac{\partial p'}{\partial t} + \mathbf{u}\cdot\nabla p'\right) + \rho c_p u_z\frac{\mathrm{d}\tilde{T}}{\mathrm{d}z} - \alpha T u_z\frac{\mathrm{d}\tilde{p}}{\mathrm{d}z}$$
$$= \nabla\cdot\left(k\nabla T'\right) + 2\mu\mathbf{G}^s : \mathbf{G}^s + \left(\mu_b - \frac{2}{3}\mu\right)(\nabla\cdot\mathbf{u})^2 + Q'. \tag{3.37}$$

This equation no longer suffers from issues associated with higher order corrections to the flow divergence and can be used along with the simplified law of mass conservation $\nabla\cdot(\tilde{\rho}\mathbf{u}) = 0$. The sum of the terms $\rho c_p u_z \mathrm{d}\tilde{T}/\mathrm{d}z - \alpha T u_z \mathrm{d}\tilde{p}/\mathrm{d}z$ in Eq. (3.37) is also of the order $\mathcal{O}(\delta\mathcal{U}\bar{\rho}\bar{g})$, however, a simpler form is not easily obtained in the general case of an unspecified equation of state. Some simplification is possible under the assumption that the fluid satisfies the perfect gas equation of state, $p = \rho RT$, where $R = k_B/m_m$ is the specific gas constant, which depends on the type of gas/fluid; $k_B$ is the Boltzmann constant and $m_m$ is the molecular mass of the fluid particles. Then $\alpha = 1/T$ and $\beta = 1/p$ and since in such a case $c_p - c_v = R = \text{const}$ and $c_p/c_v = \gamma = \text{const}$[3] we can assume $\tilde{c}_v = c_v = \text{const.}$ and $\tilde{c}_p = c_p = \text{const.}$ Therefore by virtue of $\mathrm{d}\tilde{p}/\mathrm{d}z = -\tilde{\rho}\tilde{g}$ and (3.26e) the energy equation at leading order takes the form

$$\tilde{\rho}c_p\left(\frac{\partial T'}{\partial t} + \mathbf{u}\cdot\nabla T'\right) - \left(\frac{\partial p'}{\partial t} + \mathbf{u}\cdot\nabla p'\right) + \tilde{\rho}c_p\frac{\mathrm{d}\tilde{T}}{\mathrm{d}z}u_z\left(\frac{p'}{\tilde{p}} - \frac{T'}{\tilde{T}}\right) - \tilde{\rho}c_p u_z\Delta_S$$
$$= \nabla\cdot\left(k\nabla T'\right) + 2\mu\mathbf{G}^s : \mathbf{G}^s + \left(\mu_b - \frac{2}{3}\mu\right)(\nabla\cdot\mathbf{u})^2 + Q'. \tag{3.38}$$

Such an equation does not involve the flow divergence and can be used along with $\nabla\cdot(\tilde{\rho}\mathbf{u}) = 0$, however, the lagrangian derivative of pressure needs to be kept, since it provides a contribution of the same order of magnitude as all the other terms in the above equation (which is, of course, also true for the general temperature equation (3.37)). The temperature equation (3.37) or in the case of a perfect gas (3.38) can replace the entropy equation[4] in the closed system of Eqs. (3.26a–3.26e) describing the anelastic convection.

However, in the case of a perfect gas under uniform gravity, $\mathbf{g} = -g\hat{\mathbf{e}}_z$, $g = \text{const}$, the temperature equation (3.38) can be combined with a rearranged Navier-Stokes equation

$$\frac{1}{\tilde{T}}\left[\frac{\partial\mathbf{u}}{\partial t} + (\mathbf{u}\cdot\nabla)\,\mathbf{u}\right] = -\nabla\left(\frac{p'}{\tilde{\rho}\tilde{T}}\right) - \frac{T'}{\tilde{T}^2}\mathbf{g} + \frac{\mu}{\tilde{\rho}\tilde{T}}\nabla^2\mathbf{u} + \left(\frac{\mu}{3\tilde{\rho}\tilde{T}} + \frac{\mu_b}{\tilde{\rho}\tilde{T}}\right)\nabla(\nabla\cdot\mathbf{u})$$
$$+ \frac{2}{\tilde{\rho}\tilde{T}}\nabla\mu\cdot\mathbf{G}^s + \frac{1}{\tilde{\rho}\tilde{T}}\nabla\left(\mu_b - \frac{2}{3}\mu\right)\nabla\cdot\mathbf{u}, \tag{3.39}$$

so that the buoyancy is expressed solely by the temperature fluctuation. This way both, the velocity field equation and the Poisson-type problem for pressure obtained from the Navier-Stokes equation supplied by $\nabla\cdot(\tilde{\rho}\mathbf{u}) = 0$ (by multiplying (3.39) by $\tilde{\rho}\tilde{T}$ and taking its divergence) depend on the temperature fluctuation $T'$, which depends on the lagrangian pressure derivative. However, as shown in Sect. 1.5.2 the sound waves are indeed filtered out within such an approach, despite the presence of

[3]Of course in a real fluid/gas the specific heat ratio depends on temperature since the number of degrees of freedom of fluid/gas particles increases with temperature. However, since the case of a general equation of state is presented in detail therefore whenever the perfect gas equation of state will be considered here, it will also be assumed for simplicity, that the specific heats at constant volume and pressure are uniform (and then a generalisation to nonuniform $c_v$ and $c_p$ in the case of a perfect gas state equation is fairly simple).

[4]To be precise the correct terminology for these equations is the energy equation expressed in terms of either the entropy or the temperature.

the lagrangian pressure derivative in the energy equation. Therefore the temperature formulation of the anelastic equations for a perfect gas under uniform gravity consists of the closed system of Eqs. (3.38), (3.39) and $\nabla \cdot (\tilde{\rho}\mathbf{u}) = 0$.

Finally, the anelastic mass conservation equation,

$$\nabla \cdot (\tilde{\rho}\mathbf{u}) = 0, \tag{3.40}$$

which filtrates the fast sound waves, can be cast in a somewhat different form by the use of (3.32), the hydrostatic force balance $\mathrm{d}_z\tilde{p} = -\tilde{\rho}\tilde{g}$ and the definition of $\Delta_S = -\mathrm{d}_z\tilde{T} - \tilde{\alpha}\tilde{T}\tilde{g}/\tilde{c}_p$. Namely the latter two imply

$$\frac{\mathrm{d}\tilde{T}}{\mathrm{d}z} = -\Delta_S - \frac{\tilde{\alpha}\tilde{T}\tilde{g}}{\tilde{c}_p} = -\Delta_S + \frac{\tilde{\alpha}\tilde{T}}{\tilde{c}_p\tilde{\rho}}\frac{\mathrm{d}\tilde{p}}{\mathrm{d}z}, \tag{3.41}$$

which can be introduced into (3.32) to obtain

$$\begin{aligned} \frac{1}{\tilde{\rho}}\frac{\mathrm{d}\tilde{\rho}}{\mathrm{d}z} =& \frac{\tilde{\beta}}{\tilde{c}_p}\left(\tilde{c}_p - \frac{\tilde{\alpha}^2\tilde{T}}{\tilde{\rho}\tilde{\beta}}\right)\frac{\mathrm{d}\tilde{p}}{\mathrm{d}z} + \tilde{\alpha}\Delta_S \\ =& \frac{\tilde{\beta}}{\tilde{\gamma}}\frac{\mathrm{d}\tilde{p}}{\mathrm{d}z} + \tilde{\alpha}\Delta_S, \end{aligned} \tag{3.42}$$

and we have used the thermodynamic identity (2.38) and $\gamma = c_p/c_v$. This allows to transform the mass conservation equation (3.40) into

$$\nabla \cdot \mathbf{u} + \frac{\tilde{\beta}}{\tilde{\gamma}}\frac{\mathrm{d}\tilde{p}}{\mathrm{d}z}u_z = -\tilde{\alpha}\Delta_S u_z = \mathcal{O}\left(\delta\sqrt{\frac{\tilde{g}}{L}}\right). \tag{3.43}$$

For a perfect gas $\tilde{p} = \tilde{\rho}R\tilde{T}$, thus the mass conservation equation simplifies to

$$\nabla \cdot \mathbf{u} + \frac{1}{\gamma\tilde{p}}\frac{\mathrm{d}\tilde{p}}{\mathrm{d}z}u_z = 0, \tag{3.44}$$

which is often expressed in one the following equivalent forms

$$\nabla \cdot \left(\tilde{p}^{1/\gamma}\mathbf{u}\right) = 0, \tag{3.45}$$

$$\nabla \cdot \mathbf{u} - \frac{\tilde{g}}{\tilde{C}^2}u_z = 0, \tag{3.46}$$

where $\tilde{C} = \sqrt{\gamma R\tilde{T}}$ is the speed of sound in the static reference state. In the literature the set of Eqs. (3.26a, 3.26c, 3.26d, 3.26e) for a perfect gas with either (3.44) or (3.45) or (3.46) substituted for the mass conservation equation (3.26b), i.e.

$$\tilde{\rho}\left[\frac{\partial \mathbf{u}}{\partial t} + (\mathbf{u}\cdot\nabla)\,\mathbf{u}\right] = -\nabla p' + \rho'\tilde{\mathbf{g}} - \tilde{\rho}\nabla\psi' + \mu\nabla^2\mathbf{u} + \left(\frac{\mu}{3} + \mu_b\right)\nabla(\nabla\cdot\mathbf{u})$$
$$+ 2\nabla\mu\cdot\mathbf{G}^s + \nabla\left(\mu_b - \frac{2}{3}\mu\right)\nabla\cdot\mathbf{u}, \qquad (3.47a)$$

$$\tilde{\rho}\tilde{T}\left(\frac{\partial s'}{\partial t} + \mathbf{u}\cdot\nabla s'\right) - \tilde{\rho}\tilde{c}_p u_z \Delta_S = \nabla\cdot(k\nabla T') + 2\mu\mathbf{G}^s : \mathbf{G}^s$$
$$+ \left(\mu_b - \frac{2}{3}\mu\right)(\nabla\cdot\mathbf{u})^2 + Q', \qquad (3.47b)$$

$$\nabla^2\psi' = 4\pi G\rho', \qquad \nabla\cdot\left(\tilde{p}^{1/\gamma}\mathbf{u}\right) = 0, \qquad (3.47c)$$

$$\frac{\rho'}{\tilde{\rho}} = -\frac{T'}{\tilde{T}} + \frac{p'}{\tilde{p}}, \qquad s' = -R\frac{p'}{\tilde{p}} + c_p\frac{T'}{\tilde{T}}, \qquad (3.47d)$$

is called the *pseudo-incompressible approximation* (cf. Durran 1989, 2008, Klein 2009, Klein et al. 2010, Klein and Pauluis 2012 and Vasil et al. 2013[5]). The entropy equation (3.47b) can of course be replaced by the temperature equation (3.38).

### 3.1.1 Production of the Total Entropy Within the Anelastic Approximation

It is of fundamental importance, that the derived anelastic energy equation (3.26d) is consistent with the second law of thermodynamics, i.e. in the case when the system is adiabatically insulated the production of the total entropy must be positive or null. This is not immediately obvious, since we have shown in Sect. 1.4.1, that the entropy production consists of the viscous heating (positive definite) and the volume integral of the square of the temperature gradient. However, in the process of derivation of the anelastic energy balance expressed in terms of the entropy variations (3.26d) we have neglected on the left hand side of this equation the fluctuation $T'$ with respect to the reference temperature $\tilde{T}$, leaving only $\tilde{\rho}\tilde{T}\mathrm{D}_t s$. This implies appearance of terms of the type $\nabla\tilde{T}\cdot\nabla T$ in the equation for the entropy per unit volume $\tilde{\rho}s$. To demonstrate, that the anelastic approximation does not violate the second law of thermodynamics

[5]The pseudo-incompressible approximation is particularly useful in the absence of thermal effects, as it allows for a sound-proof treatment of gravity waves without the constraint of nearly adiabatic dynamics present in the anelastic approximation. The low-Mach number theory for convective flows is currently being further developed based on a variational approach by Toby Wood and his group at the University of Newcastle, UK. The aim of this approach is to allow for relaxation of the assumption of small superadiabaticity of convection required by the anelastic approximation and demand only the smallness of the pressure fluctuation, $p'/p \ll 1$, with the density and temperature fluctuations comparable in magnitude to their reference state values.

we must show, that in the absence of any heat sources and with adiabatic insulation on boundaries the system of anelastic equations does not allow the total entropy to decrease. Let us first assume adiabatic insulation

$$-k\frac{\partial}{\partial z}\left(\tilde{T}+T'\right)\bigg|_{z=0,\ L}=0, \tag{3.48}$$

and no radiogenic heat sources

$$Q=0. \tag{3.49}$$

However, we must realize, that the assumption of adiabatic insulation implies, that there is no energy input either on the boundaries or within the fluid volume, since also $Q=0$. It follows from (3.48), that the reference state gradient $\mathrm{d}_z\tilde{T}$ is either asymptotically small of the order of $\partial_z T'=\mathcal{O}(\delta T_B/L)$ or null. This puts the system in the Boussinesq limit (cf. Sect. 3.1.3). Since the assumption of small departure from adiabaticity, $\delta \ll 1$, must always hold within the anelastic approximation, we conclude, that adiabatic insulation within the anelastic approximation can only be assumed for systems characterized by very small adiabatic gradient, $\tilde{\alpha}\tilde{T}g/\tilde{c}_p=\mathcal{O}(\delta T_B/L)$, i.e. within the Boussinesq limit. In other words, the anelastic approximation, useful and distinct from the Boussinesq approximation only at strong stratification, is constructed for systems which are not adiabatically insulated. This is, of course, a desired situation, since convective flow has to be driven by a thermal energy flux at the bottom (and/or non-zero $Q$), and naturally convective systems, such as stellar and planetary interiors or atmospheres are obviously not insulated. Nevertheless, we still must demonstrate, that the derived system of anelastic equations does not violate the second law of thermodynamics and is consistent with it, even if this effectively implies consideration of the Boussinesq limit.

The viscous heating term in (3.26d) has a standard form and has been shown to be positive definite in Sect. 1.4.1. On dividing the energy equation (3.26d) by $\tilde{T}$ one obtains the equation for the entropy per unit volume $\tilde{\rho}s$, which can be cast in the form

$$\begin{aligned}\frac{\partial}{\partial t}\left[\tilde{\rho}\left(\tilde{s}+s'\right)\right]+\nabla\cdot\left[\tilde{\rho}\mathbf{u}\left(\tilde{s}+s'\right)\right]=&\nabla\cdot\left[\frac{k}{\tilde{T}}\nabla\left(\tilde{T}+T'\right)\right]+\frac{k}{\tilde{T}^2}\frac{\mathrm{d}\tilde{T}}{\mathrm{d}z}\frac{\partial}{\partial z}\left(\tilde{T}+T'\right)\\&+2\frac{\mu}{\tilde{T}}\underset{\sim}{\mathbf{G}}^s:\underset{\sim}{\mathbf{G}}^s+\frac{\mu_b}{\tilde{T}}\left(\nabla\cdot\mathbf{u}\right)^2,\end{aligned} \tag{3.50}$$

where the anelastic continuity equation was used $\nabla\cdot(\tilde{\rho}\mathbf{u})=0$ (recall, that $\underset{\sim}{\mathbf{G}}^s$ denotes the traceless part of the symmetric velocity gradient tensor). Next, if periodic boundary conditions in the horizontal directions with some periods $L_x$ and $L_y$ are assumed, the adiabatic (3.48) and impermeability $u_z(z=0,\ L)=0$ conditions on the top and bottom boundaries allow to easily demonstrate, that the two "divergence" terms, $\nabla\cdot\left[\tilde{\rho}\mathbf{u}\left(\tilde{s}+s'\right)\right]$ and $\nabla\cdot\left[k\nabla\left(\tilde{T}+T'\right)/\tilde{T}\right]$, do not contribute to the global entropy balance

$$\int_V \nabla \cdot [\tilde{\rho}\mathbf{u}(\tilde{s}+s')] \mathrm{d}^3x = \int_{\partial V} \tilde{\rho}(\tilde{s}+s')\mathbf{u}\cdot\hat{\mathbf{n}}\mathrm{d}\Sigma = 0, \tag{3.51}$$

$$\int_V \nabla \cdot \left[\frac{k}{\tilde{T}}\nabla\left(\tilde{T}+T'\right)\right] \mathrm{d}^3x = \int_{\partial V} \frac{k}{\tilde{T}}\nabla\left(\tilde{T}+T'\right)\cdot\hat{\mathbf{n}}\mathrm{d}\Sigma = 0, \tag{3.52}$$

where $V = (0,\ L_x) \times (0,\ L_y) \times (0,\ L)$ is the total periodic fluid volume. Furthermore, the adiabatic boundary condition (3.48) can always be imposed in such a way, that

$$-k\frac{\mathrm{d}\tilde{T}}{\mathrm{d}z}\bigg|_{z=0,\ L} = 0, \qquad -k\frac{\partial T'}{\partial z}\bigg|_{z=0,\ L} = 0, \tag{3.53}$$

which simply corresponds to a rather natural choice of the reference state in this case. This, however, in conjunction with the reference state equation

$$\frac{\mathrm{d}}{\mathrm{d}z}\left(k\frac{\mathrm{d}\tilde{T}}{\mathrm{d}z}\right) = 0 \Rightarrow k\frac{\mathrm{d}\tilde{T}}{\mathrm{d}z} = \text{const}, \tag{3.54}$$

implies, that

$$k\frac{\mathrm{d}\tilde{T}}{\mathrm{d}z} = 0, \tag{3.55}$$

and hence the entire term

$$\frac{k}{\tilde{T}^2}\frac{\mathrm{d}\tilde{T}}{\mathrm{d}z}\frac{\partial}{\partial z}\left(\tilde{T}+T'\right) = 0, \tag{3.56}$$

from the entropy equation vanishes. Therefore integration of the entropy balance (3.50) over the entire periodic fluid volume $V$ leads to

$$\frac{\partial}{\partial t}\int_V \tilde{\rho}s\mathrm{d}^3x = 2\int_V \frac{\mu}{\tilde{T}}\mathbf{G}^s : \mathbf{G}^s\mathrm{d}^3x + \int_V \frac{\mu_b}{\tilde{T}}(\nabla\cdot\mathbf{u})^2\mathrm{d}^3x \geq 0. \tag{3.57}$$

This way we have demonstrated, that the anelastic approximation is consistent with the second law of thermodynamics. In other words, we have shown that the production of the total entropy in anelastic, adiabatically insulated systems, calculated from the entropy equation (3.26d) is positive definite.

### 3.1.2 Conservation of Mass and Values of the Mean Pressure at Boundaries

The solenoidal constraint $\nabla \cdot (\tilde{\rho}\mathbf{u}) = 0$, obtained by neglection of the density time variation implies sound-proof dynamics and therefore the pressure spreads infinitely fast. Consequently, the pressure fluctuation is determined by a Poisson-type, elliptic equation

$$\begin{aligned}\nabla^2 p' =& \frac{\partial}{\partial z}\left(\rho'\tilde{g}\right) - \nabla \cdot \left[\nabla \cdot (\tilde{\rho}\mathbf{u}\mathbf{u}) + \tilde{\rho}\nabla\psi'\right] - \left(\frac{4}{3}\mu + \mu_b\right)\nabla^2\left(\frac{1}{\tilde{\rho}}\frac{\mathrm{d}\tilde{\rho}}{\mathrm{d}z}u_z\right)\\ &+ 2\frac{\mathrm{d}\mu}{\mathrm{d}z}\nabla^2 u_z - \frac{\mathrm{d}}{\mathrm{d}z}\left(\frac{2}{3}\mu + 2\mu_b\right)\frac{\partial}{\partial z}\left(\frac{1}{\tilde{\rho}}\frac{\mathrm{d}\tilde{\rho}}{\mathrm{d}z}u_z\right)\\ &+ 2\frac{\mathrm{d}^2\mu}{\mathrm{d}z^2}\frac{\partial u_z}{\partial z} - 2\frac{\mathrm{d}^2}{\mathrm{d}z^2}\left(\mu_b - \frac{2}{3}\mu\right)\frac{\partial}{\partial z}\left(\frac{1}{\tilde{\rho}}\frac{\mathrm{d}\tilde{\rho}}{\mathrm{d}z}u_z\right),\end{aligned} \tag{3.58}$$

obtained by taking a divergence of the Navier-Stokes equation (3.26a), utilizing $\nabla \cdot (\tilde{\rho}\mathbf{u}) = 0$ and a simplifying assumption that $\mu = \mu(z)$ and $\mu_b = \mu_b(z)$ are functions of height only. However, neglection of the time variation of density implies also, that the total mass is *not* conserved in the dynamics described by the anelastic system of Eqs. (3.26a–3.26e), and thus conservation of the total mass must be imposed additionally. This means, that if we assume that the total mass is contained in the reference state $\tilde{\rho}$, we must impose

$$\left\langle \rho' \right\rangle = 0 \quad \text{at all times.} \tag{3.59}$$

The latter constraint must be imposed at every instant. This is achieved in a very simple way in the general case of non-uniform gravity, when because of

$$\frac{\mathrm{d}^2\left\langle \psi' \right\rangle_h}{\mathrm{d}z^2} = 4\pi G \left\langle \rho' \right\rangle_h \tag{3.60}$$

the mass conservation is simply ensured by continuity of the gravitational field at top and bottom (cf. (3.5)),

$$\left.\frac{\mathrm{d}\left\langle \psi' \right\rangle_h}{\mathrm{d}z}\right|_{z=L} = \left.\frac{\mathrm{d}\left\langle \psi' \right\rangle_h}{\mathrm{d}z}\right|_{z=0} = 0. \tag{3.61}$$

It follows, that if we average the $z$-component of the Navier-Stokes equation (3.26a) (with the viscosities $z$-dependent only) over the entire periodic domain, we get

$$\left\langle p' \right\rangle_h (z = L) - \left\langle p' \right\rangle_h (z = 0) = -L\left\langle \rho'\tilde{g} \right\rangle - L\left\langle \tilde{\rho}\frac{\partial \psi'}{\partial z} \right\rangle = -2L\left\langle \rho'\tilde{g} \right\rangle, \tag{3.62}$$

where we have used (3.7b) and (3.61) to get the final equality.

When the gravity is uniform the right hand side of the latter equation needs to vanish, since then $\tilde{g} = \text{const}$ and $\langle \rho' \rangle$ needs to vanish due to the mass conservation constraint. In such a case the gravitational potential $\psi'$ drops out of the dynamical equations (cf. the discussion below (3.6)) thus there is no need to solve for $\psi'$ and as a result the condition (3.61) is never applied. Therefore when gravity is uniform the simplest way to achieve the total mass conservation is to impose a null mean pressure fluctuation jump across the layer,

$$\langle p' \rangle_h (z = L) = \langle p' \rangle_h (z = 0) \quad \text{at all times,} \tag{3.63}$$

similarly as in the case of Boussinesq convection (cf. Sect. 2.1.4). This constitutes a boundary condition, which must be imposed on the pressure field at every moment in time, used in tandem with the elliptic equation (3.58) at $\tilde{g} = \text{const}$ and $\nabla\psi' = 0$. We note, however, that whether or not the condition (3.63) is imposed, the convective velocity field remains uninfluenced, since a shift in pressure fluctuation which is time-dependent only, corresponds to a simple gauge transformation. Nevertheless, the condition (3.63) is important in order to fully resolve the dynamics of the thermodynamic fluctuations.

An important consequence of the mass conservation constraint (3.59) and (3.26e), is that

$$-\langle \tilde{\rho}\tilde{\alpha}T' \rangle + \left\langle \tilde{\rho}\tilde{\beta}p' \right\rangle = 0 \quad \text{at all times} \tag{3.64}$$

or equivalently

$$-\left\langle \tilde{\rho}\tilde{\alpha}\tilde{T}\frac{s'}{\tilde{c}_p} \right\rangle + \left\langle \frac{\tilde{c}_v}{\tilde{c}_p}\tilde{\rho}\tilde{\beta}p' \right\rangle = 0 \quad \text{at all times} \tag{3.65}$$

must also be satisfied throughout the evolution of the system.

### 3.1.3 Boussinesq Limit

Perhaps the easiest link with the Boussinesq approximation is obtained through the energy equation expressed in terms of entropy (3.27), by assuming that the scale heights associated with density, temperature and pressure are large compared to the fluid layer thickness, i.e. $\mathrm{d}\tilde{\rho}/\mathrm{d}z \ll \bar{\rho}/L$, $\mathrm{d}\tilde{T}/\mathrm{d}z \ll \bar{T}/L$ and $\mathrm{d}\tilde{p}/\mathrm{d}z \ll \bar{p}/L$. This implies $\delta \lesssim \mathcal{O}(\epsilon)$, $\tilde{\rho} \approx \bar{\rho}$, $\tilde{T} \approx \bar{T}$, $\tilde{p} \approx \bar{p}$, $\mathrm{d}\tilde{\rho}/\mathrm{d}z \approx \mathrm{d}\tilde{\bar{\rho}}/\mathrm{d}z$, $\mathrm{d}\tilde{T}/\mathrm{d}z \approx \mathrm{d}\tilde{\bar{T}}/\mathrm{d}z$, $\mathrm{d}\tilde{p}/\mathrm{d}z \approx \mathrm{d}\tilde{\bar{p}}/\mathrm{d}z$ where $\epsilon = \Delta\tilde{\bar{\rho}}/\bar{\rho}$ and the hydrostatic contributions to thermodynamic fields, just as in the previous chapter are decomposed into the mean and a hydrostatic vertically varying correction, e.g. $\tilde{T} = \bar{T} + \tilde{\bar{T}}$. Note also, that in the Boussinesq limit $L$ becomes so small (or alternatively the averaged temperature $\bar{T}$ so large) that by the use of general formula (3.9), resulting directly from the definition of $c_p$ and thermodynamic identities, and with the aid of the hydrostatic balance we get $c_p \sim -\bar{\alpha}\bar{T}gL/(L\mathrm{d}_z\tilde{\bar{T}}) \sim \epsilon^{-1}gL/\bar{T}$. This transforms the above entropy equation (3.27) into the Boussinesq one (2.49) and the continuity equation into $\nabla \cdot \mathbf{u} \approx 0$ in a straightforward way (the term proportional to $\Delta_S \sim -\mathrm{d}\tilde{T}/\mathrm{d}z \approx -\mathrm{d}\tilde{\tilde{T}}/\mathrm{d}z \sim \mathcal{O}(\bar{T}\epsilon/L)$ remains in the entropy equation (3.27) since it is of the same order $\mathcal{O}(\epsilon^{1/2})$ as the rest of the terms in that equation, but in the continuity equation $u_z \mathrm{d}\tilde{\rho}/\mathrm{d}z \sim \mathcal{O}(\bar{\rho}\sqrt{g/L}\epsilon^{3/2})$ is negligible compared to $\bar{\rho}\nabla \cdot \mathbf{u} \sim \mathcal{O}(\bar{\rho}\sqrt{g/L}\epsilon^{1/2})$). Moreover, as demonstrated in the previous chapter (cf. discussion below (2.19)) the assumption of large scale heights implies also $\left|\tilde{\beta}p'\right| \ll \left|\tilde{\alpha}T'\right|$, which allows to simplify the buoyancy force in the Navier-Stokes equation (3.26a) to $-\tilde{\alpha}T'\mathbf{g}$ thus obtaining the Boussinesq momentum balance (2.26). A consequence of $\delta \lesssim \mathcal{O}(\epsilon)$ is that the departure from adiabatic state in the Boussinesq limit is always small but in the sense, that the total temperature (and density, pressure) variation is very weak in comparison with its mean value; the actual difference $-(L/\bar{T})(\mathrm{d}\tilde{\tilde{T}}/\mathrm{d}z + g\bar{\alpha}\bar{T}/\bar{c}_p)$ can be even smaller, but it is not a necessary requirement; in other words under the Boussinesq approximation both contributions from $(L/\bar{T})(\mathrm{d}\tilde{\tilde{T}}/\mathrm{d}z)$ and $(L/\bar{T})(g\bar{\alpha}\bar{T}/\bar{c}_p)$ are weak, of the order $\mathcal{O}(\epsilon)$, but it is allowed that $g\bar{\alpha}\bar{T}/\bar{c}_p \ll |\mathrm{d}\tilde{\tilde{T}}/\mathrm{d}z|$ (cf. Sect. 3.7.2).

Taking the Boussinesq limit of the temperature equation (3.37) is achieved by neglection of the entire lagrangian pressure derivative on the grounds of $\left|\tilde{\beta}p'\right| \ll \left|\tilde{\alpha}T'\right|$ and

$$\begin{aligned} \rho c_p u_z \frac{\mathrm{d}\tilde{T}}{\mathrm{d}z} - \alpha T u_z \frac{\mathrm{d}\tilde{p}}{\mathrm{d}z} &= \bar{\rho}\bar{c}_p u_z \frac{\mathrm{d}\tilde{\tilde{T}}}{\mathrm{d}z} - \bar{\alpha}\bar{T} u_z \frac{\mathrm{d}\tilde{\tilde{p}}}{\mathrm{d}z} + \mathcal{O}\left(\epsilon^2 \bar{c}_p \mathcal{U} \bar{\rho}\bar{T}/L\right) \\ &\approx - \bar{\rho} c_p u_z \Delta_S + \mathcal{O}\left(\epsilon^2 \bar{c}_p \mathcal{U} \bar{\rho}\bar{T}/L\right). \end{aligned} \tag{3.66}$$

## 3.2 The Reference State

We will focus for a moment on the static reference state, which by assumption is close to an adiabatic state, to give an idea of the possible formulations within the anelastic approximation. There are basically two options - either, as in all the previous sections, the reference state is not adiabatic but its departure from adiabaticity is small by assumption and given by (3.11) or the reference state itself can be chosen to be adiabatic and then the flow is driven by the boundary conditions which only slightly depart from those corresponding to the adiabatic state. The latter case requires an alternative definition of the parameter $\delta$ involving then the boundary conditions. Both formulations are equivalent and can be transformed into each other. The case of the adiabatic reference state is postponed until the end of the current section and we will start with the case, when the hydrostatic reference state slightly departs from adiabatic and satisfies the boundary conditions, which drive convection. From the mathematical point of view this corresponds simply to a well-established trick in the theory of differential equations of subtracting a stationary state which satisfies the non-homogeneous boundary conditions, to allow for homogeneous boundary conditions for the remainder (fluctuation). The hydrostatic state equations (3.7a)–(3.7d) for a perfect gas in the absence of the radiative heat sources and at uniform gravity take the form[6]

$$\frac{\mathrm{d}\tilde{p}}{\mathrm{d}z} = -\tilde{\rho} g, \quad \frac{\mathrm{d}}{\mathrm{d}z}\left(k\frac{\mathrm{d}\tilde{T}}{\mathrm{d}z}\right) = 0, \quad \tilde{p} = \tilde{\rho} R \tilde{T}, \quad \tilde{s} = c_v \ln\frac{\tilde{p}}{\tilde{\rho}^{\gamma}} + \text{const}. \tag{3.67}$$

Let us consider an illustrative example when the thermal conductivity $k$ is assumed uniform, in which case the reference state is polytropic with the reference temperature $\tilde{T}$ being a simple linear function of $z$ and

$$\tilde{T} = T_B\left(1 - \theta\frac{z}{L}\right), \quad \tilde{\rho} = \rho_B\left(1 - \theta\frac{z}{L}\right)^m, \quad \tilde{p} = \frac{gL\rho_B}{\theta(m+1)}\left(1 - \theta\frac{z}{L}\right)^{m+1}, \tag{3.68a}$$

$$\tilde{s} = c_p\frac{m+1-\gamma m}{\gamma}\ln\left(1 - \theta\frac{z}{L}\right) + \text{const}, \tag{3.68b}$$

$$m = \frac{gL}{R\Delta T} - 1. \tag{3.68c}$$

where $\theta = \Delta T/T_B$ is the magnitude of the basic temperature gradient and a measure of compressibility of the fluid; $T_B$ and $\rho_B$ denote here the values of temperature and density at the bottom of the layer in the reference state respectively, $\Delta T = T_B - T_T > 0$ is the temperature jump across the layer in the reference state and $0 \le \theta < 1$. We note, that for the case when the temperature is held constant at boundaries the values $T_B$ and $T_T$ simply constitute the thermodynamic boundary conditions, however, in the cases of specified heat flux at boundaries or isentropic boundaries, the actual top and bottom values of the temperature in the convective state are generally shifted with respect to $T_T$ and $T_B$ by corrections of the order $\delta$, namely $T'(x, y, z = 0, t)$ and $T'(x, y, z = L, t)$. On the other hand the bottom value of density in the reference state $\rho_B$ is established by the total mass of the fluid in the considered domain and thus in general the total bottom density value is also altered by order $\delta$ corrections in a convective state.

Since in the adiabatic state $\tilde{\rho} \sim \tilde{T}^{1/(\gamma-1)}$, here the departure from adiabaticity is manifested by the small difference between the adiabatic exponent $1/(\gamma - 1)$ and the polytropic index $m$, i.e. $\delta \sim 1/(\gamma - 1) - m > 0$. More precisely, in this case

$$\delta = \left(\frac{1}{\gamma - 1} - m\right)\frac{\gamma - 1}{\gamma}\ln\frac{1}{1-\theta} = \frac{L}{T_B}\left(\frac{\Delta T}{L} - \frac{g}{c_p}\right)\frac{1}{\theta}\ln\frac{1}{1-\theta} > 0; \tag{3.69}$$

note, that the non-dimensional expression

[6]Note, that the expression for the entropy of an ideal gas results directly from the definition of the specific heat at constant volume $c_v = T(\partial_T s)_\rho$, the Maxwell relation $\rho^2(\partial_\rho s)_T = (\partial_T p)_\rho$ and the equation of state $p = \rho R T$.

$$\frac{L}{T_B}\left(\frac{\Delta T}{L}-\frac{g}{c_p}\right) \tag{3.70}$$

is often utilized as an alternative definition of the small anelastic parameter $\delta$.

Let us expand the reference state about the adiabatic neglecting terms of the order $\mathcal{O}(c_p^2\Delta_S^2/g^2)$ and higher; this yields

$$\tilde{T} = T_B\left(1-\frac{gz}{c_pT_B}\right)-\Delta_S z, \tag{3.71a}$$

$$\begin{aligned}\tilde{\rho} \approx & \rho_B\left(1-\frac{gz}{c_pT_B}-\frac{\Delta_S}{T_B}z\right)^{\frac{1}{\gamma-1}-\frac{\Delta_S c_p}{g(\gamma-1)}} \approx \rho_B\left(1-\frac{gz}{c_pT_B}\right)^{\frac{1}{\gamma-1}} \\ & -\frac{\rho_B c_p\Delta_S}{g(\gamma-1)}\left[\frac{gz}{c_pT_B}\left(1-\frac{gz}{c_pT_B}\right)^{\frac{2-\gamma}{\gamma-1}}+\left(1-\frac{gz}{c_pT_B}\right)^{\frac{1}{\gamma-1}}\ln\left(1-\frac{gz}{c_pT_B}\right)\right],\end{aligned} \tag{3.71b}$$

$$\begin{aligned}\tilde{p} \approx & \rho_B RT_B\left(1-\frac{gz}{c_pT_B}-\frac{\Delta_S}{T_B}z\right)^{\frac{\gamma}{\gamma-1}-\frac{\Delta_S c_p}{g(\gamma-1)}} \approx \rho_B RT_B\left(1-\frac{gz}{c_pT_B}\right)^{\frac{\gamma}{\gamma-1}} \\ & -\frac{\rho_B T_B c_v c_p\Delta_S}{g}\left[\frac{\gamma gz}{c_pT_B}\left(1-\frac{gz}{c_pT_B}\right)^{\frac{1}{\gamma-1}}+\left(1-\frac{gz}{c_pT_B}\right)^{\frac{\gamma}{\gamma-1}}\ln\left(1-\frac{gz}{c_pT_B}\right)\right],\end{aligned} \tag{3.71c}$$

$$\tilde{s} \approx \frac{c_p^2\Delta_S}{g}\ln\left(1-\frac{gz}{c_pT_B}\right)+\text{const.} \tag{3.71d}$$

The order $\mathcal{O}(c_p\Delta_S/g)$ corrections are vital, since they allow to satisfy the boundary conditions, which drive the flow. It is this order correction to the constant adiabatic entropy profile in (3.71d) which is responsible for non-zero entropy gradient and driving in the energy equation (3.26d), i.e. $\tilde{\rho}\tilde{T}u_z\mathrm{d}\tilde{s}/\mathrm{d}z \neq 0$. It is, therefore, important to realize, that the $\mathcal{O}(c_p\Delta_S/g)$ corrections to the adiabatic profile in the reference state need to be established precisely, therefore the reference state must be obtained from the full equations, not those approximated, with some $\mathcal{O}(c_p\Delta_S/g)$ order terms already neglected.

As mentioned, an often used alternative formulation is based on taking the reference state hydrostatic and adiabatic and application of non-homogeneous boundary conditions to the fluctuations. In such a case the Navier-Stokes and continuity equations (3.26a)–(3.26b) remain unchanged but in the energy equation (3.26d) the basic entropy gradient vanishes (there is no term proportional to $\Delta_S$ in (3.27)), so that it takes on a simpler form

$$\rho_{ad} T_{ad} \left( \frac{\partial s'}{\partial t} + \mathbf{u} \cdot \nabla s' \right) = \nabla \cdot \left( k \nabla T' \right) + 2\mu \mathbf{G}^s : \mathbf{G}^s + \left( \mu_b - \frac{2}{3}\mu \right) (\nabla \cdot \mathbf{u})^2 + Q', \tag{3.72}$$

where the subscript $ad$ denotes the static, adiabatic reference state and the boundary conditions on $T'$ are non-zero in this case. In the above example of a perfect gas with uniform thermal conductivity and gravity and no radiative heat sources the adiabatic static state can take the form

$$T_{ad} = T_{ad\,B} \left( 1 - \frac{gz}{c_p T_{ad\,B}} \right), \quad \rho_{ad} = \rho_{ad\,B} \left( 1 - \frac{gz}{c_p T_{ad\,B}} \right)^{\frac{1}{\gamma - 1}}, \tag{3.73a}$$

$$p_{ad} = \rho_{ad\,B} R T_{ad\,B} \left( 1 - \frac{gz}{c_p T_B} \right)^{\frac{\gamma}{\gamma-1}}, \quad s_{ad} = \text{const.}, \tag{3.73b}$$

which is of the same form as (3.71a–3.71d) but without the $\mathcal{O}(c_p \Delta_S / g)$ corrections, and satisfies the hydrostatic dynamical equations (the quantities $\rho_{ad\,B}$ and $T_{ad\,B}$ are simply the bottom values of density and temperature in the adiabatic state). The small parameter measuring the departure from adiabaticity introduced by the boundary conditions, either with fixed temperature or fixed heat flux, which imply $\Delta T = T_B - T_T > gL/c_p$ is conveniently defined as

$$\delta_{ad} = \frac{1}{T_B} \left( \Delta T - \frac{gL}{c_p} \right), \tag{3.74}$$

and in the current case of constant thermal conductivity when the temperature profile is linear it is simply equivalent to $\delta_{ad} = \delta = L\Delta_S / T_B$. Note, that the total mass (per horizontal unit surface) contained in the fluid domain can be expressed in two ways

$$\int_0^L \tilde{\rho} \mathrm{d}z = \frac{R\rho_B T_B}{g} \left[ 1 - \left( 1 - \frac{\Delta T}{T_B} \right)^{\frac{gL}{R\Delta T}} \right], \tag{3.75a}$$

$$\int_0^L \rho_{ad} \mathrm{d}z = \frac{R\rho_{ad\,B} T_B}{g} \left[ 1 - \left( 1 - \frac{gL}{c_p T_B} \right)^{\frac{\gamma}{\gamma-1}} \right], \tag{3.75b}$$

where $T_{ad\,B} = T_B$ has been assumed and thus a full correspondence between the two formulations with the hydrostatic conduction and adiabatic reference states is obtained by assuming

$$\rho_{ad\,B} = \rho_B \frac{1 - \left( 1 - \frac{\Delta T}{T_B} \right)^{\frac{gL}{R\Delta T}}}{1 - \left( 1 - \frac{gL}{c_p T_B} \right)^{\frac{\gamma}{\gamma-1}}} = \rho_B + \mathcal{O}\left( \delta \rho_B \right), \tag{3.76}$$

which means that the bottom values of the density in the hydrostatic conduction reference state and in the hydrostatic adiabatic state are slightly shifted by an order $\mathcal{O}(\delta\rho_B)$ value. This way the total mass of the fluid is contained in the reference states, $\tilde{\rho}$ and $\rho_{ad}$, and therefore the mass conservation implies that the spatial average over the entire fluid domain of the density fluctuation is always zero, both in the case of fluctuation about the hydrostatic conduction reference state, $\langle\rho - \tilde{\rho}\rangle = 0$ and in the case of fluctuation about the hydrostatic adiabatic reference state, $\langle\rho - \rho_{ad}\rangle = 0$. Recall, that in practice this is obtained by imposing condition (3.63) on the mean pressure fluctuation.

To illustrate full correspondence between the conduction reference state formulation and the formulation with the adiabatic reference state let us consider a perfect gas, characterized by constant dynamic viscosities $\mu = \text{const}$, $\mu_b = \text{const}$, thermal conductivity $k = \text{const}$, specific heats $c_v = \text{const}$, $c_p = \text{const}$ and gravity $g = \text{const}$ and the total mass $M_f$, which is driven by keeping the temperature fixed at the boundaries and $\Delta T = T_B - T_T > gL/c_p$. Of course the system physically responds to the driving regardless of our choice of mathematical description, i.e. no matter which formulation we pick, the final results for the velocity field and the total temperature $T(\mathbf{x}, t)$, total pressure $p(\mathbf{x}, t)$, total density $\rho(\mathbf{x}, t)$, and the total entropy $s(\mathbf{x}, t)$ must be the same. The total thermodynamic variables are expressed in the following way

$$T(\mathbf{x}, t) = \tilde{T}(z) + T'(\mathbf{x}, t) = T_{ad}(z) + T_S(\mathbf{x}, t), \tag{3.77a}$$

$$\rho(\mathbf{x}, t) = \tilde{\rho}(z) + \rho'(\mathbf{x}, t) = \rho_{ad}(z) + \rho_S(\mathbf{x}, t), \tag{3.77b}$$

$$p(\mathbf{x}, t) = \tilde{p}(z) + p'(\mathbf{x}, t) = p_{ad}(z) + p_S(\mathbf{x}, t), \tag{3.77c}$$

$$s(\mathbf{x}, t) = \tilde{s}(z) + s'(\mathbf{x}, t) = \text{const} + s_S(\mathbf{x}, t), \tag{3.77d}$$

where the subscript $S$ denotes the superadiabatic fluctuation about the adiabatic state. The latter expressions define transformations between the two formulations of the type

$$T_S(\mathbf{x}, t) = T'(\mathbf{x}, t) + \tilde{T}(z) - T_{ad}(z) = T'(\mathbf{x}, t) - \Delta_S z, \tag{3.78}$$

where (3.71a) and (3.73a) have been used and $T_{ad\,B} = T_B$. The transformation between the entropy fluctuations $s_S(\mathbf{x}, t) = s'(\mathbf{x}, t) + \tilde{s}(z) + \text{const}$ is also rather simple, since the inhomogeneous correction involves only the vertical variation of $\tilde{s}(z)$, cf. (3.71d). On the other hand the transformations for the density and pressure fluctuations, which result from (3.77b, 3.77c), are slightly more complicated, because they must involve the bottom values of the density in both formulations, i.e. $\rho_B$ and $\rho_{ad\,B}$. Therefore if we assume, that the total mass of the fluid is contained in the references states (which is a natural assumption, see Sect. 3.1.2), and utilize the relation (3.76), the transformations for the density and pressure fluctuations between the two anelastic formulations about the conduction and adiabatic reference states become well defined. In particular for the density fluctuations we provide the explicit transformation formula (cf. (3.71b) and (3.73a))

$$\begin{aligned}\rho_S(\mathbf{x},t) &= \rho'(\mathbf{x},t) + \tilde{\rho}(z) - \rho_{ad}(z)\\ &= \rho'(\mathbf{x},t) + \rho_{ad\,B}\left(1-\frac{gz}{c_pT_B}\right)^{\frac{1}{\gamma-1}}\left\{\frac{\rho_B}{\rho_{ad\,B}} - 1\right.\\ &\left.\quad -\frac{c_p\Delta_S}{g\,(\gamma-1)}\left[\frac{\frac{gz}{c_pT_B}}{1-\frac{gz}{c_pT_B}} + \ln\left(1-\frac{gz}{c_pT_B}\right)\right]\right\} + \mathcal{O}\left(\rho_{ad\,B}\delta^2\right),\end{aligned} \tag{3.79}$$

where by the use of $\delta = L\Delta_S/T_B = \theta - gL/c_pT_B$ and (3.76) the ratio $\rho_B/\rho_{ad\,B}$ is given by

$$\frac{\rho_B}{\rho_{ad\,B}} = 1 - \frac{c_p\Delta_S}{g\,(\gamma-1)}\frac{\gamma\left(1-\frac{gL}{c_pT_B}\right)^{\frac{\gamma}{\gamma-1}}}{1-\left(1-\frac{gL}{c_pT_B}\right)^{\frac{\gamma}{\gamma-1}}}\left[\frac{\frac{gL}{c_pT_B}}{1-\frac{gL}{c_pT_B}} + \ln\left(1-\frac{gL}{c_pT_B}\right)\right] + \mathcal{O}\left(\delta^2\right). \tag{3.80}$$

Of course the correction to $\rho'(\mathbf{x},t)$ in Eq. (3.79) is of the same order of magnitude as the density fluctuation, i.e. $\mathcal{O}(\rho_{ad\,B}\delta)$. Similar transformation formula can be obtained for the pressure fluctuation. We conclude, that the results obtained with one formulation, say for fluctuations about a conduction reference state, can be easily transformed into fluctuations about the adiabatic state in the same physical setting using (3.77a–3.77d). In practice, it is often necessary to compare results obtained from numerical simulations utilizing the two different formulations. The above recipe allows to do this, but of course the physical situation modelled with the two approaches *must* be the same in order for the results to correspond directly to each other. Therefore such comparisons must be done with great care. In the considered example the two formulations necessarily produce the same results, as long as the aforementioned set of physical parameters, i.e. $\mu$, $\mu_b$, $k$, $c_p$, $c_v$, $g$, the total mass of the fluid $M_f$ and the driving bottom-top temperature difference $\Delta T$ is the same in both approaches. We elaborate on the issue of how to compare the results of the conduction reference state formulation with the results of the formulation with the adiabatic reference state in Sect. 3.7.1.

Finally we note, that in numerical modelling of anelastic convection an often undertaken approach is to utilize time-dependent basic states. All the variables such as $\rho$, $T$, $p$ and $\mathbf{u}$ are horizontally averaged at each time step, so that the basic state $\langle\rho\rangle_h(z,t)$, $\langle T\rangle_h(z,t)$, $\langle p\rangle_h(z,t)$ and $\langle\mathbf{u}\rangle_h(z,t)$ depends only on height and time. Often a time average is applied as well,

$$\langle\rho\rangle_{h,t} = \frac{1}{t}\int_0^t \langle\rho\rangle_h(z,s)\mathrm{d}s, \tag{3.81}$$

which implies that the time dependence of the basic state is slow. Such a state naturally satisfies all the boundary conditions responsible for driving the flow; if the fluid flow is thermally driven this implies that the temperature fluctuation satisfies homogeneous boundary conditions. If, however, convection is driven by fixed entropy at boundaries, the entropy fluctuation has to vanish there. The small parameter $\delta$ is then naturally defined by the boundary conditions, as in (3.74).

## 3.3 Simplifications Through Entropy Formulations

The aim of this section is to further simplify the full system of dynamical anelastic equations and express them solely in terms of two thermodynamic variables, the entropy $s'$ and the pressure $p'$ with the latter appearing only in the Navier-Stokes equation under the $\nabla$ operator, thus being easily removable by taking its curl. By making use of Eqs. (3.26e) which result from the equation of state, one can eliminate the density and temperature fluctuations and express them by the entropy and pressure fluctuations

$$\rho' = -\frac{\tilde{\alpha}\tilde{T}\tilde{\rho}}{\tilde{c}_p}s' + \tilde{\rho}\tilde{\beta}\frac{\tilde{c}_v}{\tilde{c}_p}p', \tag{3.82a}$$

$$T' = \frac{\tilde{T}}{\tilde{c}_p}s' + \frac{\tilde{\alpha}\tilde{T}}{\tilde{c}_p\tilde{\rho}}p', \tag{3.82b}$$

where the thermodynamic identity

$$c_p - c_v = \frac{\alpha^2 T}{\beta\rho}, \tag{3.83}$$

justified in (2.38) was used in obtaining the above expression for $\rho'$. Introduction of (3.82b) into the energy balance (3.26d) leads to

$$\begin{aligned}\tilde{\rho}\tilde{T}\left[\frac{\partial s'}{\partial t} + \mathbf{u}\cdot\nabla\left(s' + \tilde{s}\right)\right] =& \nabla\cdot\left[k\nabla\left(\frac{\tilde{T}}{\tilde{c}_p}s'\right)\right] + \nabla\cdot\left[k\nabla\left(\frac{\tilde{\alpha}\tilde{T}}{\tilde{c}_p}\frac{p'}{\tilde{\rho}}\right)\right]\\ &+ 2\mu\mathbf{G}^s : \mathbf{G}^s + \left(\mu_b - \frac{2}{3}\mu\right)(\nabla\cdot\mathbf{u})^2 + Q'. \end{aligned}\tag{3.84}$$

On the other hand the pressure term together with the buoyancy force in the Navier-Stokes equation (3.26a), divided by $\tilde{\rho}$, with the aid of (3.82a) can be expressed in the following way

$$\begin{aligned}-\frac{1}{\tilde{\rho}}\nabla p' + \frac{\rho'}{\tilde{\rho}}\tilde{\mathbf{g}} - \nabla\psi' &= -\frac{1}{\tilde{\rho}}\nabla p' + \left(\frac{\tilde{\alpha}\tilde{T}}{\tilde{c}_p}s' - \tilde{\beta}\frac{\tilde{c}_v}{\tilde{c}_p}p'\right)\tilde{g}\hat{\mathbf{e}}_z - \nabla\psi'\\ &= -\nabla\left(\frac{p'}{\tilde{\rho}} + \psi'\right) + \frac{\tilde{\alpha}\tilde{T}}{\tilde{c}_p}s'\tilde{g}\hat{\mathbf{e}}_z - \left(\tilde{g}\tilde{\beta}\frac{\tilde{c}_v}{\tilde{c}_p} + \frac{1}{\tilde{\rho}^2}\frac{\mathrm{d}\tilde{\rho}}{\mathrm{d}z}\right)p'\hat{\mathbf{e}}_z. \end{aligned}\tag{3.85}$$

Next, since the $z$-derivative of the static density distribution is given by (3.32) and, again, making use of $c_p - c_v = \alpha^2 T/\beta\rho$ (cf. (3.83)), one obtains

$$
\begin{aligned}
-\frac{1}{\tilde{\rho}}\nabla p' + \frac{\rho'}{\tilde{\rho}}\tilde{\mathbf{g}} - \nabla\psi' &= -\nabla\left(\frac{p'}{\tilde{\rho}} + \psi'\right) + \frac{\tilde{\alpha}\tilde{T}}{\tilde{c}_p}s'\tilde{g}\hat{\mathbf{e}}_z - \tilde{\alpha}\frac{p'}{\tilde{\rho}}\Delta_S\hat{\mathbf{e}}_z \\
&= -\nabla\left(\frac{p'}{\tilde{\rho}} + \psi'\right) + \frac{\tilde{\alpha}\tilde{T}}{\tilde{c}_p}s'\tilde{g}\hat{\mathbf{e}}_z + \mathcal{O}\left(\bar{g}\delta^2\right), \quad (3.86)
\end{aligned}
$$

since $\Delta_S = \mathcal{O}(\bar{T}\delta/L)$ and $p'/\tilde{p} = \mathcal{O}(\delta)$.

This allows to rewrite the system of dynamic equations under the anelastic approximation in the form

$$
\begin{aligned}
\frac{\partial\mathbf{u}}{\partial t} + (\mathbf{u}\cdot\nabla)\,\mathbf{u} = &-\nabla\left(\frac{p'}{\tilde{\rho}} + \psi'\right) + \frac{\tilde{\alpha}\tilde{T}}{\tilde{c}_p}s'\tilde{g}\hat{\mathbf{e}}_z + \frac{\mu}{\tilde{\rho}}\nabla^2\mathbf{u} + \left(\frac{\mu}{3\tilde{\rho}} + \frac{\mu_b}{\tilde{\rho}}\right)\nabla(\nabla\cdot\mathbf{u}) \\
&+ \frac{2}{\tilde{\rho}}\nabla\mu\cdot\mathbf{G}^s + \frac{1}{\tilde{\rho}}\nabla\left(\mu_b - \frac{2}{3}\mu\right)\nabla\cdot\mathbf{u}, \quad (3.87a)
\end{aligned}
$$

$$
\nabla^2\psi' = 4\pi G\frac{\tilde{\rho}}{\tilde{c}_p}\left(\tilde{\beta}\tilde{c}_v p' - \tilde{\alpha}\tilde{T}s'\right) \quad (3.87b)
$$

$$
\nabla\cdot(\tilde{\rho}\mathbf{u}) = 0, \quad (3.87c)
$$

$$
\begin{aligned}
\tilde{\rho}\tilde{T}\left(\frac{\partial s'}{\partial t} + \mathbf{u}\cdot\nabla s'\right) - \tilde{\rho}\tilde{c}_p u_z\Delta_S = &\nabla\cdot\left[k\nabla\left(\frac{\tilde{T}}{\tilde{c}_p}s'\right)\right] + \nabla\cdot\left[k\nabla\left(\frac{\tilde{\alpha}\tilde{T}}{\tilde{c}_p}\frac{p'}{\tilde{\rho}}\right)\right] \\
&+ 2\mu\mathbf{G}^s : \mathbf{G}^s + \left(\mu_b - \frac{2}{3}\mu\right)(\nabla\cdot\mathbf{u})^2 + Q', \quad (3.87d)
\end{aligned}
$$

which constitute a closed problem for two thermodynamic variables $s'$ and $p'$ and the velocity field $\mathbf{u}$. However, up to now we have merely substituted for the density and temperature fluctuations from (3.82a, 3.82b), but the pressure fluctuation still appears in the energy equation and in the equation for the gravitational potential. Therefore we will now assume, that the gravitational acceleration is constant (cf. the discussion below (3.6))

$$
\text{assume:}\quad g \approx \tilde{g} \approx \text{const}, \quad \left|\nabla\psi'\right| \ll \frac{\rho'}{\tilde{\rho}}g, \quad (3.88)
$$

which allows to eliminate the fluctuation $\mathbf{g}' = -\nabla\psi'$ from the momentum balance, as it is negligibly small in comparison with the term $\tilde{\alpha}\tilde{T}s'g\hat{\mathbf{e}}_z/\tilde{c}_p$, and hence the Eq. (3.87b) can also be removed.

Next, multiplying the Eq. (3.87a) by $\tilde{\rho}$ and taking its divergence, by the use of the continuity equation (3.87c) we obtain a stationary Poisson-type problem for $p'/\tilde{\rho}$, at the leading order in $\delta$,

$$\nabla\cdot\left(\tilde{\rho}\nabla\frac{p'}{\tilde{\rho}}\right)=\nabla\cdot\left[\nabla\cdot\left(2\mu\mathbf{G}^s-\tilde{\rho}\mathbf{uu}\right)\right]+\nabla^2\left[\left(\mu_b-\frac{2}{3}\mu\right)\nabla\cdot\mathbf{u}\right]+\frac{\partial}{\partial z}\left(\frac{\tilde{\alpha}\tilde{T}\tilde{\rho}g}{\tilde{c}_p}s'\right), \tag{3.89}$$

where

$$\nabla\cdot\left[\nabla\cdot\left(2\mu\mathbf{G}^s-\tilde{\rho}\mathbf{uu}\right)\right]=\partial_i\partial_j\left(2\mu G^s_{ij}-\tilde{\rho}u_iu_j\right), \tag{3.90}$$

and with the aid of the estimate of the static density vertical variation in (3.33) we get

$$\nabla\cdot\left(\tilde{\rho}\nabla\frac{p'}{\tilde{\rho}}\right)=\tilde{\rho}\nabla^2\frac{p'}{\tilde{\rho}}-\frac{\tilde{c}_v\tilde{\beta}\tilde{\rho}^2g}{\tilde{c}_p}\frac{\partial}{\partial z}\frac{p'}{\tilde{\rho}}. \tag{3.91}$$

The Poisson-type, stationary problem for the pressure perturbation is a manifestation of the fact, that under the anelastic approximation all the terms $\mathcal{O}(Ma^2)$ times smaller than the leading order are neglected and thus pressure spreads infinitely fast. Note, that up to now no additional assumptions have been made except for $g=$ const and the standard anelastic assumption of small system departure from the adiabatic state. If we assume, that the fluid satisfies the equation of state of a perfect gas (with constant specific heats) the above system of dynamical equations reduces to

$$\begin{aligned}\frac{\partial\mathbf{u}}{\partial t}+(\mathbf{u}\cdot\nabla)\,\mathbf{u}=&-\nabla\frac{p'}{\tilde{\rho}}+\frac{gs'}{c_p}\hat{\mathbf{e}}_z+\frac{\mu}{\tilde{\rho}}\nabla^2\mathbf{u}+\left(\frac{\mu}{3\tilde{\rho}}+\frac{\mu_b}{\tilde{\rho}}\right)\nabla\left(\nabla\cdot\mathbf{u}\right)\\&+\frac{2}{\tilde{\rho}}\nabla\mu\cdot\mathbf{G}^s+\frac{1}{\tilde{\rho}}\nabla\left(\mu_b-\frac{2}{3}\mu\right)\nabla\cdot\mathbf{u},\end{aligned} \tag{3.92a}$$

$$\nabla\cdot(\tilde{\rho}\mathbf{u})=0, \tag{3.92b}$$

$$\begin{aligned}\tilde{\rho}\tilde{T}\left(\frac{\partial s'}{\partial t}+\mathbf{u}\cdot\nabla s'\right)-\tilde{\rho}c_pu_z\Delta_S=&\nabla\cdot\left[\kappa\tilde{\rho}\nabla\left(\tilde{T}s'\right)\right]+\nabla\cdot\left[\kappa\tilde{\rho}\nabla\left(\frac{p'}{\tilde{\rho}}\right)\right]\\&+2\mu\mathbf{G}^s:\mathbf{G}^s+\left(\mu_b-\frac{2}{3}\mu\right)(\nabla\cdot\mathbf{u})^2+Q',\end{aligned} \tag{3.92c}$$

where we have introduced the thermal diffusivity

$$\kappa=\frac{k}{\tilde{\rho}c_p}. \tag{3.93}$$

The Poisson-type problem for the pressure is given by (3.89), or more explicitly

$$\nabla^2 \frac{p'}{\bar{\rho}} - \frac{g}{c_p(\gamma-1)\tilde{T}} \frac{\partial}{\partial z} \frac{p'}{\bar{\rho}} = \frac{1}{\bar{\rho}} \nabla \cdot [\nabla \cdot (2\mu \mathbf{G}^s - \tilde{\rho}\mathbf{u}\mathbf{u})] + \frac{1}{\bar{\rho}} \nabla^2 \left[ \left( \mu_b - \frac{2}{3}\mu \right) \nabla \cdot \mathbf{u} \right] + \frac{1}{c_p \bar{\rho}} \frac{\partial}{\partial z} (\tilde{\rho} g s'), \tag{3.94}$$

with $\gamma = c_p/c_v = \text{const}$ being the specific heat ratio.

### 3.3.1 *Boundary Conditions for the Entropy*

Formulation in terms of the entropy density per unit mass $s'$ requires specification of boundary conditions for this variable. These naturally depend on a particular problem and may differ from one application to another. However, on general grounds we can say, that by the use of (3.14b) the fixed temperature boundary conditions $T'|_{z=0,L} = 0$ correspond to

$$s' + \frac{\tilde{\alpha}}{\tilde{\rho}} p' = 0 \quad \text{at } z = 0,\ L \tag{3.95}$$

whereas fixed heat flux at boundaries $\partial_z T'|_{z=0,L} = 0$ is equivalent to

$$\frac{\partial}{\partial z} \left( \frac{\tilde{T}}{\tilde{c}_p} s' + \frac{\tilde{\alpha}\tilde{T}}{\tilde{c}_p \tilde{\rho}} p' \right) = 0 \quad \text{at } z = 0,\ L. \tag{3.96}$$

Either one of the boundary conditions, i.e. fixed temperature or fixed heat flux is supplied by the general mass conservation law $\partial_t(\tilde{\rho} + \rho') + \nabla \cdot [(\tilde{\rho} + \rho')\mathbf{u}] = 0$ which implies $\langle \tilde{\rho} + \rho' \rangle = \text{const}$. Under standard initial conditions not introducing additional mass into the system it follows that $\langle \rho' \rangle = 0$, and consequently

$$\left\langle \frac{\tilde{\alpha}\tilde{T}\tilde{\rho}}{\tilde{c}_p} s' \right\rangle = \left\langle \frac{\tilde{c}_v \tilde{\beta}\tilde{\rho}}{\tilde{c}_p} p' \right\rangle. \tag{3.97}$$

Since there are two boundary conditions required for the entropy field $s'$ and one for the pressure $p'$ the three relations: two boundary conditions at $z = 0,\ L$ either (3.95) or (3.96) and the relation between spatial means (3.97) constitute a necessary and sufficient set of conditions, allowing to fully determine the fields. However, this set of conditions is rather cumbersome in terms of applicability in numerical simulations, because it not only couples the entropy and pressure but also involves computation of means in order to fully determine the entropy and pressure fields.

#### 3.3.1.1 The Case of Isentropic Boundary Conditions

Turbulent convection is often modelled with application of isentropic boundary conditions. Although keeping the entropy fixed at boundaries does not seem physical, since physically one can not control the pressure and hence also the entropy at boundaries, it is a common practice due to simplicity achieved when the entropy formulations are used in tandem with fixed entropy boundary conditions. In such a case $s'|_{z=0,L} = 0$ corresponds to

$$-\frac{p'}{c_p\tilde{\rho}} + T' = 0 \quad \text{at } z = 0,\ L \tag{3.98}$$

whereas the conservation of the total mass, $\langle \rho' \rangle = 0$, implies

$$\langle \tilde{\rho} s' \rangle = \frac{c_v}{R}\left\langle \frac{p'}{\tilde{T}} \right\rangle, \tag{3.99}$$

cf. Eqs. (3.14a, 3.14b).

The following calculation reveals some interesting features of convective flows. First we note, that by the mass conservation law the horizontally averaged vertical velocity must satisfy

$$\frac{\partial \langle u_z \rangle_h}{\partial z} = -\frac{\langle u_z \rangle_h}{D_\rho}, \tag{3.100}$$

where $D_\rho = -\tilde{\rho}/\mathrm{d}_z\tilde{\rho}$ is the density scale height, therefore impermeability conditions at the boundaries at $z = 0,\ L$ imply

$$\langle u_z \rangle_h = 0. \tag{3.101}$$

A horizontal average of the $z$-component of the stationary Navier-Stokes equation (3.26a) with constant gravity yields

$$\frac{\partial}{\partial z}\langle \tilde{\rho} u_z^2 \rangle_h = -\frac{\partial \langle p' \rangle_h}{\partial z} - g\langle \rho' \rangle_h, \tag{3.102}$$

therefore the mean pressure fluctuation $\langle p' \rangle_h$ in convection satisfies a non-hydrostatic balance, influenced by inertia. Next, integration of the latter equality over $z$ from 0 to $L$ shows

$$\Delta\langle p' \rangle_h = \langle p' \rangle_{h,B} - \langle p' \rangle_{h,T} = 0, \tag{3.103}$$

which means that the mean pressure fluctuation is the same at both, top and bottom boundaries. Consequently the boundary conditions (3.98) can be expressed in the following way

$$\langle T' \rangle_{h,B} = \frac{\langle p' \rangle_{h,B}}{c_p\tilde{\rho}_B}, \tag{3.104a}$$

$$\langle T' \rangle_{h,T} = \frac{\langle p' \rangle_{h,B}}{c_p\tilde{\rho}_T} = \frac{\tilde{\rho}_B}{\tilde{\rho}_T}\langle T' \rangle_{h,B}, \tag{3.104b}$$

and of course at leading order in $\delta$ the values of the reference state density at top and bottom $\tilde{\rho}_T$ and $\tilde{\rho}_B$ could be simply replaced by the top and bottom values of the total density $\rho(z=0)$ and $\rho(z=L)$, which in general differ by order $\mathcal{O}(\delta)$ corrections. Since density decreases with height, $\tilde{\rho}_B > \tilde{\rho}_T$, it is clear from (3.104b), that when the boundaries are isentropic, the mean temperature fluctuation is of the same sign at the top and bottom boundaries and its magnitude is significantly greater at the top (note that similar calculation could be done for the case of isothermal boundaries, with analogous results for the magnitudes and signs of the mean entropy fluctuation at boundaries).

### 3.3.2 *Constant Thermal Diffusivity Formulation for an Ideal Gas with Non-vanishing Heat Sink*

[7]Holding the assumption, that the fluid satisfies the equation of state of a perfect gas (with constant specific heats) and $g = \text{const}$ a further simplification can be achieved by elimination of the pressure term from the energy equation. The pressure term in that equation can be easily expressed in the form

$$\nabla\cdot\left[\kappa\tilde{\rho}\nabla\left(\frac{p'}{\tilde{\rho}}\right)\right]=\kappa\nabla\cdot\left[\tilde{\rho}\nabla\left(\frac{p'}{\tilde{\rho}}\right)\right]+\tilde{\rho}\frac{\mathrm{d}\kappa}{\mathrm{d}z}\frac{\partial}{\partial z}\left(\frac{p'}{\tilde{\rho}}\right). \tag{3.105}$$

Therefore by virtue of (3.89) the energy equation can be rewritten in the form

$$\begin{aligned}\tilde{\rho}\tilde{T}\left(\frac{\partial s'}{\partial t}+\mathbf{u}\cdot\nabla s'\right)-\tilde{\rho}c_p u_z\Delta_S =&\nabla\cdot\left[\kappa\tilde{\rho}\nabla\left(\tilde{T}s'\right)\right]+\kappa\frac{g}{c_p}\frac{\partial}{\partial z}\left(\tilde{\rho}s'\right)\\ &+\tilde{\rho}\frac{\mathrm{d}\kappa}{\mathrm{d}z}\frac{\partial}{\partial z}\frac{p'}{\tilde{\rho}}+\mathcal{J}+Q',\end{aligned} \tag{3.106}$$

where the term

$$\begin{aligned}\mathcal{J} =&\kappa\nabla\cdot\left[\nabla\cdot\left(2\mu\mathbf{G}^s-\tilde{\rho}\mathbf{u}\mathbf{u}\right)\right]+\kappa\nabla^2\left[\left(\mu_b-\frac{2}{3}\mu\right)\nabla\cdot\mathbf{u}\right]\\ &+2\mu\mathbf{G}^s:\mathbf{G}^s+\left(\mu_b-\frac{2}{3}\mu\right)(\nabla\cdot\mathbf{u})^2,\end{aligned} \tag{3.107}$$

gathers all terms nonlinear in the velocity field and all associated with viscous diffusion. Next, the two terms on the right hand side of (3.106) involving the entropy derivatives, on the basis of the definition of the parameter $\delta$ in (3.11) yielding

$$\frac{\mathrm{d}\tilde{T}}{\mathrm{d}z}=-\frac{g}{c_p}+\mathcal{O}\left(\delta\frac{\bar{T}}{L}\right), \tag{3.108}$$

[7]This section follows the derivation of Mizerski (2017).

can be manipulated to give at leading order

$$\begin{aligned}
\nabla\cdot\left[\kappa\tilde{\rho}\nabla\left(\tilde{T}s'\right)\right]+\kappa\frac{1}{c_p}\frac{\partial}{\partial z}\left(\tilde{\rho}gs'\right) &= \frac{\partial}{\partial z}\left(\kappa\tilde{\rho}\frac{\mathrm{d}\tilde{T}}{\mathrm{d}z}s'\right)+\nabla\cdot\left(\kappa\tilde{\rho}\tilde{T}\nabla s'\right)+\kappa\frac{g}{c_p}\frac{\partial}{\partial z}\left(\tilde{\rho}s'\right)\\
&= -\frac{g}{c_p}\frac{\mathrm{d}\kappa}{\mathrm{d}z}\tilde{\rho}s'+\nabla\cdot\left(\kappa\tilde{\rho}\tilde{T}\nabla s'\right)+\mathcal{O}\left(\delta^{5/2}\bar{\rho}g\sqrt{gL}\right).
\end{aligned} \tag{3.109}$$

Finally the energy equation can be cast in the following form

$$\tilde{\rho}\tilde{T}\left(\frac{\partial s'}{\partial t}+\mathbf{u}\cdot\nabla s'\right)-\tilde{\rho}c_p u_z\Delta_S=\nabla\cdot\left(\kappa\tilde{\rho}\tilde{T}\nabla s'\right)+\tilde{\rho}\frac{\mathrm{d}\kappa}{\mathrm{d}z}\left[\frac{\partial}{\partial z}\frac{p'}{\tilde{\rho}}-\frac{g}{c_p}s'\right]+\mathcal{J}+Q'. \tag{3.110}$$

It is clear now, that the assumption of uniform thermal diffusivity $\kappa$ allows to simplify the energy equation by removing the entire term proportional to the $z$-derivative of $\kappa$, which is the only term involving the pressure fluctuation and then the full system of anelastic equations reads

$$\begin{aligned}
\frac{\partial\mathbf{u}}{\partial t}+(\mathbf{u}\cdot\nabla)\,\mathbf{u} &= -\nabla\frac{p'}{\tilde{\rho}}+\frac{gs'}{c_p}\hat{\mathbf{e}}_z+\frac{\mu}{\tilde{\rho}}\nabla^2\mathbf{u}+\left(\frac{\mu}{3\tilde{\rho}}+\frac{\mu_b}{\tilde{\rho}}\right)\nabla\left(\nabla\cdot\mathbf{u}\right)\\
&\quad+\frac{2}{\tilde{\rho}}\nabla\mu\cdot\mathbf{G}^s+\frac{1}{\tilde{\rho}}\nabla\left(\mu_b-\frac{2}{3}\mu\right)\nabla\cdot\mathbf{u},
\end{aligned} \tag{3.111a}$$

$$\nabla\cdot(\tilde{\rho}\mathbf{u})=0, \tag{3.111b}$$

$$\tilde{\rho}\tilde{T}\left(\frac{\partial s'}{\partial t}+\mathbf{u}\cdot\nabla s'\right)-\tilde{\rho}c_p u_z\Delta_S=\kappa\nabla\cdot\left(\tilde{\rho}\tilde{T}\nabla s'\right)+\mathcal{J}+Q', \tag{3.111c}$$

where $\mathcal{J}$ is given in (3.107).

This constitutes a closed system of equations for an ideal gas, obtained by no additional assumptions other than $g=\text{const}$ and $\kappa=\text{const}$, expressed solely in terms of the velocity field and the pressure and entropy fluctuations. If the pressure distribution is required, it can be calculated from the Poisson-type problem for pressure fluctuation provided in (3.94). However, the pressure fluctuation is entirely eliminated from the energy equation and it appears only in the momentum balance under the gradient operator. Therefore the pressure problem, which as mentioned involves some quite significant complications for computational implementations, can be easily avoided by taking curl of the Navier-Stokes equation and eliminating the pressure fluctuations from the full set of the dynamical equations. The significant advantage is that physically one can not control pressure at the boundaries and therefore boundary conditions on pressure are computationally cumbersome and often have to involve application of spatial averages. Hence, the elimination of pressure is desired. Such an approach typically involves introduction of some potentials, such as e.g. the toroidal (say $\mathcal{T}$) and poloidal (say $\mathcal{P}$) potentials

$$\tilde{\rho}\mathbf{u} = \nabla \times \left(\mathcal{T}\hat{\mathbf{e}}_z\right) + \nabla \times \nabla \times \left(\mathcal{P}\hat{\mathbf{e}}_z\right) \tag{3.112}$$

or the vector potential $\tilde{\rho}\mathbf{u} = \nabla \times \mathbf{A}$ (with some gauge conditions for $\mathcal{T}$ and $\mathcal{P}$ or for $\mathbf{A}$) to satisfy the solenoidal constraint $\nabla \cdot (\tilde{\rho}\mathbf{u}) = 0$, thus a reduction of the number of variables. It is often accompanied by an arbitrary *ad hoc* but greatly simplifying assumption that the entropy is constant and known on the boundaries, even though physically one cannot control the entropy nor pressure on boundaries. The boundary conditions for the potentials ($\mathcal{T}$ and $\mathcal{P}$ or $\mathbf{A}$) then need to be derived on the basis of the physical conditions assumed for the velocity field.

We have started the derivation of the final simplified energy equation (3.111c) from the Eq. (3.26d) which is equivalent to (3.27). Although the latter two are already simplified with respect to the full energy equation (3.1d), we have demonstrated in Sect. 3.1.1 that this simplification does not affect the sign of total entropy production,

$$\frac{\partial}{\partial t}\int \tilde{\rho}s'\mathrm{d}^3x, \tag{3.113}$$

which in accordance with the second law of thermodynamics is positive, if the boundaries are assumed adiabatically insulating. In the process of derivation of (3.111c) we have neglected some terms of the order $\mathcal{O}(\delta^{5/2})$, however, careful considerations show, that all the neglected terms in the energy equation can be gathered to yield a correction of the form

$$\begin{aligned} \kappa\frac{\partial}{\partial z}\left[\frac{\Delta_S}{\tilde{T}}\left(p' + \tilde{\rho}\tilde{T}s'\right)\right] &= \kappa\frac{\partial}{\partial z}\left[\frac{c_p\Delta_S}{g}\left(\frac{\gamma-1}{\gamma}\frac{p'}{\tilde{p}} + \frac{s'}{c_p}\right)\tilde{\rho}g\right] \\ &= \mathcal{O}\left(\delta^{5/2}\bar{\rho}g\sqrt{gL}\right). \end{aligned} \tag{3.114}$$

These terms, divided by $\tilde{T}$ and integrated over the entire fluid volume provide only a small, $\mathcal{O}(\delta^{5/2})$ order correction to the total entropy production which is of the order $\mathcal{O}(\delta^{3/2})$. Therefore the total entropy production in (3.111c) is necessarily positive.

#### 3.3.2.1 The Reference State at Constant Thermal Diffusivity

Let us focus now on the possible forms of the hydrostatic reference state in the case, when the thermal diffusivity $\kappa = k/\tilde{\rho}c_p$ is uniform.[8] We consider an ideal gas at uniform gravity $g = \text{const}$. The reference state equations (3.7a–3.7d) in this case

[8]We stress, that the above entropy formulation is valid only when the parameter $\kappa = k/\tilde{\rho}c_p$ is uniform, which is the thermal diffusivity defined with the basic density profile $\tilde{\rho}$, not the 'full' thermal diffusivity $k/c_p(\tilde{\rho} + \rho')$. This corresponds to an assumption of a particular vertical profile of the thermal conduction coefficient $k \sim \tilde{\rho}$.

simplify to

$$\frac{\mathrm{d}\tilde{p}}{\mathrm{d}z} = -\tilde{\rho}g, \qquad \frac{\mathrm{d}}{\mathrm{d}z}\left(\tilde{\rho}\frac{\mathrm{d}\tilde{T}}{\mathrm{d}z}\right) = -\frac{\tilde{Q}}{\kappa c_p}, \qquad \tilde{p} = \tilde{\rho}R\tilde{T}. \tag{3.115}$$

Of course, once $\tilde{p}$, $\tilde{\rho}$ and $\tilde{T}$ are known the basic state entropy can be found from

$$\tilde{s} = c_v \ln \frac{\tilde{p}}{\tilde{\rho}^{\gamma}} + \text{const.} \tag{3.116}$$

Elimination of the pressure from Eqs. (3.115) leaves

$$\frac{\mathrm{d}\tilde{T}}{\mathrm{d}z} + \frac{1}{\tilde{\rho}}\frac{\mathrm{d}\tilde{\rho}}{\mathrm{d}z}\tilde{T} + \frac{g}{R} = 0, \qquad \frac{\mathrm{d}}{\mathrm{d}z}\left(\tilde{\rho}\frac{\mathrm{d}\tilde{T}}{\mathrm{d}z}\right) = -\frac{\tilde{Q}}{\kappa c_p}. \tag{3.117}$$

The above reference state equations must be satisfied simultaneously with the fundamental assumption $\delta \ll 1$, which implies

$$-\frac{L}{\bar{T}}\left(\frac{\mathrm{d}\tilde{T}}{\mathrm{d}z} + \frac{g}{c_p}\right) = \mathcal{O}(\delta) \ll 1. \tag{3.118}$$

Note, that under the current assumptions the adiabatic gradient $g/c_p$ is uniform, therefore the latter equality can only be satisfied when either both, the constant adiabatic gradient and vertically varying gradient of the reference temperature profile are small compared to the mean temperature of the fluid divided by the layer's thickness $\bar{T}/L$ or the temperature gradient in the reference state is uniform at leading order. The former case implies only a weak stratification, thus effectively reduces the model to the Boussinesq approximation and it is only the latter case which is of interest. When the reference state gradient is (at least approximately) uniform the energy equation becomes $\tilde{Q} = -\kappa c_p \mathrm{d}_z\tilde{T}\mathrm{d}_z\tilde{\rho}$ and since both the gradients $\mathrm{d}_z\tilde{\rho}$ and $\mathrm{d}_z\tilde{T}$ are negative this necessarily implies heat sinks, $\tilde{Q} < 0$, i.e. cooling processes in the fluid volume. Note, that the nearly adiabatic temperature gradient $\mathrm{d}_z\tilde{T} = -g/c_p + \mathcal{O}(\delta\Delta T/L)$ together with the hydrostatic force balance $\mathrm{d}_z\tilde{p} = -\tilde{\rho}g$ and the state equation $\tilde{p} = \tilde{\rho}R\tilde{T}$ imply the reference state in the form (3.68a–3.68c) and consequently the volumetric heat sink is given by

$$\begin{aligned}\tilde{Q} &= -\kappa c_p \frac{\mathrm{d}\tilde{T}}{\mathrm{d}z}\frac{\mathrm{d}\tilde{\rho}}{\mathrm{d}z} = -\frac{g^2\kappa}{\gamma R}\frac{\tilde{\rho}}{\tilde{T}} + \mathcal{O}\left(\delta^{3/2}\rho_B\sqrt{g^3L}\right) \\ &= -\frac{g^2\kappa\rho_B}{\gamma R T_B}\left(1-\theta\frac{z}{L}\right)^{m-1} + \mathcal{O}\left(\delta^{3/2}\rho_B\sqrt{g^3L}\right).\end{aligned} \tag{3.119}$$

In other words the mathematical description of anelastic convection formulated in terms of the entropy fluctuation in (3.111a–3.111c), although certainly attractive from the point of view of modelling, becomes applicable only when the modelled system is being cooled at every height by some processes, with cooling rate in the form (3.119). Such processes can involve natural cooling, e.g. through thermal radiation or chemical reactions or laboratory induced cooling, e.g. through the laser cooling process. The latter is a very effective mechanism of cooling of gases based on interactions of the gas molecules with a unidirectional ensemble of laser rays and the Doppler effect, which result in homogenization of the velocity distribution of particles and thus temperature decrease (cf. Phillips 1998). Let us provide a simple example of a natural cooling process through thermal radiation. As in Sect. 2.2.6 we adopt the model of Goody (1956) and Goody and Yung (1989), but we do not assume that both the boundaries have radiational properties of a black body, in particular the intensity of radiation from boundaries can be set to zero (as in the case of vacuum).

The heat per unit volume emitted from the system in a time unit by thermal radiation can be expressed by the radiative energy flux,

$$Q = -\nabla \cdot \mathbf{j}_{rad}, \tag{3.120}$$

which under the so-called double-stream Milne-Eddington approximation has to satisfy the following equation

$$\nabla \frac{1}{\alpha_a} \nabla \cdot \mathbf{j}_{rad} - 3\alpha_a \mathbf{j}_{rad} = 4\sigma_{rad} \nabla \left(T^4\right), \tag{3.121}$$

where $\alpha_a$ is the coefficient of absorption of radiation per unit volume and $\sigma_{rad}$ is the Stefan-Boltzmann constant; the absorption coefficient is a function of temperature and pressure, $\alpha_a = \alpha_a(T, p)$. Let us assume, that the boundaries are kept isothermal. The general radiative boundary conditions derived from the radiative transfer equation (cf. pp. 429–430 in Goody 1956), supplied by the thermal boundary conditions, yield for the reference state (which is $z$-dependent only)

$$\left.\left(\frac{\mathrm{d}\tilde{j}_{rad\,z}}{\mathrm{d}z} - 2\alpha_a \tilde{j}_{rad\,z}\right)\right|_{z=0} = 4\alpha_a \beta_B, \quad \left.\left(\frac{\mathrm{d}\tilde{j}_{rad\,z}}{\mathrm{d}z} + 2\alpha_a \tilde{j}_{rad\,z}\right)\right|_{z=L} = 4\alpha_a \beta_T, \tag{3.122a}$$

$$\tilde{T}(z=0) = T_B, \quad \tilde{T}(z=L) = T_T, \tag{3.122b}$$

where $\beta_B = B(T_B) - I_B^+$ is the difference between the Planck intensity of black body radiation at temperature $T_B$ denoted by $B(T_B) = \sigma_{rad} T_B^4$ and the upward radiation intensity from the bottom boundary $I_B^+$; analogously $\beta_T = B(T_T) - I_T^-$ with $I_T^-$ denoting the downward radiation intensity from the top boundary and $B(T_T) = \sigma_{rad} T_T^4$. Naturally, in the case when both boundaries are black bodies we get $\beta_T = \beta_B = 0$. Note, that when the radiative flux is $z$-dependent only the Eq. (3.121) implies

$$\tilde{j}_{rad\,x} = \tilde{j}_{rad\,y} = 0. \tag{3.123}$$

To provide a simple example of a possible form of the reference state let us assume that the absorption coefficient $\alpha_a$ is constant and consider the limit of a transparent fluid, when the mean free path of photons is much larger than $L$ thus the absorption coefficient is very small, $\alpha_a \ll L^{-1}$. In such a case the smallest term $3\alpha_a \mathbf{j}_{rad}$ in the radiative flux equation (3.121) can be neglected and the simplified equation reads

$$\frac{\mathrm{d}^2 \tilde{j}_{rad\,z}}{\mathrm{d}z^2} = 16\alpha_a \sigma_{rad} \tilde{T}^3 \frac{\mathrm{d}\tilde{T}}{\mathrm{d}z}. \tag{3.124}$$

The latter has to be solved together with the energy balance in the hydrostatic reference state at constant $\kappa = k/c_p\tilde{\rho}$, cf. the second equation in (3.117) and (3.120)

$$\frac{\mathrm{d}}{\mathrm{d}z}\left(\tilde{\rho}\frac{\mathrm{d}\tilde{T}}{\mathrm{d}z}\right) = \frac{1}{\kappa c_p}\frac{\mathrm{d}\tilde{j}_{rad\,z}}{\mathrm{d}z}, \tag{3.125}$$

subject to the boundary conditions (3.122a, 3.122b). Since in the case at hand the adiabatic gradient $g/c_p$ is constant the temperature profile must be linear in $z$,

$$\tilde{T} = T_B\left(1 - \theta\frac{z}{L}\right), \tag{3.126}$$

where $\theta = \Delta T/T_B$. Consequently the first equation in (3.117) and the perfect gas law $\tilde{p} = \tilde{\rho}R\tilde{T}$ imply the polytropic form of the reference state provided in (3.68a–3.68c). Therefor by the use of (3.125) we get

$$\frac{\mathrm{d}\tilde{j}_{rad\,z}}{\mathrm{d}z} = \frac{m\kappa c_p \rho_B T_B \theta^2}{L^2}\left(1 - \theta\frac{z}{L}\right)^{m-1}, \tag{3.127}$$

and finally (3.124) implies

$$\frac{m\,(m-1)\,\kappa c_p \rho_B \theta^2}{L^2 T_B^{m-2}} \tilde{T}^{m-2} = 16\alpha_a \sigma_{rad} \tilde{T}^3. \tag{3.128}$$

The latter can only be satisfied when $m = 5$ (and then by $\delta \ll 1$ we must have $\gamma = 6/5 + \mathcal{O}(\delta)$) and when the following relation between $\rho_B$, $T_B$ and $\theta$ is satisfied

$$\kappa c_p \rho_B \theta^2 = \frac{4}{5}\alpha_a \sigma_{rad} T_B^3 L^2. \tag{3.129}$$

The solution for the radiative flux, obtained from (3.127) and (3.122a) reads

$$\tilde{j}_{rad\,z} = \frac{\kappa c_p \rho_B T_B \theta}{L}\left[\frac{5\theta}{2\alpha_a L} + 1 - \left(1 - \theta\frac{z}{L}\right)^5\right] - 2\beta_B$$
$$= 4\sigma_{rad} T_B^4 \left\{\frac{1}{2} + \frac{\alpha_a L}{5\theta}\left[1 - \left(1 - \theta\frac{z}{L}\right)^5\right]\right\} - 2\beta_B \tag{3.130}$$

where on top of the relation (3.129) between the system parameters also

$$\sigma_{rad} T_B^4 \left\{1 + (1-\theta)^4 + \frac{2\alpha_a L}{5\theta}\left[1 - (1-\theta)^5\right]\right\} = \beta_T + \beta_B, \tag{3.131}$$

must be satisfied; the latter, however, does not involve $\rho_B$. We note, that when radiation from the bottom boundary can be neglected, $I_B^+ \approx 0$, and the top boundary can be assumed to have the radiational properties of a black body, $\beta_T = 0$, and in the limit of strong stratification when $T_B/T_T \gg 1$ thus $1 - \theta \ll 1$ the term proportional to $\alpha_a L/\theta$ can be neglected since $\alpha_a L$ was assumed small, and then the relation (3.131) is naturally satisfied.

It should be stressed again, that due to the requirement, that the temperature gradient in the reference state must be close to adiabatic, this entropy formulation with constant thermal diffusivity can be applied only to a class of systems with non-vanishing volume cooling, e.g. in the form (3.130) with (3.129) and (3.131).

### *3.3.3 Braginsky and Roberts' Formulation for Turbulent Convection*

The turbulent flow is produced by superadiabatic gradient and hence in case of a fully developed convectively driven turbulence it is natural to assume that the heat flux at large scales, produced by small-scale turbulent fluctuations is proportional to the entropy, not the temperature gradient. Note, that the turbulent fluctuations in this case are not the same as the thermodynamic fluctuations denoted by primes. The turbulent fluctuations are obtained by separating the thermodynamic fluctuations into large-scale means and small-scale corrections, the latter being the turbulent fluctuations. Their action on the means is often modelled by introduction of turbulent transport coefficients, such as turbulent viscosity and thermal diffusivity. In this subsection we will utilize the concept of turbulent transport coefficients and therefore consider only the mean thermodynamic fluctuations without making a distinction in notation between the mean and full thermodynamic fluctuations. In other words the primed variables will simply denote the mean thermodynamic fluctuations.

Braginsky and Roberts (1995) have proposed a simple ansatz for a fully developed, well-mixed turbulence in convective flow, that the molecular heat transport in such a case is much smaller than that generated by small scale turbulence. Therefore in the evolution of large scale turbulent components (that is the mean quantities) the dominant contribution to diffusive heat flux comes from turbulent diffusivity and can be modeled by $-\tilde{\rho}\tilde{T}\boldsymbol{\kappa}_t \cdot \nabla s' = -\tilde{T}\boldsymbol{k}_t \cdot \nabla s'/\tilde{c}_p$, where $\boldsymbol{\kappa}_t$ and $\boldsymbol{k}_t = \tilde{\rho}\tilde{c}_p\boldsymbol{\kappa}_t$ are now the turbulent diffusivity and conductivity respectively, which are typically non-isotropic and hence for generality should be kept in tensorial form. This allows to neglect the term $\nabla \cdot \left(k\nabla T'\right)$ describing the molecular heat transport in the general energy equation (3.27) with respect to the flux associated with turbulent diffusivity, typically at least few orders of magnitude larger. Therefore we can write down the set of dynamical equations in a form much resembling that from the previous section devoted to constant molecular thermal diffusivity formulation, but with turbulent transport tensors $\boldsymbol{\kappa}_t$, $\boldsymbol{\mu}_t$ and $\boldsymbol{\mu}_{bt}$

$$\frac{\partial \mathbf{u}}{\partial t} + (\mathbf{u} \cdot \nabla)\,\mathbf{u} = -\nabla\frac{p'}{\tilde{\rho}} + \frac{gs'}{c_p}\hat{\mathbf{e}}_z + \nabla \cdot \left[2\boldsymbol{\mu}_t \cdot \mathbf{G}^s + \left(\boldsymbol{\mu}_{bt} - \frac{2}{3}\boldsymbol{\mu}_t\right)\nabla \cdot \mathbf{u}\right] \tag{3.132a}$$

$$\nabla \cdot (\tilde{\rho}\mathbf{u}) = 0, \tag{3.132b}$$

$$\begin{aligned} &\tilde{\rho}\tilde{T}\left(\frac{\partial s'}{\partial t} + \mathbf{u} \cdot \nabla s'\right) - \tilde{\rho}c_p u_z \Delta_S \\ &\quad = \nabla \cdot \left(\tilde{\rho}\tilde{T}\boldsymbol{\kappa}_t \cdot \nabla s'\right) + \left[2\boldsymbol{\mu}_t \cdot \mathbf{G}^s + \left(\boldsymbol{\mu}_{bt} - \frac{2}{3}\boldsymbol{\mu}_t\right)\nabla \cdot \mathbf{u}\right] : \mathbf{G} + Q'. \end{aligned} \tag{3.132c}$$

In such a way one obtains a closed system of equations expressed solely in terms of the velocity field, pressure and the entropy fluctuations, however, applicable only to a fully developed turbulence in convective flows. The pressure problem is easily avoidable by taking curl of the Navier-Stokes equation.

Such a formulation, however, suffers from having the capability to violate the second law of thermodynamics, i.e. the production of the total entropy in adiabatically insulated system may turn out negative. In the case of no-slip boundary conditions the flow near the boundaries can not be turbulent and therefore the heat flux at boundaries reduces to the molecular one. Consequently the adiabatic insulation implies

$$-k\frac{\partial}{\partial z}\left(\tilde{T} + T'\right)\Big|_{z=0,\,L} = 0. \tag{3.133}$$

In the absence of heat sources, $Q = 0$, very similar manipulations to those done in Sect. 3.1.1 (cf. Eqs. (3.51) and (3.55)) allow to derive from (3.132c) the following formula for the production of the total entropy

$$\begin{aligned} \frac{\partial}{\partial t}\int_V \tilde{\rho}s\mathrm{d}^3x &= \int_{\partial V} \tilde{\rho}\left(\boldsymbol{\kappa}_t \cdot \nabla s'\right) \cdot \hat{\mathbf{n}}\mathrm{d}\Sigma \\ &\quad + \int_V \frac{1}{\tilde{T}}\left[2\boldsymbol{\mu}_t \cdot \mathbf{G}^s + \left(\boldsymbol{\mu}_{bt} - \frac{2}{3}\boldsymbol{\mu}_t\right)\nabla \cdot \mathbf{u}\right] : \mathbf{G}\mathrm{d}^3x, \end{aligned} \tag{3.134}$$

where we have made a natural assumption, that the molecular heat conduction coefficient $k$ is non-zero everywhere in the fluid volume. Since the boundary conditions (3.133) involve the temperature only, the term

$$\int_{\partial V} \tilde{\rho} \left(\boldsymbol{\kappa}_t \cdot \nabla s'\right) \cdot \hat{\mathbf{n}} \mathrm{d}\Sigma \tag{3.135}$$

is in general neither zero nor positive definite and hence the entropy production can, in principle, be negative. The possible lack of consistency with the second law of thermodynamics, resulting from application of the concept of turbulent transport coefficients, and therefore poor control over the sign of the production of the total entropy is a significant disadvantage of this formulation.

#### 3.3.3.1 The Hydrostatic Reference State

In this case the hydrostatic reference state can either be adiabatic or close to adiabatic, but as in the previous cases has to satisfy the standard static ($\partial_t \equiv 0$ and $\mathbf{u} \equiv 0$) equations, i.e. with molecular, not turbulent, diffusion coefficients, and in particular

$$\frac{\mathrm{d}}{\mathrm{d}z}\left(k \frac{\mathrm{d}\tilde{T}}{\mathrm{d}z}\right) = -\tilde{Q}. \tag{3.136}$$

The turbulent diffusion coefficients, as previously the molecular ones, now have to satisfy (for all $i = 1, 2, 3$ and $j = 1, 2, 3$)

$$\left(\mu_{b_t}\right)_{ij} \lesssim (\mu_t)_{ij} \sim \delta^{1/2} \bar{\rho} \sqrt{\bar{g}L} L. \tag{3.137}$$

$$(\kappa_t)_{ij} \sim \delta^{1/2} \sqrt{\bar{g}L} L, \tag{3.138}$$

for consistency of the anelastic formulation. The molecular transport coefficients $k$, $\mu$ and $\mu_b$ are by assumption negligibly small compared to the turbulent ones. This, of course, requires the heating term $\tilde{Q}$ to be small in order to balance the weak molecular diffusion in the hydrostatic state, as in (3.136).

## 3.4 Energetic Properties of Anelastic Systems

Just as in the Boussinesq case we now proceed to describe the mean physical and in particular energetic properties of anelastic systems. For simplicity and clarity it will be assumed that the system is periodic in horizontal directions, there are no radiative heat sources, thus $Q = 0$ and the acceleration of gravity is constant $g = \text{const}$. The evolution of compressible convection under the anelastic approximation is described by the set of the Navier-Stokes (3.87a), mass conservation and energy equations (3.26b)–(3.26e), i.e.

$$\frac{\partial \mathbf{u}}{\partial t} + (\mathbf{u}\cdot\nabla)\mathbf{u} = -\nabla\frac{p'}{\tilde{\rho}} + \frac{\tilde{\alpha}\tilde{T}}{\tilde{c}_p}s' g\hat{\mathbf{e}}_z + \frac{\mu}{\tilde{\rho}}\nabla^2\mathbf{u} + \left(\frac{\mu}{3\tilde{\rho}} + \frac{\mu_b}{\tilde{\rho}}\right)\nabla(\nabla\cdot\mathbf{u})$$
$$+ \frac{2}{\tilde{\rho}}\nabla\mu\cdot\mathbf{G}^s + \frac{1}{\tilde{\rho}}\nabla\left(\mu_b - \frac{2}{3}\mu\right)\nabla\cdot\mathbf{u}, \tag{3.139a}$$

$$\nabla\cdot(\tilde{\rho}\mathbf{u}) = 0, \tag{3.139b}$$

$$\tilde{\rho}\tilde{T}\left[\frac{\partial s'}{\partial t} + \mathbf{u}\cdot\nabla(\tilde{s}+s')\right] = \nabla\cdot(k\nabla T) + 2\mu\mathbf{G}^s : \mathbf{G}^s + \left(\mu_b - \frac{2}{3}\mu\right)(\nabla\cdot\mathbf{u})^2, \tag{3.139c}$$

$$\frac{\rho'}{\tilde{\rho}} = -\tilde{\alpha}T' + \tilde{\beta}p', \quad s' = -\tilde{\alpha}\frac{p'}{\tilde{\rho}} + \tilde{c}_p\frac{T'}{\tilde{T}}. \tag{3.139d}$$

Understanding the energy transfer and production in convective flow is the key to understanding the physics of compressible convection. Therefore we derive now a few exact relations which allow to describe some general aspects of the dynamics of developed compressible convection. By multiplying the Navier-Stokes equation (3.139a) by $\tilde{\rho}\mathbf{u}$ and averaging over the entire periodic volume (periodicity referring to the '$x$' and '$y$' directions) we obtain the following relation

$$\frac{\partial}{\partial t}\left\langle\frac{1}{2}\tilde{\rho}\mathbf{u}^2\right\rangle = \left\langle\frac{g\tilde{\alpha}\tilde{T}\tilde{\rho}}{\tilde{c}_p}u_z s'\right\rangle - 2\left\langle\mu\mathbf{G}^s : \mathbf{G}^s\right\rangle + \left\langle\left(\frac{2}{3}\mu - \mu_b\right)(\nabla\cdot\mathbf{u})^2\right\rangle, \tag{3.140}$$

where the impermeable and either no-slip or stress-free boundary conditions, as in (2.65a) and (2.65b) were used. The latter relation states, that changes in the total kinetic energy are due to the total work per unit volume of the buoyancy force averaged over the horizontal directions and the total viscous dissipation in the fluid volume; in a (statistically) stationary state the work of the buoyancy force and the viscous dissipation are equal

$$\left\langle\frac{g\tilde{\alpha}\tilde{T}\tilde{\rho}}{\tilde{c}_p}u_z s'\right\rangle = 2\left\langle\mu\mathbf{G}^s : \mathbf{G}^s\right\rangle - \left\langle\left(\frac{2}{3}\mu - \mu_b\right)(\nabla\cdot\mathbf{u})^2\right\rangle. \tag{3.141}$$

Next we derive expressions for the total, superadiabatic heat flux in the system at every $z$. First we recall, that the impermeability conditions at boundaries together with the horizontally averaged mass conservation law imply

$$\langle u_z\rangle_h = 0, \tag{3.142}$$

cf. Eq. (3.101). Now, averaging the heat equation (3.139c) over a horizontal plane and integrating from 0 to $z$ leads to,

$$\frac{\partial}{\partial t}\int_0^z \left\langle \tilde{\rho}\tilde{T}s'\right\rangle_h \mathrm{d}z = \int_0^z \tilde{\rho}\frac{\mathrm{d}\tilde{T}}{\mathrm{d}z}\left\langle u_z s'\right\rangle_h \mathrm{d}z - \tilde{\rho}\tilde{T}\left\langle u_z s'\right\rangle_h - k\frac{\partial\left\langle T\right\rangle_h}{\partial z}\bigg|_{z=0} + k\frac{\partial\left\langle T\right\rangle_h}{\partial z}$$
$$+2\int_0^z \left\langle \mu\mathbf{G}^s : \mathbf{G}^s\right\rangle_h \mathrm{d}z - \int_0^z \left\langle\left(\frac{2}{3}\mu - \mu_b\right)(\nabla\cdot\mathbf{u})^2\right\rangle_h \mathrm{d}z. \tag{3.143}$$

In a stationary state the left hand side of the latter equation vanishes. Thus setting the upper limit of the vertical integration in (3.143) to $z = L$ (i.e. integrating over the entire fluid volume), applying the boundary conditions of impermeability $u_z(z = 0,\ L) = 0$ and utilizing (3.140) allows to obtain

$$L\left\langle\frac{g\tilde{\alpha}\tilde{T}\tilde{\rho}}{\tilde{c}_p}u_z s'\right\rangle + L\left\langle\tilde{\rho}\frac{\mathrm{d}\tilde{T}}{\mathrm{d}z}u_z s'\right\rangle = -k\frac{\partial\left\langle T\right\rangle_h}{\partial z}\bigg|_{z=L} + k\frac{\partial\left\langle T\right\rangle_h}{\partial z}\bigg|_{z=0}, \tag{3.144}$$

and since $\mathrm{d}\tilde{T}/\mathrm{d}z = -g\tilde{\alpha}\tilde{T}/\tilde{c}_p + \mathcal{O}(\delta\tilde{T}/L)$ it is clear that the two terms on the left hand side cancel at leading order. This leads to the expectable conclusion, that in a stationary state

$$-k\frac{\partial\left\langle T\right\rangle_h}{\partial z}\bigg|_{z=0} = -k\frac{\partial\left\langle T\right\rangle_h}{\partial z}\bigg|_{z=L}, \tag{3.145}$$

hence the total, horizontally averaged heat flux entering the system at the bottom equals the flux, which leaves the system at the top. Furthermore, on denoting the total, horizontally averaged heat flux by $F_{total}(z)$, the expression (3.143) can be used to derive a formula for the heat flux $F_{total}(z = 0)$ which enters the system at the bottom in a stationary state

$$\begin{aligned}F_{total}(z=0) &= -k\frac{\mathrm{d}}{\mathrm{d}z}\left(\tilde{T} + \left\langle T'\right\rangle_h\right)\bigg|_{z=0}\\ &= -k\frac{\mathrm{d}}{\mathrm{d}z}\left(\tilde{T} + \left\langle T'\right\rangle_h\right)\\ &\quad + \tilde{\rho}\tilde{T}\left\langle u_z s'\right\rangle_h - \int_0^z \tilde{\rho}\frac{\mathrm{d}\tilde{T}}{\mathrm{d}z}\left\langle u_z s'\right\rangle_h \mathrm{d}z\\ &\quad - 2\int_0^z \left\langle \mu\mathbf{G}^s : \mathbf{G}^s\right\rangle_h \mathrm{d}z + \int_0^z \left\langle\left(\frac{2}{3}\mu - \mu_b\right)(\nabla\cdot\mathbf{u})^2\right\rangle_h \mathrm{d}z,\end{aligned} \tag{3.146}$$

which we will refer to as *the first formula for total heat flux*. Clearly, the heat flux

$$F_{total}(z) = -k\frac{\mathrm{d}}{\mathrm{d}z}\left(\tilde{T} + \left\langle T'\right\rangle_h\right) + \tilde{\rho}\tilde{T}\left\langle u_z s'\right\rangle_h, \tag{3.147}$$

which consists of the conductive $-k\mathrm{d}_z\left(\tilde{T}+\langle T\rangle_h\right)$ and advective $\tilde{\rho}\tilde{T}\left\langle u_z s'\right\rangle_h$ parts, contrary to the Boussinesq case is not constant at every $z$, because the work of the buoyancy force, likewise the viscous entropy production substantially modify the total flux for $0<z<1$, i.e.

$$\begin{aligned}F_{total}(z)=&F_{total}(z=0)-\int_0^z\frac{\tilde{\alpha}\tilde{T}g\tilde{\rho}}{\tilde{c}_p}\left\langle u_z s'\right\rangle_h\mathrm{d}z\\&+2\int_0^z\left\langle\mu\mathbf{G}^s:\mathbf{G}^s\right\rangle_h\mathrm{d}z-\int_0^z\left\langle\left(\frac{2}{3}\mu-\mu_b\right)(\nabla\cdot\mathbf{u})^2\right\rangle_h\mathrm{d}z,\end{aligned}\tag{3.148}$$

where we have used $\mathrm{d}_z\tilde{T}=-\tilde{\alpha}\tilde{T}g/\tilde{c}_p+\mathcal{O}(\delta g/\tilde{c}_p)$. Note, that $Q=0$ implies $k\mathrm{d}_z\tilde{T}=$ const, hence the terms $-\left.k\mathrm{d}_z\tilde{T}\right|_{z=0}$ and $-k\mathrm{d}_z\tilde{T}$ are equal and can be cancelled on the both sides of (3.146) leaving an expression for the convective heat flux only

$$F_{conv.}=F_{total}+k\mathrm{d}_z\tilde{T}=-k\left.\frac{\mathrm{d}}{\mathrm{d}z}\left\langle T'\right\rangle_h\right|_{z=0}=-k\frac{\mathrm{d}}{\mathrm{d}z}\left\langle T'\right\rangle_h+\tilde{\rho}\tilde{T}\left\langle u_z s'\right\rangle_h,\tag{3.149a}$$

$$F_{conv.}(z=0)=F_{total}(z=0)+k\mathrm{d}_z\tilde{T}=-k\left.\frac{\mathrm{d}}{\mathrm{d}z}\left\langle T'\right\rangle_h\right|_{z=0}.\tag{3.149b}$$

Moreover, the formula (3.146) can be used to derive *the first formula for superadiabatic heat flux*

$$\begin{aligned}F_S(z=0)=&-k\left.\frac{\mathrm{d}}{\mathrm{d}z}\left(\tilde{T}+\left\langle T'\right\rangle_h-T_{ad}\right)\right|_{z=0}\\=&-k\frac{\mathrm{d}}{\mathrm{d}z}\left(\tilde{T}+\left\langle T'\right\rangle_h-T_{ad}\right)-\left(k\frac{\mathrm{d}T_{ad}}{\mathrm{d}z}-k\left.\frac{\mathrm{d}T_{ad}}{\mathrm{d}z}\right|_{z=0}\right)\\&+\tilde{\rho}\tilde{T}\left\langle u_z s'\right\rangle_h-\int_0^z\tilde{\rho}\frac{\mathrm{d}\tilde{T}}{\mathrm{d}z}\left\langle u_z s'\right\rangle_h\mathrm{d}z\\&-2\int_0^z\left\langle\mu\mathbf{G}^s:\mathbf{G}^s\right\rangle_h\mathrm{d}z+\int_0^z\left\langle\left(\frac{2}{3}\mu-\mu_b\right)(\nabla\cdot\mathbf{u})^2\right\rangle_h\mathrm{d}z,\end{aligned}\tag{3.150}$$

Note, that since the basic state profile is assumed close to the adiabatic the term $-(k\mathrm{d}_zT_{ad}-k\mathrm{d}_zT_{ad}|_{z=0})$ which describes the heat flux jump along the adiabat is of the order $\mathcal{O}(\delta k\tilde{T}/L)=\mathcal{O}(\delta^{3/2}\tilde{\rho}\tilde{c}_p\sqrt{\tilde{g}L\tilde{T}})$, which is consistent with the order of magnitude of the entire formula (3.150). The adiabatic temperature profile is in general curvilinear, however if e.g. one considers the case of an ideal gas with uniform gravity $g=$ const, then the adiabatic gradient becomes uniform and the term $-(k\mathrm{d}_zT_{ad}-k\mathrm{d}_zT_{ad}|_{z=0})$ vanishes; in such a case setting $z=L$ in (3.150) gives $-k\,\mathrm{d}_z(\tilde{T}+\left\langle T'\right\rangle_h-T_{ad})\big|_{z=0}=-k\,\mathrm{d}_z(\tilde{T}+\left\langle T'\right\rangle_h-T_{ad})\big|_{z=L}$, so that the superadiabatic heat flux entering the system at $z = 0$ is equal to the superadiabatic heat flux leaving the system at $z = L$. Most generally, however, when the adiabatic gradient cannot be assumed uniform the vertical variation of the heat flux conducted down the adiabat, $-(k\mathrm{d}_z T_{ad}|_{z=L} - k\mathrm{d}_z T_{ad}|_{z=0})$ contributes to the superadiabatic heat flux balance at the top and bottom boundaries.

Next we consider the balance of entropy per unit volume, that is the heat equation (3.139c) divided by $\tilde{T}$. If we average the entropy equation over the horizontal planes and integrate it from bottom to arbitrary height $z$, then for a stationary state and impermeable and either stress-free or no-slip boundaries we obtain *the second formula for the total heat flux*

$$\begin{aligned}
F_{total}(z=0) =& -k\frac{\mathrm{d}}{\mathrm{d}z}\left(\tilde{T} + \left\langle T'\right\rangle_h\right)\Big|_{z=0} \\
=& -k\frac{\tilde{T}_B}{\tilde{T}}\frac{\mathrm{d}}{\mathrm{d}z}\left(\tilde{T} + \left\langle T'\right\rangle_h\right) - k\frac{\mathrm{d}\tilde{T}}{\mathrm{d}z}\left[1 - \frac{\tilde{T}_B}{\tilde{T}}\right] \\
& - \tilde{T}_B k\frac{\mathrm{d}\tilde{T}}{\mathrm{d}z}\left(\frac{\left\langle T'\right\rangle_h}{\tilde{T}^2} - \frac{\left\langle T'\right\rangle_h\big|_{z=0}}{\tilde{T}_B^2}\right) - 2\tilde{T}_B k\frac{\mathrm{d}\tilde{T}}{\mathrm{d}z}\int_0^z \frac{\left\langle T'\right\rangle_h}{\tilde{T}^3}\frac{\mathrm{d}\tilde{T}}{\mathrm{d}z}\mathrm{d}z \\
& + \tilde{\rho}\tilde{T}_B\left\langle u_z s'\right\rangle_h \\
& - 2\tilde{T}_B\int_0^z\left\langle\frac{\mu}{\tilde{T}}\mathbf{G}^s : \mathbf{G}^s\right\rangle_h \mathrm{d}z + \tilde{T}_B\int_0^z\left\langle\frac{2\mu - 3\mu_b}{3\tilde{T}}\left(\nabla\cdot\mathbf{u}\right)^2\right\rangle_h \mathrm{d}z.
\end{aligned} \tag{3.151}$$

The latter implies

$$\begin{aligned}
F_S(z=0) =& -k\frac{\mathrm{d}}{\mathrm{d}z}\left(\tilde{T} + \left\langle T'\right\rangle_h - T_{ad}\right)\Big|_{z=0} \\
=& -k\frac{\tilde{T}_B}{\tilde{T}}\frac{\mathrm{d}}{\mathrm{d}z}\left(\tilde{T} + \left\langle T'\right\rangle_h - T_{ad}\right) \\
& - \left[k\frac{\mathrm{d}\left(\tilde{T} - T_{ad}\right)}{\mathrm{d}z}\Big|_{z=0} - \frac{\tilde{T}_B}{\tilde{T}}k\frac{\mathrm{d}\left(\tilde{T} - T_{ad}\right)}{\mathrm{d}z}\right] \\
& - \tilde{T}_B k\frac{\mathrm{d}\tilde{T}}{\mathrm{d}z}\left(\frac{\left\langle T'\right\rangle_h}{\tilde{T}^2} - \frac{\left\langle T'\right\rangle_h\big|_{z=0}}{\tilde{T}_B^2}\right) - 2\tilde{T}_B k\frac{\mathrm{d}\tilde{T}}{\mathrm{d}z}\int_0^z \frac{\left\langle T'\right\rangle_h}{\tilde{T}^3}\frac{\mathrm{d}\tilde{T}}{\mathrm{d}z}\mathrm{d}z \\
& + \tilde{\rho}\tilde{T}_B\left\langle u_z s'\right\rangle_h \\
& - 2\tilde{T}_B\int_0^z\left\langle\frac{\mu}{\tilde{T}}\mathbf{G}^s : \mathbf{G}^s\right\rangle_h \mathrm{d}z + \tilde{T}_B\int_0^z\left\langle\frac{2\mu - 3\mu_b}{3\tilde{T}}\left(\nabla\cdot\mathbf{u}\right)^2\right\rangle_h \mathrm{d}z.
\end{aligned} \tag{3.152}$$

which will be referred to as *the second formula for superadiabatic heat flux*. In the above we have utilized the fact, that for $Q = 0$ one obtains $k\mathrm{d}_z\tilde{T} = \mathrm{const}$. Note, that the distinction between the bottom value of the reference temperature $\tilde{T}_B$ and the bottom value of the total temperature $\tilde{T}_B + T'(\mathbf{x}, t)|_{z=0}$ is kept here for clarity, since the boundaries are not necessarily isothermal (recall, that the corrections from the temperature fluctuation $T'(\mathbf{x}, t)|_{z=0}$ are of the order $\mathcal{O}(\delta)$). However, this distinction is not really necessary in the Eq. (3.152) for the superadiabatic flux, since all the terms in this expression are of the order $\mathcal{O}(\delta^{3/2})$, thus $\mathcal{O}(\delta)$ corrections to $\tilde{T}_B$ are irrelevant, as they produce negligible $\mathcal{O}(\delta^{5/2})$ corrections to the superadiabatic heat flux.

Equation (3.151) taken at $z = L$, i.e. the heat equation divided by $\tilde{T}$ integrated over the entire volume from $z$ to $L$, with the aid of (3.145) gives the following relation between the convective heat flux in the system and the viscous dissipation with corrections proportional to the basic state heat flux

$$F_{conv.}(z=0)\left(\tilde{\Gamma}-1\right) = 2\tilde{T}_B L\left\langle\frac{\mu}{\tilde{T}}\mathbf{G}^s : \mathbf{G}^s\right\rangle - \tilde{T}_B L\left\langle\frac{2\mu - 3\mu_b}{3\tilde{T}}\left(\nabla\cdot\mathbf{u}\right)^2\right\rangle + k\frac{\mathrm{d}\tilde{T}}{\mathrm{d}z}\left[\frac{\tilde{\Gamma}^2\left\langle T'\right\rangle_h\big|_{z=L} - \left\langle T'\right\rangle_h\big|_{z=0}}{\tilde{T}_B} + 2\tilde{T}_B\int_0^L\frac{\left\langle T'\right\rangle_h}{\tilde{T}^3}\frac{\mathrm{d}\tilde{T}}{\mathrm{d}z}\mathrm{d}z\right], \tag{3.153}$$

where we have introduced the bottom to top temperature ratio

$$\tilde{\Gamma} = \frac{\tilde{T}_B}{\tilde{T}_T} = \frac{1}{1-\tilde{\theta}} > 1, \tag{3.154}$$

and $\tilde{\theta} = \Delta\tilde{T}/\tilde{T}_B = (\tilde{T}_B - \tilde{T}_T)/\tilde{T}_B$. Of course, in the case, when both boundaries are held at constant temperature we have $\tilde{T}_B = T_B$, $\tilde{T}_T = T_T$, $\Delta\tilde{T} = \Delta T$ and the term proportional to $\Gamma^2\left\langle T'\right\rangle_h\big|_{z=L} - \left\langle T'\right\rangle_h\big|_{z=0}$ vanishes.

One of the possible approaches is to define the Nusselt number simply as the ratio of the convective heat flux to the total flux conducted by the basic state and the Rayleigh number with the use of the basic state gradient $-k\mathrm{d}_z\tilde{T}/\bar{k}$; this leads to a simple relation between the Nusselt and Rayleigh numbers via (3.153). We will, however, consider now a simplified case of a fluid governed by the ideal gas equation of state with constant specific heats $c_p$ and $c_v$, constant gravity $g$, viscosity $\mu$ and thermal conductivity $k$. This will allow for fairly simple relations between the Nusselt and Rayleigh numbers defined in a similar manner as in the Boussinesq case. We separate the two inherently distinct problems of isothermal (thermally perfectly conducting) and fixed heat flux (thermally insulating) boundaries.

### 3.4.1 *Isothermal Boundaries, Perfect Gas, Uniform Fluid Properties, g = const.*

Under the current assumptions the Eq. (3.152) taken at $z = L$ supplied by (3.145) provides an expression for the superadiabatic heat flux in the form

$$
\begin{aligned}
\left[F_S\left(z=0\right)-k\Delta_S\right]\left(\Gamma-1\right) = {}& 2T_B L\mu\left\langle\frac{1}{\tilde{T}}\mathbf{G}^s:\mathbf{G}^s\right\rangle \\
& - T_B L\left(\frac{2}{3}\mu-\mu_b\right)\left\langle\frac{(\nabla\cdot\mathbf{u})^2}{\tilde{T}}\right\rangle \\
& + 2T_B Lk\left(\frac{\Delta T}{L}\right)^2\left\langle\frac{T'}{\tilde{T}^3}\right\rangle,
\end{aligned}
\tag{3.155}
$$

On defining the Nusselt number $Nu$ and the Rayleigh number $Ra$ in the following way

$$
Nu = \frac{F_S\left(z=0\right)}{k\Delta_S} = \frac{-k\,\frac{\mathrm{d}}{\mathrm{d}z}\left(\tilde{T}+\left\langle T'\right\rangle_h - T_{ad}\right)\Big|_{z=0}}{k\Delta_S},
\tag{3.156}
$$

$$
Ra = \frac{g\,\Delta_S\,L^4\rho_B^2 c_p}{T_B\mu k},
\tag{3.157}
$$

where $k\Delta_S = k(\Delta T/L - g/c_p)$ is the superadiabatic conductive heat flux in the hydrostatic basic state and $1/T_B = \tilde{\alpha}_B$, the relation (3.155) can be rewritten to yield

$$
\begin{aligned}
\frac{k^2}{\rho_B^2 c_p^2 L^4}Ra\left(Nu-1\right)\left(\Gamma-1\right) = {}& 2\Delta T\left\langle\frac{1}{\tilde{T}}\mathbf{G}^s:\mathbf{G}^s\right\rangle \\
& -\Delta T\left(\frac{2}{3}-\frac{\mu_b}{\mu}\right)\left\langle\frac{(\nabla\cdot\mathbf{u})^2}{\tilde{T}}\right\rangle \\
& +2c_p Pr^{-1}L\left(\frac{\Delta T}{L}\right)^3\left\langle\frac{T'}{\tilde{T}^3}\right\rangle,
\end{aligned}
\tag{3.158}
$$

where we have used $\Delta T/L = g/c_p + \mathcal{O}(\delta\tilde{T}/L)$. In the above

$$
Pr = \frac{\mu c_p}{k},
\tag{3.159}
$$

is the Prandtl number. We note, that since in the case at hand the basic state is given by the Eqs. (3.68a–3.68c) we get

$$
\Delta_S = \frac{\Delta T}{L} - \frac{g}{c_p} = -\frac{T_B}{c_p}\frac{\mathrm{d}\tilde{s}}{\mathrm{d}z}\bigg|_{z=0} = \frac{\Delta T\Delta\tilde{s}}{c_p L\ln\Gamma},
\tag{3.160}
$$

where

$$\Delta\tilde{s} = \tilde{s}_B - \tilde{s}_T = c_p \frac{m + 1 - \gamma m}{\gamma} \ln \Gamma \tag{3.161}$$

is the entropy jump across the fluid layer in the basic hydrostatic state. Thus we arrive at an alternative expression for the Rayleigh number

$$Ra = \frac{g \Delta T \Delta\tilde{s} L^3 \rho_B^2}{\ln \Gamma T_B \mu k}. \tag{3.162}$$

Another useful relation comes from averaging the first formula for total heat flux (3.146) over the vertical fluid gap $0 \leq z \leq L$, which under the current assumptions can be cast in the following form

$$\begin{aligned} F_{conv.}(z=0) =& 2\left\langle \tilde{\rho}\tilde{T} u_z s' \right\rangle - 2\mu L \left\langle \left( \frac{\tilde{T}}{\Delta T} - 1 \right) \mathbf{G}^s : \mathbf{G}^s \right\rangle \\ &+ \left( \frac{2}{3}\mu - \mu_b \right) L \left\langle \left( \frac{\tilde{T}}{\Delta T} - 1 \right) (\nabla \cdot \mathbf{u})^2 \right\rangle, \end{aligned} \tag{3.163}$$

where

$$\frac{\tilde{T}}{\Delta T} - 1 = \frac{1}{\Gamma - 1} - \frac{z}{L}. \tag{3.164}$$

An alternative expression for the convective flux at the bottom is provided by (3.153), which in the case at hand implies

$$\begin{aligned} F_{conv.}(z=0)(\Gamma - 1) =& 2T_B L \left\langle \frac{\mu}{\tilde{T}} \mathbf{G}^s : \mathbf{G}^s \right\rangle - T_B L \left\langle \frac{2\mu - 3\mu_b}{3\tilde{T}} (\nabla \cdot \mathbf{u})^2 \right\rangle \\ &+ 2T_B L k \left( \frac{\Delta T}{L} \right)^2 \left\langle \frac{T'}{\tilde{T}^3} \right\rangle, \end{aligned} \tag{3.165}$$

### 3.4.2 Boundaries Held at Constant Heat Flux, Perfect Gas, Uniform Fluid Properties, g = const.

When the heat flux is held constant at the boundaries we have

$$\frac{\mathrm{d}}{\mathrm{d}z} \left\langle T' \right\rangle_h \Big|_{z=0} = \frac{\mathrm{d}}{\mathrm{d}z} \left\langle T' \right\rangle_h \Big|_{z=L} = 0, \tag{3.166}$$

and the basic state gradient is uniform, as in the previous case $\tilde{T} = \tilde{T}_B - \Delta\tilde{T} z/L$, and $\Delta\tilde{T} = \tilde{T}_B - \tilde{T}_T$. This means that the definition of the Nusselt number (3.156) utilized in the case of isothermal boundaries becomes useless in the current case, since the expression $-k\, \mathrm{d}_z(\tilde{T} + \langle T' \rangle_h - T_{ad})\Big|_{z=0} / k\Delta_S$ is exactly unity (cf. Eq. (2.80) and the comment below for a similar result in the Boussinesq case). Therefore an alternative definition of the Nusselt number is required. Contrary to the Boussinesq case the total convective heat flux (3.149a) is non-zero away from the boundaries (cf. (2.81) for the Boussinesq case), because of the substantial influence of viscous heating and the work done by buoyancy effects. Nevertheless, it seems reasonable to define the Nusselt number only by the mean advective contribution to the total convective flux (3.149a),

$$Nu_Q = \frac{\langle \tilde{\rho}\tilde{T} u_z s' \rangle}{k\Delta_S} \tag{3.167}$$

since the mean conductive contribution is expected to be much smaller in strongly developed convection. Averaging the first formula for total heat flux (3.146) over $0 \leq z \leq L$ and taking into account (3.166) leads to

$$\begin{aligned} 0 = -k\frac{\Delta \langle T' \rangle_h}{L} + 2\langle \tilde{\rho}\tilde{T} u_z s' \rangle - 2\mu L \left\langle \left( \frac{\tilde{T}}{\Delta T} - 1 \right) \mathbf{G}^s : \mathbf{G}^s \right\rangle \\ + \left( \frac{2}{3}\mu - \mu_b \right) L \left\langle \left( \frac{\tilde{T}}{\Delta T} - 1 \right) (\nabla \cdot \mathbf{u})^2 \right\rangle, \end{aligned} \tag{3.168}$$

where[9]

$$\Delta \langle T' \rangle_h = \langle T' \rangle_h \big|_{z=L} - \langle T' \rangle_h \big|_{z=0}, \tag{3.169}$$

is the temperature fluctuation jump across the layer. Therefore with the use of the Nusselt number definition we get the following relation

$$\begin{aligned} \frac{k^2}{\tilde{\rho}_B^2 c_p^2 L^4} Ra \left( Nu_Q - \frac{k\frac{\Delta \langle T' \rangle_h}{2L}}{k\Delta_S} \right) \frac{\Gamma}{\Gamma - 1} = \left\langle \left( \frac{\tilde{T}}{\Delta T} - 1 \right) \mathbf{G}^s : \mathbf{G}^s \right\rangle \\ - \left( \frac{1}{3} - \frac{\mu_b}{2\mu} \right) \left\langle \left( \frac{\tilde{T}}{\Delta T} - 1 \right) (\nabla \cdot \mathbf{u})^2 \right\rangle. \end{aligned} \tag{3.170}$$

As mentioned above, the term $k\Delta \langle T' \rangle_h / 2Lk\Delta_S$ describing the ratio of the mean conductive heat flux resulting from convection to the superadiabatic heat flux in the basic state is expected to be negligible with respect to $Nu_Q$ at high $Ra$, when convection is strongly turbulent and the Nusselt number could be much greater than

[9] Although it does not play an important role in the following analysis, we can observe that in the case at hand it is expected, that convection tends to equalize the total top and bottom temperatures, which implies $\Delta \langle T' \rangle_h > 0$.

unity. If, however this is not the case and the mean conductive flux is significant it is more convenient to include it in the definition of the Nusselt number, since the values of the temperature at the top and bottom boundaries, $\left.\langle T'\rangle_h\right|_{z=L}$ and $\left.\langle T'\rangle_h\right|_{z=0}$ are not known. Thus in such a case it is natural to define the Nusselt number in the following way

$$Nu'_Q = \frac{\left\langle \tilde{\rho}\tilde{T} u_z s'\right\rangle - k\frac{\Delta\langle T'\rangle_h}{2L}}{k\Delta_S}. \tag{3.171}$$

Note however, that the constant heat flux boundary conditions (3.166) imply, that $F_{conv.}(z=0) = -k\, \mathrm{d}_z \left.\langle T'\rangle_h\right|_{z=0} = 0$, thus (3.153) allows to express the mean fluctuation temperature at one boundary (say at the top $\left.\langle T'\rangle_h\right|_{z=L}$) by the value of the mean fluctuation temperature at the other boundary (at the bottom $\left.\langle T'\rangle_h\right|_{z=0}$) and mean quantities.

## 3.5 Linear Stability Analysis (ideal Gas, $Q = 0$, Isothermal, Stress-Free and Impermeable Boundaries, Constant $\nu$, $k$, g and $c_p$)

It is of interest to study the properties of marginal (linear) anelastic convection near threshold (in a domain of large horizontal extent), to demonstrate the influence of density stratification on the critical Rayleigh number, convective flow and the temperature distribution at convection threshold. This will be achieved via formal expansions in parameter $\theta = \Delta T/T_B$, which is a measure of compressibility/stratification of the fluid. Since the calculations in the compressible case are cumbersome, we limit ourselves here only to the simplest case of stress-free, impermeable and isothermal boundaries, thus we assume

$$u_z\,(z=0,\ L) = 0, \qquad \left.\frac{\partial \mathbf{u}_h}{\partial z}\right|_{z=0,\ L} = 0 \tag{3.172}$$

and

$$T'\,(z=0,\ L) = 0. \tag{3.173}$$

At convection threshold and slightly above it the magnitude of perturbations to the hydrostatic basic state can be safely assumed small, $\mathbf{u}/\delta^{1/2}\sqrt{gL} \ll 1$ and $T'/\delta\tilde{T} \ll 1$, etc. which allows to linearize the full set of dynamical equations (3.139a–3.139d) to get

$$\frac{\partial \mathbf{u}}{\partial t} = -\nabla \frac{p'}{\tilde{\rho}} + \frac{g}{c_p} s' \hat{\mathbf{e}}_z + \nu \nabla^2 \mathbf{u} + \frac{\nu}{3} \nabla (\nabla \cdot \mathbf{u}) + \frac{\nu}{\tilde{\rho}} \frac{\mathrm{d}\tilde{\rho}}{\mathrm{d}z} \left( \nabla u_z + \frac{\partial \mathbf{u}}{\partial z} \right) + \frac{2}{3} \nu \left( \frac{1}{\tilde{\rho}} \frac{\mathrm{d}\tilde{\rho}}{\mathrm{d}z} \right)^2 u_z \hat{\mathbf{e}}_z, \tag{3.174a}$$

$$\nabla \cdot (\tilde{\rho} \mathbf{u}) = 0, \tag{3.174b}$$

$$\tilde{\rho} \tilde{T} \left( \frac{\partial s'}{\partial t} + \frac{\mathrm{d}\tilde{s}}{\mathrm{d}z} u_z \right) = k \nabla^2 T, \tag{3.174c}$$

$$\frac{\rho'}{\tilde{\rho}} = -\frac{T'}{\tilde{T}} + \frac{p'}{\tilde{p}}, \qquad \tilde{T} s' = -\frac{p'}{\tilde{\rho}} + c_p T', \tag{3.174d}$$

where for simplicity we have assumed that the fluid satisfies the equation of state of an ideal gas $p = \rho R T$ and the heat capacities $c_p$ and $c_v$ are constants. Moreover, it has been assumed, that the kinematic viscosity $\nu = \mu / \tilde{\rho}$ and the thermal conductivity coefficient $k$ are uniform and that the bulk viscosity vanishes, $\mu_b = 0$. Under the current assumptions the hydrostatic reference state is given by Eqs. (3.68a–3.68c), hence we can compute the inverse density scale height, which appears explicitly in the Eq. (3.174a),

$$\frac{1}{\tilde{\rho}} \frac{\mathrm{d}\tilde{\rho}}{\mathrm{d}z} = -\frac{1}{L} \frac{m\theta}{1 - \theta \frac{z}{L}}, \tag{3.175}$$

and $\mathrm{d}_z \tilde{s} = -c_p \Delta_S / \tilde{T}$ by the use of the general formula (3.10).

To derive one equation for the $z$-dependent amplitude of the vertical velocity $\hat{u}_z$, analogous to (2.104a), we first decompose the perturbation fields into Fourier modes

$$\mathbf{u}(x, y, z, t) = \Re\mathfrak{e}\ \hat{\mathbf{u}}(z) \mathrm{e}^{\sigma t} \mathrm{e}^{\mathrm{i}(\mathcal{K}_x x + \mathcal{K}_y y)}, \tag{3.176a}$$

$$T'(x, y, z, t) = \Re\mathfrak{e}\ \hat{T}(z) \mathrm{e}^{\sigma t} \mathrm{e}^{\mathrm{i}(\mathcal{K}_x x + \mathcal{K}_y y)}, \tag{3.176b}$$

$$s'(x, y, z, t) = \Re\mathfrak{e}\ \hat{s}(z) \mathrm{e}^{\sigma t} \mathrm{e}^{\mathrm{i}(\mathcal{K}_x x + \mathcal{K}_y y)}, \tag{3.176c}$$

$$p'(x, y, z, t) = \Re\mathfrak{e}\ \hat{p}(z) \mathrm{e}^{\sigma t} \mathrm{e}^{\mathrm{i}(\mathcal{K}_x x + \mathcal{K}_y y)}, \tag{3.176d}$$

where $\sigma$ is the growth rate, and adopt the following procedure. We introduce the above form of the perturbation fields into the Eqs. (3.174a–3.174d), and take a $z$-component of the double curl of the Eq. (3.174a), to eliminate the pressure and relate the entropy amplitude $\hat{s}$ to $\hat{u}_z$. Next we take a divergence of the Eq. (3.174a) to obtain an equation for the pressure amplitude. With the aid of (3.174b) this yields

$$-\sigma \mathfrak{D}^2 \hat{u}_z = \frac{g}{c_p} \mathcal{K}^2 \hat{s} - \nu \mathfrak{D}^2 \mathfrak{D}^2 \hat{u}_z + \frac{2}{3} \nu \mathcal{K}^2 \frac{m+3}{m} \left( \frac{1}{\tilde{\rho}} \frac{\mathrm{d}\tilde{\rho}}{\mathrm{d}z} \right)^2 \hat{u}_z, \tag{3.177a}$$

$$\nabla^2 \frac{\hat{p}}{\tilde{\rho}} = \sigma \frac{1}{\tilde{\rho}} \frac{\mathrm{d}\tilde{\rho}}{\mathrm{d}z} \hat{u}_z + \frac{g}{c_p} \frac{\mathrm{d}\hat{s}}{\mathrm{d}z} - \frac{4}{3} \nu \mathfrak{D}^2 \left( \frac{1}{\tilde{\rho}} \frac{\mathrm{d}\tilde{\rho}}{\mathrm{d}z} \hat{u}_z \right) + \nu \frac{1}{\tilde{\rho}} \frac{\mathrm{d}\tilde{\rho}}{\mathrm{d}z} \mathfrak{D}^2 \hat{u}_z$$
$$- \frac{2\nu}{m} \left( \frac{1}{\tilde{\rho}} \frac{\mathrm{d}\tilde{\rho}}{\mathrm{d}z} \right)^2 \frac{\mathrm{d}\hat{u}_z}{\mathrm{d}z} - \frac{2\nu}{m} \left( \frac{1}{\tilde{\rho}} \frac{\mathrm{d}\tilde{\rho}}{\mathrm{d}z} \right)^3 \hat{u}_z, \tag{3.177b}$$

$$\sigma \tilde{\rho} \tilde{T} \hat{s} - c_p \Delta_S \tilde{\rho} \hat{u}_z = k \nabla^2 \hat{T}, \tag{3.177c}$$

$$\frac{\hat{p}}{\tilde{\rho}} = c_p \hat{T} - \tilde{T} \hat{s}, \tag{3.177d}$$

where

$$\nabla^2 = -\mathcal{K}^2 + \frac{\mathrm{d}^2}{\mathrm{d}z^2}, \qquad \mathfrak{D}^2 \hat{u}_z = \nabla^2 \hat{u}_z + \frac{\mathrm{d}}{\mathrm{d}z} \left( \frac{1}{\tilde{\rho}} \frac{\mathrm{d}\tilde{\rho}}{\mathrm{d}z} \hat{u}_z \right), \tag{3.178}$$

and we have used (3.175) to write $\mathrm{d}_z(\mathrm{d}_z\tilde{\rho}/\tilde{\rho}) = -(1/m)(\mathrm{d}_z\tilde{\rho}/\tilde{\rho})^2$.

We now introduce $\hat{p}/\tilde{\rho}$ from (3.177d) into (3.177b), substitute for the laplacian of temperature $\hat{T}$ from (3.177c) and for the entropy $\hat{s}$ from (3.177a) to obtain the final equation for one variable $\hat{u}_z$, which reads

$$- \frac{\sigma^2}{\kappa_B \nu} \frac{\tilde{\rho}}{\rho_B} \tilde{T} \mathfrak{D}^2 \hat{u}_z + \sigma \left[ \frac{1}{\kappa_B} \frac{\tilde{\rho}}{\rho_B} \tilde{T} \mathfrak{D}^2 \mathfrak{D}^2 \hat{u}_z + \frac{1}{\nu} \nabla^2 \left( \tilde{T} \mathfrak{D}^2 \hat{u}_z \right) + \frac{g}{c_p \nu} \frac{\mathrm{d}}{\mathrm{d}z} \mathfrak{D}^2 \hat{u}_z \right]$$
$$- \sigma \mathcal{K}^2 \hat{u}_z \left( \frac{1}{\tilde{\rho}} \frac{\mathrm{d}\tilde{\rho}}{\mathrm{d}z} \right) \left[ \frac{2}{3} \frac{m+3}{m} \frac{1}{\kappa} \frac{\tilde{\rho}}{\rho_B} \tilde{T} \left( \frac{1}{\tilde{\rho}} \frac{\mathrm{d}\tilde{\rho}}{\mathrm{d}z} \right) + \frac{g}{c_p \nu} \right]$$
$$= T_B \frac{\mathcal{K}^2}{L^4} Ra \frac{\tilde{\rho}}{\rho_B} \hat{u}_z + \nabla^2 \left( \tilde{T} \mathfrak{D}^2 \mathfrak{D}^2 \hat{u}_z \right) + \frac{g}{c_p} \frac{\mathrm{d}}{\mathrm{d}z} \mathfrak{D}^2 \mathfrak{D}^2 \hat{u}_z$$
$$- \frac{4}{3} \frac{g}{c_p} \mathcal{K}^2 \mathfrak{D}^2 \left( \frac{1}{\tilde{\rho}} \frac{\mathrm{d}\tilde{\rho}}{\mathrm{d}z} \hat{u}_z \right) + \frac{g}{c_p} \mathcal{K}^2 \frac{1}{\tilde{\rho}} \frac{\mathrm{d}\tilde{\rho}}{\mathrm{d}z} \mathfrak{D}^2 \hat{u}_z$$
$$- \frac{2}{3} \mathcal{K}^2 \frac{m+3}{m} \nabla^2 \left[ \tilde{T} \left( \frac{1}{\tilde{\rho}} \frac{\mathrm{d}\tilde{\rho}}{\mathrm{d}z} \right)^2 \hat{u}_z \right] - \frac{2}{3} \frac{g}{c_p} \mathcal{K}^2 \frac{m+3}{m} \frac{\mathrm{d}}{\mathrm{d}z} \left[ \left( \frac{1}{\tilde{\rho}} \frac{\mathrm{d}\tilde{\rho}}{\mathrm{d}z} \right)^2 \hat{u}_z \right]$$
$$- \frac{2}{m} \frac{g}{c_p} \mathcal{K}^2 \left[ \left( \frac{1}{\tilde{\rho}} \frac{\mathrm{d}\tilde{\rho}}{\mathrm{d}z} \right)^2 \frac{\mathrm{d}\hat{u}_z}{\mathrm{d}z} + \left( \frac{1}{\tilde{\rho}} \frac{\mathrm{d}\tilde{\rho}}{\mathrm{d}z} \right)^3 \hat{u}_z \right], \tag{3.179}$$

where

$$\kappa_B = \frac{k}{\rho_B c_p}. \tag{3.180}$$

We will now assume, that the *Principle of the exchange of stabilities* holds in the case at hand, later to verify this hypothesis *a posteriori*. Therefore the growth rate $\sigma$ at the onset of convection is now assumed to be purely real and the system becomes convectively unstable as soon as there appears at least one mode with $\sigma > 0$, whereas exactly at threshold $\sigma = 0$. This allows to set the entire left hand side of (3.179) to zero in the marginal state and on multiplying this equation by $L^6 T_B^{-1}$ and introducing non-dimensional coordinates $\mathbf{x}^\sharp = \mathbf{x}/L$ the equation for $\hat{u}_z$ reduces to

$$\begin{aligned}
0 = &\mathcal{K}^{\sharp 2} Ra \left(1 - \theta z^\sharp\right)^m \hat{u}_z + \nabla^{\sharp 2}\left[\left(1 - \theta z^\sharp\right) \mathfrak{D}^{\sharp 2} \mathfrak{D}^{\sharp 2} \hat{u}_z\right] + \theta \frac{\mathrm{d}}{\mathrm{d}z^\sharp} \mathfrak{D}^{\sharp 2} \mathfrak{D}^{\sharp 2} \hat{u}_z \\
&+ \frac{4}{3} m \theta^2 \mathcal{K}^{\sharp 2} \mathfrak{D}^{\sharp 2} \left(\frac{\hat{u}_z}{1 - \theta z^\sharp}\right) - m \theta^2 \mathcal{K}^{\sharp 2} \frac{1}{1 - \theta z^\sharp} \mathfrak{D}^{\sharp 2} \hat{u}_z \\
&- \frac{2}{3} m (m + 3) \theta^2 \mathcal{K}^{\sharp 2} \nabla^{\sharp 2} \left(\frac{\hat{u}_z}{1 - \theta z^\sharp}\right) - \frac{2}{3} m (m + 3) \theta^3 \mathcal{K}^{\sharp 2} \frac{\mathrm{d}}{\mathrm{d}z^\sharp} \left[\frac{\hat{u}_z}{\left(1 - \theta z^\sharp\right)^2}\right] \\
&- 2 m \theta^3 \mathcal{K}^{\sharp 2} \frac{1}{\left(1 - \theta z^\sharp\right)^2} \left(\frac{\mathrm{d}\hat{u}_z}{\mathrm{d}z^\sharp} - m\theta \frac{\hat{u}_z}{1 - \theta z^\sharp}\right),
\end{aligned} \tag{3.181}$$

where $\mathcal{K}^\sharp = \mathcal{K} L$ and

$$Ra = \frac{g \Delta_S L^4}{T_B \kappa_B \nu}, \tag{3.182}$$

$$\nabla^{\sharp 2} = -\mathcal{K}^{\sharp 2} + \frac{\mathrm{d}^2}{\mathrm{d}z^{\sharp 2}}, \qquad \mathfrak{D}^{\sharp 2} \hat{u}_z = \nabla^{\sharp 2} \hat{u}_z - m\theta \frac{\mathrm{d}}{\mathrm{d}z^\sharp} \left(\frac{\hat{u}_z}{1 - \theta z^\sharp}\right). \tag{3.183}$$

This equation is subject to boundary conditions, that is the impermeability of boundaries

$$\hat{u}_z \left(z^\sharp = 0,\ 1\right) = 0, \tag{3.184}$$

and the stress-free boundary condition (3.172) for the horizontal velocity components, which by taking the $z$-derivative of the mass conservation equation $\nabla_h^\sharp \cdot \hat{\mathbf{u}}_h + \partial_{z^\sharp} \hat{u}_z = m\theta \hat{u}_z / (1 - \theta z^\sharp)$ can be easily transformed to boundary conditions involving only $\hat{u}_z$

$$\left.\frac{\partial^2 \hat{u}_z}{\partial z^{\sharp 2}}\right|_{z^\sharp = 0} = m\theta \left.\frac{\partial \hat{u}_z}{\partial z^\sharp}\right|_{z^\sharp = 0}, \tag{3.185a}$$

$$\left.\frac{\partial^2 \hat{u}_z}{\partial z^{\sharp 2}}\right|_{z^\sharp = 1} = \frac{m\theta}{1 - \theta} \left.\frac{\partial \hat{u}_z}{\partial z^\sharp}\right|_{z^\sharp = 1}. \tag{3.185b}$$

This problem of the sixth order differential equation (3.181) with four boundary conditions (3.184) and (3.185a, 3.185b) must be supplied with the temperature equation (3.177c), which now reads

$$\nabla^{\sharp 2} \frac{\hat{T}}{T_B} = -\frac{\nu}{g L^2} Ra \left(1 - \theta z^\sharp\right)^m \hat{u}_z, \tag{3.186}$$

with two boundary conditions

$$\hat{T}\left(z^{\sharp}=0,1\right)=0. \tag{3.187}$$

The full problem defined by Eqs. (3.181) and (3.186)[10] with the boundary conditions (3.184), (3.185a, 3.185b) and (3.187) is rather cumbersome and perhaps even unsolvable analytically. Therefore to get some insight into the effect of density stratification on the convection threshold we will assume now, that the stratification is weak through imposing

$$\theta = 1 - \frac{1}{\Gamma} \ll 1. \tag{3.188}$$

This allows to expand all the dependent variables in perturbation series, which under additional simplifying assumption, that at threshold the convective flow takes the form of two-dimensional rolls leads to the following form of the velocity and temperature fields

$$u_z = \left(\hat{u}_{0z} + \theta\hat{u}_{1z} + \theta^2\hat{u}_{2z} + \cdots\right)\cos\left(\mathcal{K}x\right), \tag{3.189a}$$

$$u_x = \left(\hat{u}_{0x} + \theta\hat{u}_{1x} + \theta^2\hat{u}_{2x} + \cdots\right)\sin\left(\mathcal{K}x\right), \tag{3.189b}$$

$$T' = \left(\hat{T}_0 + \theta\hat{T}_1 + \theta^2\hat{T}_2 + \cdots\right)\cos\left(\mathcal{K}x\right). \tag{3.189c}$$

The expansion of dependent variables must be accompanied by expansion of the critical Rayleigh number and the critical wave number, since these parameters are expected to be influenced by stratification,

$$Ra_{crit} = Ra_0 + \theta Ra_1 + \theta^2 Ra_2 + \cdots, \quad \mathcal{K}^{\sharp}_{crit} = \mathcal{K}^{\sharp}_0 + \theta\mathcal{K}^{\sharp}_1 + \theta^2\mathcal{K}^{\sharp 2}_2 + \cdots. \tag{3.190}$$

Introducing the above expansions into the equations and gathering terms at the leading order $\theta^0$ one obtains

$$-\mathcal{K}^{\sharp 2}_0 Ra_0\hat{u}_{0z} = \left(-\mathcal{K}^{\sharp 2}_0 + \frac{\mathrm{d}^2}{\mathrm{d}z^{\sharp 2}}\right)^3 \hat{u}_{0z}, \tag{3.191a}$$

$$\left(-\mathcal{K}^{\sharp 2}_0 + \frac{\mathrm{d}^2}{\mathrm{d}z^{\sharp 2}}\right)\frac{\hat{T}_0}{T_B} = -\frac{\nu}{gL^2}Ra_0\hat{u}_{0z}, \tag{3.191b}$$

which is, of course, an equivalent set of equations to that obtained in the Boussinesq case (2.105) and (2.104b). It is, however, not immediately obvious, that the boundary conditions for the velocity $\hat{u}_{0z}$ can be expressed in the same form as (2.110), so as to extract only the $\sin\pi z^{\sharp}$ type solutions out of the class of six independent solutions of the sixth order equation (3.191a). This is because the anelastic linear problem (3.177a–3.177d), in contrast to the Boussinesq case, involves the entropy and pressure

[10]Still with some aid of (3.177a–3.177d) to obtain self-consistency, i.e. eliminate the solution of the homogeneous temperature problem $\nabla^{\sharp 2}\hat{T}/T_B = 0$.

fields, for which the boundary conditions are not specified. In particular the entropy appears in the buoyancy force, thus complicating derivation of a full set of boundary conditions solely in terms of the vertical velocity component $\hat{u}_z$. However, the state equation (3.177d), with the aid of the basic state formulae (3.68a) can be easily rearranged into the following non-dimensional form

$$\theta \frac{\hat{p}}{gL\rho_B \left(1 - \theta z^\sharp\right)^m} = \frac{\hat{T}}{T_B} - \left(1 - \theta z^\sharp\right) \frac{\hat{s}}{c_p}, \tag{3.192}$$

where the fact, that $m = 1/(\gamma - 1) + \mathcal{O}(\delta)$ was utilized. This implies, that at the leading order in $\theta$ the standard Boussinesq consonance of the temperature and entropy fluctuations occurs, i.e. $\hat{s}_0 = c_p \hat{T}_0 / T_B$ and therefore the leading order problem here corresponds directly to the threshold of Boussinesq convection. It follows that the leading order solution corresponds to that from Sect. 2.2.1 and reads

$$\hat{u}_{0z} = A \sin \pi z^\sharp, \qquad \hat{T}_0 = T_B \frac{\nu}{gL^2} A \frac{9}{2} \pi^2 \sin\left(\pi z^\sharp\right), \tag{3.193}$$

$$Ra_0 = \frac{27}{4} \pi^4, \qquad \mathcal{K}_0^2 = \frac{\pi^2}{2}. \tag{3.194}$$

Gathering now all the terms next to the first power of $\theta$ in thc Eqs. (3.181) and (3.186) and in the boundary conditions (3.184), (3.185a, 3.185b) and (3.187) and utilizing (3.193), allows to write down the set of equations determining the first order corrections to the Boussinesq solutions (3.193) resulting from compressibility that is $\hat{u}_1$, $\hat{T}_1$, $Ra_1$ and $\mathcal{K}_1$ in the following simple form

$$\frac{27}{8}\pi^6 \hat{u}_{1z} + \left(-\frac{\pi^2}{2} + \frac{\mathrm{d}^2}{\mathrm{d}z^{\sharp 2}}\right)^3 \hat{u}_{1z} = \frac{27}{8}\pi^6 \left( (m-1) z^\sharp - \frac{Ra_1}{Ra_0} \right) \sin \pi z^\sharp + \frac{9}{4}\pi^5 (2m+1) \cos \pi z^\sharp, \tag{3.195a}$$

$$\left(-\mathcal{K}_0^{\sharp 2} + \frac{\mathrm{d}^2}{\mathrm{d}z^{\sharp 2}}\right) \frac{\hat{T}_1}{T_B} = \frac{\nu}{gL^2} \frac{27}{4} \pi^4 \left[ -\hat{u}_{1z} + \sin \pi z^\sharp \left( m z^\sharp - \frac{Ra_1}{Ra_0} + \frac{4}{3\pi^2} \mathcal{K}_0 \mathcal{K}_1 \right) \right], \tag{3.195b}$$

with the boundary conditions

$$\left.\frac{\partial^2 \hat{u}_{1z}}{\partial z^{\sharp 2}}\right|_{z^\sharp = 0} = m\pi A, \qquad \left.\frac{\partial^2 \hat{u}_{1z}}{\partial z^{\sharp 2}}\right|_{z^\sharp = 1} = -m\pi A, \tag{3.196a}$$

$$\hat{u}_z\left(z^\sharp = 0, 1\right) = 0, \qquad \hat{T}\left(z^\sharp = 0, 1\right) = 0. \tag{3.196b}$$

The next step is to consider the solvability condition for the latter set of equations, i.e. apply the Fredholm Alternative Theorem (cf. Korn and Korn 1961). This requires the knowledge of the solution to the homogeneous problem from (3.195a) with adequate boundary conditions, which must be orthogonal to the non-homogeneity on the right hand side of (3.195a). The homogeneous problem, however, corresponds directly to the leading order one, i.e. the Boussinesq problem, both in terms of the equations and boundary conditions, thus its solution is simply $\sin \pi z^{\sharp}$ and the solvability condition yields

$$\begin{aligned} 0 = & \frac{27}{8}\pi^6 \int_0^1 \mathrm{d}z \left( (m-1)\, z^{\sharp} - \frac{Ra_1}{Ra_0} \right) \sin^2 \pi z^{\sharp} \\ & + \frac{9}{4}\pi^5 (2m+1) \int_0^1 \mathrm{d}z \cos \pi z^{\sharp} \sin \pi z^{\sharp}. \end{aligned} \tag{3.197}$$

This implies, that the first order correction to the critical Rayleigh number at threshold takes the form

$$Ra_1 = \frac{1}{2}(m-1)\, Ra_0 = \frac{27}{8}\pi^4 (m-1). \tag{3.198}$$

The latter result means, that when $m > 1$, which roughly corresponds to fluids characterized by $\gamma < 2$ (see discussion below (3.68a–3.68c)[11]), the critical Rayleigh number in the case at hand

$$Ra_{crit} \approx \frac{27}{4}\pi^4 \left[ 1 + \frac{1}{2}\theta (m-1) \right], \tag{3.199}$$

is increased by the presence of stratification. As a result the total heat per unit mass accumulated in the entire fluid layer (between top and bottom boundaries) in the marginal state, which according to (1.60) equals

$$-\int_0^L c_p \left( \frac{\mathrm{d}\tilde{T}}{\mathrm{d}z} + \frac{g}{c_p} \right) \mathrm{d}z = \frac{c_p \kappa_B \nu T_B}{g L^3} Ra_{crit}, \tag{3.200}$$

(and is equal to the total heat per unit mass released in the marginal state by a fluid parcel rising from the bottom to the top of the fluid layer), where $Ra_{crit}$ is given in (3.199) is also greater in the stratified case, than in the Boussinesq one.

This, however, is by no means a rule and not every anelastic system is characterized by a greater critical Rayleigh number than its Boussinesq analogue. In fact our choice of constant $\nu$ and $k$, thus $z$-dependent $\mu = \tilde{\rho}\nu$ and $\kappa = k/\tilde{\rho} c_p$ allowed to simplify the algebraic manipulations, but also influenced the stability characteristics of the system. Mizerski and Tobias (2011) have studied the marginal anelastic convection in the limit of rapid background rotation $E \ll 1$ (with $Pr = \nu/\kappa_B$ significantly exceeding the Boussinesq transitional value $\mathscr{P}r$ between stationary and oscillatory marginal

[11] The definition of the parameter $\delta$ in (3.11) and the basic state formulae (3.68a–3.68c) imply, that $m \approx 1/(\gamma - 1)$ up to terms as small as $\mathcal{O}(\delta)$.

convection) and weak stratification[12] $\theta \ll 1$, and they have shown, that when $\nu$ and $\kappa$ are assumed constant instead of $\nu$ and $k$, the presence of stratification decreases the critical Rayleigh number to $Ra_{crit} = 3(\pi/\sqrt{2}E)^{4/3}\,[1 - \theta\gamma/2(\gamma - 1)]$. Moreover, the efficiency of heat transport near the marginal state was also shown to depend strongly on whether the thermal conductivity coefficient $k$ or the thermal diffusivity $\kappa$ are assumed constant. In particular they have demonstrated for rapidly rotating convective systems, that when $k = $ const and $\gamma < 2$ the Nusselt number is decreased by the presence of compressibility and the opposite is true, when $\kappa =$ const. However, the specific heat ratio can achieve values greater than two, as e.g. in the case of dry air at high pressure ($\sim 200$ atm) and low temperature (200 K), as reported by Perry et al. (1997) and Kamari et al. (2014). Therefore it is also possible, that in the case of constant $\nu$ and $k$ studied above the critical Rayleigh number (3.199) may be less than the Boussinesq value $27\pi^4/4$, when the polytropic index $m$ falls below unity.

Once the correction $Ra_1$ to the critical Rayleigh number is known, the correction to the critical wave number $\mathcal{K}_1^\sharp$ can be established. Due to the nature of the analysis in the asymptotic regime $\theta \ll 1$ and the perturbation expansions it is, in fact, rather simple to predict, that the Rayleigh number at convection threshold can only depend on $\mathcal{K}^\sharp$ through the $\theta^2$ correction of the type $+\text{Const}\theta^2\mathcal{K}_1^{\sharp 2}$, where Const $> 0$ and thus minimization of $Ra(\mathcal{K}^\sharp)$ over all possible wave numbers leads to $\mathcal{K}_{1\,crit}^\sharp = 0$; no correction proportional to $\mathcal{K}_1^\sharp$ in the first power is possible, since the Rayleigh number at threshold can not depend on the sign of the wave number. This indeed, can be verified by consideration of the $\theta^2$ order terms in the $\hat{u}_z$ Eq. (3.181), even without solving yet for the $\hat{u}_{1z}$ correction to the vertical velocity.[13] From the solvability condition at the order $\theta^2$ all the terms proportional to first and second powers of $\mathcal{K}_1^\sharp$ and first power of $\mathcal{K}_2^\sharp$ can be gathered, which leads to cancelling of all the terms proportional to $\mathcal{K}_1^\sharp$ and $\mathcal{K}_2^\sharp$ in the first powers and the correction $Ra_2$ to the Rayleigh number at threshold takes the form $Ra_2 = 4\mathcal{K}_1^{\sharp 2}/3 + \text{const}$, where const is entirely independent of the wave number. This allows to verify, that the minimal value of the Rayleigh number at convection threshold, in other words the critical Rayleigh number for weakly stratified, anelastic convection with constant $\nu$ and $k$ is achieved by modes with $\mathcal{K}_1^\sharp = \mathcal{K}_{1\,crit}^\sharp = 0$.

We can now solve the first order problem (3.195a–3.196b) with $\mathcal{K}_1$ set to zero in (3.195b). The homogeneous equation for the vertical velocity correction $\hat{u}_{1z}$,

$$\frac{27}{8}\pi^6\hat{u}_{1z}^{HE} + \left(-\frac{\pi^2}{2} + \frac{\mathrm{d}^2}{\mathrm{d}z^{\sharp 2}}\right)^3 \hat{u}_{1z}^{HE} = 0, \tag{3.201}$$

possesses a class of solutions

[12] Note the different sign in definition of $\theta$ in Mizerski and Tobias (2011).

[13] This still requires the assumption formulated above, that the *Principle of the exchange of stability* is valid, which will be later verified for consistency at least up to the order $\theta$, by a straightforward calculation of the growth rate with its order $\theta$ correction.

$$\begin{aligned}\hat{u}_{1z}^{HE} &= C_1 \sin\left(q_1 z^\sharp\right)\cosh\left(q_2 z^\sharp\right) + C_2 \cos\left(q_1 z^\sharp\right)\sinh\left(q_2 z^\sharp\right) \\ &+ C_3 \cos\left(q_1 z^\sharp\right)\cosh\left(q_2 z^\sharp\right) + C_4 \sin\left(q_1 z^\sharp\right)\sinh\left(q_2 z^\sharp\right) \\ &+ C_5 \cos\left(\pi z^\sharp\right) + C_6 \sin\left(\pi z^\sharp\right), \end{aligned} \tag{3.202}$$

where $C_j$ with $j = 1, \ldots, 6$ are constants and

$$q_1 = \frac{\pi}{2\sqrt{2}}\sqrt{\sqrt{52} - 5}, \quad q_2 = \frac{\pi}{2\sqrt{2}}\sqrt{\sqrt{52} + 5}. \tag{3.203}$$

The term $C_6 \sin\left(\pi z^\sharp\right)$ is exactly of the same type as the leading order solution, and hence the constant $C_6$ can be set to zero without loss of generality. The solution of the homogeneous problem (3.202) must be accompanied by the specific solution of the non-homogeneous equation (3.195a), which can be written in the form

$$A\left[\frac{1}{24}(19m - 7)\, z^\sharp \sin\left(\pi z^\sharp\right) - \frac{\pi}{8}(m - 1)\, z^\sharp\left(z^\sharp - 1\right)\cos\left(\pi z^\sharp\right)\right]. \tag{3.204}$$

One can now solve for the general form of the temperature from (3.195b) and apply the boundary conditions (3.196a, 3.196b) to obtain the final formulae for the order $\theta$ corrections to the vertical velocity and temperature. Once the form of $\hat{u}_z$ is known, the mass conservation equation can be used to calculate the horizontal velocity component

$$\mathcal{K}_0^\sharp \hat{u}_{0x} = -\frac{\mathrm{d}\hat{u}_{0z}}{\mathrm{d}z^\sharp}, \tag{3.205a}$$

$$\mathcal{K}_0^\sharp \hat{u}_{1x} = -\frac{\mathrm{d}\hat{u}_{1z}}{\mathrm{d}z^\sharp} + m\theta\hat{u}_{0z}. \tag{3.205b}$$

The full solution in the dimensional coordinates, together with the leading order terms invoked here again for the sake of completeness, takes the following form

$$\hat{u}_{0z} = A \sin\left(\pi\frac{z}{L}\right), \tag{3.206}$$

$$\begin{aligned}\hat{u}_{1z} = A\frac{2}{27\pi}\Big\{&(m+2)\left[\cos\left(\pi\frac{z}{L}\right) - \cos\left(q_1\frac{z}{L}\right)\cosh\left(q_2\frac{z}{L}\right)\right] \\ &- \sqrt{3}(m-4)\sin\left(q_1\frac{z}{L}\right)\sinh\left(q_2\frac{z}{L}\right) \\ &+ C_1 \sin\left(q_1\frac{z}{L}\right)\cosh\left(q_2\frac{z}{L}\right) + C_2\cos\left(q_1\frac{z}{L}\right)\sinh\left(q_2\frac{z}{L}\right)\Big\} \\ + \frac{A}{8}\Big[&\frac{1}{3}(19m-7)\frac{z}{L}\sin\left(\pi\frac{z}{L}\right) - \pi(m-1)\frac{z}{L}\left(\frac{z}{L} - 1\right)\cos\left(\pi\frac{z}{L}\right)\Big]\end{aligned} \tag{3.207}$$

$$\hat{u}_{0x} = -A\sqrt{2}\cos\left(\pi\frac{z}{L}\right), \tag{3.208}$$

$$\begin{aligned}
\hat{u}_{1x} = &- A\frac{2\sqrt{2}}{27\pi^2}\Bigg\{ [q_1C_1 + q_2C_2]\cos\left(q_1\frac{z}{L}\right)\cosh\left(q_2\frac{z}{L}\right) \\
&+ [q_2C_1 - q_1C_2]\sin\left(q_1\frac{z}{L}\right)\sinh\left(q_2\frac{z}{L}\right) \\
&+ \left[q_1(m+2) - q_2\sqrt{3}(m-4)\right]\sin\left(q_1\frac{z}{L}\right)\cosh\left(q_2\frac{z}{L}\right) \\
&- \left[q_2(m+2) + q_1\sqrt{3}(m-4)\right]\cos\left(q_1\frac{z}{L}\right)\sinh\left(q_2\frac{z}{L}\right)\Bigg\} \\
&- \frac{A\sqrt{2}}{8}\Bigg\{\left[\pi(m-1)\frac{z}{L}\left(\frac{z}{L}-1\right) - \frac{61m+95}{27\pi}\right]\sin\left(\pi\frac{z}{L}\right) \\
&+ \left[\frac{1}{3}(13m-1)\frac{z}{L} + m - 1\right]\cos\left(\pi\frac{z}{L}\right)\Bigg\}
\end{aligned} \tag{3.209}$$

$$\hat{T}_0 = T_B\frac{\nu}{gL^2}A\frac{9}{2}\pi^2\sin\left(\pi\frac{z}{L}\right), \tag{3.210}$$

$$\begin{aligned}
\hat{T}_1 = T_B\frac{\nu}{gL^2}A\frac{\pi}{3}\Bigg\{ &-(m-7)\cos\left(q_1\frac{z}{L}\right)\cosh\left(q_2\frac{z}{L}\right) \\
&+ \sqrt{3}(m-1)\sin\left(q_1\frac{z}{L}\right)\sinh\left(q_2\frac{z}{L}\right) \\
&+ C_3\sin\left(q_1\frac{z}{L}\right)\cosh\left(q_2\frac{z}{L}\right) + C_4\cos\left(q_1\frac{z}{L}\right)\sinh\left(q_2\frac{z}{L}\right) \\
&+ \frac{9\pi}{2}\left[\frac{3}{8}(m-5)\frac{z}{L} + (m-1)\right]\sin\left(\pi\frac{z}{L}\right) \\
&- \left[\frac{27}{16}\pi^2(m-1)\frac{z}{L}\left(\frac{z}{L}-1\right) - m + 7\right]\cos\left(\pi\frac{z}{L}\right)\Bigg\},
\end{aligned} \tag{3.211}$$

$$Ra_{crit} = \frac{27}{4}\pi^4\left[1 + \frac{1}{2}\theta(m-1)\right] + \mathcal{O}\left(\theta^2\right), \tag{3.212}$$

$$\mathcal{K}_{crit} = \mathcal{K}_0 + \mathcal{O}\left(\theta^2\right), \tag{3.213}$$

where

$$C_1 = \frac{(m+2)\sin q_1 + \sqrt{3}(m-4)\sinh q_2}{\cosh q_2 - \cos q_1}, \tag{3.214}$$

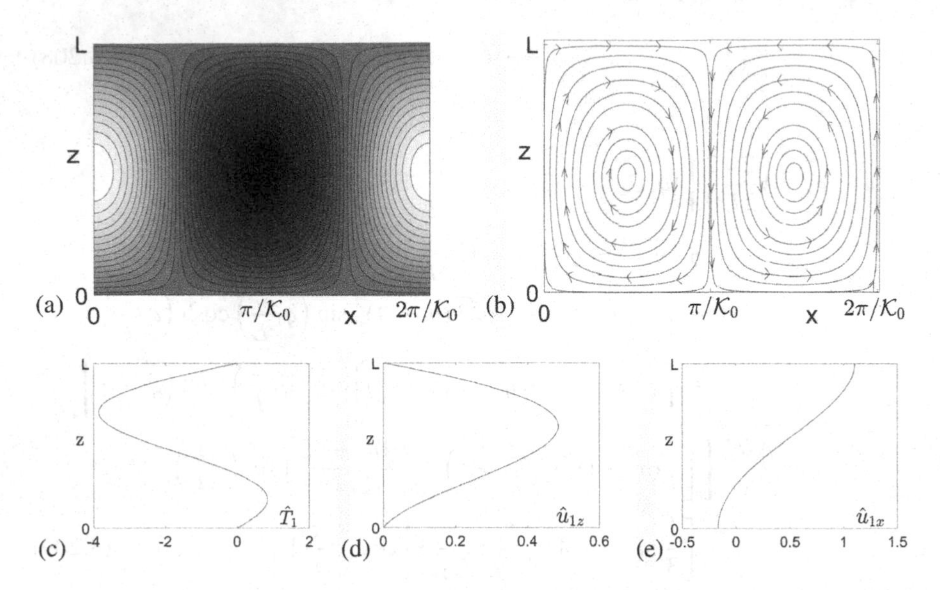


**Fig. 3.1** Visual representation of the linear solutions (3.206–3.211) at the threshold of anelastic convection for an ideal gas with isothermal, stress-free and impermeable boundaries, no heat sources $Q = 0$, constant $\nu$, $k$, $\mathbf{g}$ and $c_p$ and $\mu_b = 0$. Figure **a** presents a greyscale map and isolines of the temperature $T' = (\hat{T}_0 + \theta\hat{T}_1)\cos\mathcal{K}_0 x$ and the streamlines of the flow at threshold, given by $u_x = (\hat{u}_{0x} + \theta\hat{u}_{1x})\sin\mathcal{K}_0 x$ and $u_z = (\hat{u}_{0z} + \theta\hat{u}_{1z})\cos\mathcal{K}_0 x$ are shown on figure **b**. To clearly visualize the effect of stratification on linear solutions the stratification parameter was chosen $\theta = 0.5$. The vertical dependencies of the order $\theta$ corrections to temperature, $\hat{T}_1(z)$, vertical velocity, $\hat{u}_{1z}(z)$ and horizontal velocity, $\hat{u}_{1x}(z)$ are plotted on figures **c**, **d** and **e**

$$C_2 = \frac{(m+2)\sinh q_2 - \sqrt{3}\,(m-4)\sin q_1}{\cosh q_2 - \cos q_1}, \tag{3.215}$$

$$C_3 = \frac{(m-7)\sin q_1 - \sqrt{3}\,(m-1)\sinh q_2}{\cosh q_2 - \cos q_1}, \tag{3.216}$$

$$C_4 = \frac{(m-7)\sinh q_2 + \sqrt{3}\,(m-1)\sin q_1}{\cosh q_2 - \cos q_1}. \tag{3.217}$$

The solutions are depicted on Fig. 3.1. Note, that the centre of the convective cells is shifted downwards with respect to the symmetric Boussinesq solutions and the horizontal velocities are significantly stronger near the top than near the bottom.

The planform of the solutions was of course arbitrarily assumed here as two-dimensional rolls. The weakly nonlinear analysis and the pattern selection problem are very cumbersome in the case of anelastic convection, since even the solutions of the linear problem (3.181) are not known for arbitrary values of $\theta$ and the double asymptotic calculations involving small departure from threshold and small $\theta$ become increasingly difficult at higher orders. Therefore the problem of planform selection by anelastic convection near threshold remains almost entirely unsolved as of yet.

### 3.5.1 The Growth Rate of the Convective Instability in the Limit $\theta \ll 1$

Let us now derive the first order correction to the growth rate, which results from vertical density stratification of the fluid layer, $\theta \ll 1$. We define the non-dimensional growth rate in a similar manner as in the Boussinesq case

$$\sigma^{\sharp} = \frac{\sigma L^2}{\kappa_B}, \tag{3.218}$$

with the use of the thermal diffusion time at the bottom $L^2/\kappa_B = L^2 c_p \rho_B / k$. The non-dimensional form of the full $\hat{u}_z$-equation (3.179), which constitutes the eigen problem for $\hat{u}_z$ and $\sigma^{\sharp}$ takes the form

$$\begin{aligned}
&-\frac{\sigma^{\sharp 2}}{Pr}\left(1-\theta z^{\sharp}\right)^{m+1}\mathfrak{D}^{\sharp 2}\hat{u}_z \\
&+\sigma^{\sharp}\left[\left(1-\theta z^{\sharp}\right)^{m+1}\mathfrak{D}^{\sharp 2}\mathfrak{D}^{\sharp 2}\hat{u}_z+\frac{1}{Pr}\nabla^{\sharp 2}\left[\left(1-\theta z^{\sharp}\right)\mathfrak{D}^{\sharp 2}\hat{u}_z\right]+\frac{\theta}{Pr}\frac{\mathrm{d}}{\mathrm{d}z^{\sharp}}\mathfrak{D}^{\sharp 2}\hat{u}_z\right] \\
&-\sigma^{\sharp}\mathcal{K}^{\sharp 2}\hat{u}_z\frac{m\theta^2}{1-\theta z^{\sharp}}\left[\frac{2}{3}(m+3)\left(1-\theta z^{\sharp}\right)^{m}-\frac{1}{Pr}\right] \\
&\quad=\mathcal{K}^{\sharp 2}Ra\left(1-\theta z^{\sharp}\right)^{m}\hat{u}_z+\nabla^{\sharp 2}\left[\left(1-\theta z^{\sharp}\right)\mathfrak{D}^{\sharp 2}\mathfrak{D}^{\sharp 2}\hat{u}_z\right]+\theta\frac{\mathrm{d}}{\mathrm{d}z^{\sharp}}\mathfrak{D}^{\sharp 2}\mathfrak{D}^{\sharp 2}\hat{u}_z \\
&\quad+m\theta^2\mathcal{K}^{\sharp 2}\left[\frac{4}{3}\mathfrak{D}^{\sharp 2}\left(\frac{\hat{u}_z}{1-\theta z^{\sharp}}\right)-\frac{1}{1-\theta z^{\sharp}}\mathfrak{D}^{\sharp 2}\hat{u}_z\right] \\
&\quad-\frac{2}{3}m(m+3)\theta^2\mathcal{K}^{\sharp 2}\left\{\nabla^{\sharp 2}\left(\frac{\hat{u}_z}{1-\theta z^{\sharp}}\right)+\theta\frac{\mathrm{d}}{\mathrm{d}z^{\sharp}}\left[\frac{\hat{u}_z}{\left(1-\theta z^{\sharp}\right)^2}\right]\right\} \\
&\quad-2m\theta^3\mathcal{K}^{\sharp 2}\frac{1}{\left(1-\theta z^{\sharp}\right)^2}\left[\frac{\mathrm{d}\hat{u}_z}{\mathrm{d}z^{\sharp}}-\frac{m\theta}{1-\theta z^{\sharp}}\hat{u}_z\right],
\end{aligned} \tag{3.219}$$

where the Prandtl number

$$Pr = \frac{\nu}{\kappa_B}, \tag{3.220}$$

is also defined with the use of the thermal diffusivity at the bottom. Exactly at convection threshold the growth rate is zero, therefore we introduce the parameter

$$\eta = \frac{Ra - Ra_{crit}}{Ra_{crit}} \ll 1, \tag{3.221}$$

which measures the departure from critical state, where $Ra_{crit} = Ra_0 + \theta Ra_1$ is defined in (3.212). In other words, the state we have in mind at present, is a flow, developed via the convective instability at a Rayleigh number slightly above the threshold value $Ra = Ra_{crit}(1+\eta)$. The eigen mode in such a state in general differs from the threshold solution given in (3.206) and (3.207), hence we introduce the following double asymptotic expansion in small parameters $\theta$ and $\eta$

$$\hat{u}_z = \hat{u}_{0z} + \theta\hat{u}_{1z} + \eta\hat{v}_{0z} + \eta\theta\hat{v}_{1z} + \mathcal{O}\left(\theta^2,\, \eta^2\right), \tag{3.222}$$

where the corrections denoted by $\hat{v}_{0z}$ and $\hat{v}_{1z}$ result solely from the departure from threshold and are absent when $Ra = Ra_{crit}$. Substituting the above formula (3.222) into the eigen problem (3.219) and balancing the terms at the order $\theta^0$ (bearing in mind that $\sigma \sim \eta$) we obtain

$$-\frac{\sigma_0^{\sharp 2}}{Pr}\nabla^{\sharp 2}\hat{u}_{0z} + \sigma_0^{\sharp}\frac{1+Pr}{Pr}\nabla^{\sharp 4}\hat{u}_{0z} - \mathcal{K}^{\sharp 2} Ra_0\eta\hat{u}_{0z}$$
$$= \mathcal{K}^{\sharp 2} Ra_0\hat{v}_{0z} + \left(-\mathcal{K}^{\sharp 2} + \frac{\mathrm{d}^2}{\mathrm{d}z^{\sharp 2}}\right)^3 \hat{v}_{0z}, \tag{3.223}$$

and hence

$$\left[\frac{\sigma_0^{\sharp 2}}{Pr}\left(\mathcal{K}^{\sharp 2} + \pi^2\right) + \sigma_0^{\sharp}\frac{1+Pr}{Pr}\left(\mathcal{K}^{\sharp 2} + \pi^2\right)^2 - \mathcal{K}^{\sharp 2} Ra_0\eta\right] A\sin\pi z^{\sharp}$$
$$= \mathcal{K}^{\sharp 2} Ra_0\hat{v}_{0z} + \left(-\mathcal{K}^{\sharp 2} + \frac{\mathrm{d}^2}{\mathrm{d}z^{\sharp 2}}\right)^3 \hat{v}_{0z}, \tag{3.224}$$

where we have used the asymptotic form of the growth rate

$$\sigma^{\sharp} = \sigma_0^{\sharp} + \theta\sigma_1^{\sharp} + \mathcal{O}\left(\eta\theta^2\right). \tag{3.225}$$

The solvability condition for this equation (cf. Korn and Korn 1961) yields the same solution for the leading order term in the growth rate expansion $\sigma_0^{\sharp}$ as that from the Boussinesq case (2.167). Thus the maximal growth rate is achieved at $\mathcal{K}^{\sharp} = \mathcal{K}_0^{\sharp} = \pi/\sqrt{2}$ and is of the order $\mathcal{O}(\eta)$.[14]

We now proceed to the next order and gather all the terms of the order $\theta^1$ in the Eq. (3.219), which yields

---

[14]Note, that this means that the entire left hand side of the Eq. (3.224) vanishes, and therefore the solution for the order $\eta$ correction to vertical velocity $\hat{v}_{0z}$ simply reproduces the leading order term $\hat{u}_{0z}$; consequently $\hat{v}_{0z}$ can be assumed zero without loss of generality.

$$
\begin{aligned}
&-\frac{\sigma_0^{\sharp 2}}{Pr}\nabla^{\sharp 2}\hat{u}_{1z} - 2\frac{\sigma_0^{\sharp}\sigma_1^{\sharp}}{Pr}\nabla^{\sharp 2}\hat{u}_{0z} + m\frac{\sigma_0^{\sharp 2}}{Pr}\frac{\mathrm{d}\hat{u}_{0z}}{\mathrm{d}z^{\sharp}} + (m+1)\frac{\sigma_0^{\sharp 2}}{Pr}z^{\sharp}\nabla^{\sharp 2}\hat{u}_{0z} \\
&+\sigma_1^{\sharp}\frac{1+Pr}{Pr}\nabla^{\sharp 4}\hat{u}_{0z} + \sigma_0^{\sharp}\frac{1+Pr}{Pr}\nabla^{\sharp 4}\hat{u}_{1z} - \sigma_0^{\sharp}\left(m + \frac{1+Pr}{Pr}\right)z^{\sharp}\nabla^{\sharp 4}\hat{u}_{0z} \\
&-\left(2m + \frac{m+1}{Pr}\right)\sigma_0^{\sharp}\frac{\mathrm{d}}{\mathrm{d}z^{\sharp}}\nabla^{\sharp 2}\hat{u}_{0z} \\
&-\mathcal{K}_0^{\sharp 2}Ra_1\eta\hat{u}_{0z} - \mathcal{K}_0^{\sharp 2}Ra_0\eta\left(\hat{u}_{1z} - mz^{\sharp}\hat{u}_{0z}\right) \\
&\qquad = \mathcal{K}_0^{\sharp 2}Ra_0\hat{v}_{1z} + \left(-\mathcal{K}_0^{\sharp 2} + \frac{\mathrm{d}^2}{\mathrm{d}z^{\sharp 2}}\right)^3\hat{v}_{1z}.
\end{aligned}
\tag{3.226}
$$

It can be anticipated, that just as the leading order term $\sigma_0^{\sharp} \sim \eta$ also the correction $\sigma_1$ is proportional to $\eta$, and therefore all the terms proportional to $\sigma_0^2$ and $\sigma_0\sigma_1$ are of the order $\mathcal{O}(\eta^2)$. Neglecting all the terms $\mathcal{O}(\eta^2)$ in the Eq. (3.226) and leaving only terms of the order $\mathcal{O}(\eta)$ leads to

$$
\begin{aligned}
&\mathcal{K}_0^{\sharp 2}Ra_0\hat{v}_{1z} + \left(-\mathcal{K}_0^{\sharp 2} + \frac{\mathrm{d}^2}{\mathrm{d}z^{\sharp 2}}\right)^3\hat{v}_{1z} \\
&\qquad = -\frac{9}{4}\pi^4\left[\frac{3}{4}\pi^2(m-1)\eta - \sigma_1^{\sharp}\frac{1+Pr}{Pr}\right]A\sin\pi z^{\sharp} \\
&\qquad -\frac{27}{8}\pi^6\eta\left(1 - \frac{m}{1+Pr}\right)Az^{\sharp}\sin\pi z^{\sharp} \\
&\qquad +\frac{9}{4}\pi^5\eta\frac{Pr}{1+Pr}\left(2m + \frac{m+1}{Pr}\right)A\cos\pi z^{\sharp} \\
&\qquad -\frac{27}{8}\pi^6\eta\hat{u}_{1z} + \frac{3}{2}\pi^2\eta\left(-\mathcal{K}_0^{\sharp 2} + \frac{\mathrm{d}^2}{\mathrm{d}z^{\sharp 2}}\right)^2\hat{u}_{1z}.
\end{aligned}
\tag{3.227}
$$

The solvability condition for the above equation can now be applied. Integration by parts of the terms in the last row of (3.227) and application of boundary conditions yields

$$
\int_0^1\left[-\frac{27}{8}\pi^6\eta\hat{u}_{1z} + \frac{3}{2}\pi^2\eta\left(-\mathcal{K}_0^{\sharp 2} + \frac{\mathrm{d}^2}{\mathrm{d}z^{\sharp 2}}\right)^2\hat{u}_{1z}\right]\sin\pi z^{\sharp}\mathrm{d}z^{\sharp} = 0, \tag{3.228}
$$

hence by the use of

$$
\int_0^1\sin^2\pi z^{\sharp}\mathrm{d}z^{\sharp} = 1/2, \quad \int_0^1 z^{\sharp}\sin^2\pi z^{\sharp}\mathrm{d}z^{\sharp} = 1/4, \quad \int_0^1\cos\pi z^{\sharp}\sin\pi z^{\sharp}\mathrm{d}z^{\sharp} = 0
$$

one obtains the following equation for the correction to the growth rate $\sigma_1$

$$\frac{3}{4}\pi^2 (m-1)\,\eta - \sigma_1^\sharp \frac{1+Pr}{Pr} + \frac{3}{4}\pi^2 \eta \left(1 - \frac{m}{1+Pr}\right) = 0. \tag{3.229}$$

The solution of the latter equation

$$\sigma_1^\sharp = \frac{3}{4}\pi^2 m \left(\frac{Pr}{1+Pr}\right)^2 \eta > 0 \tag{3.230}$$

is positive, thus the convective instability in the anelastic case with $\nu$ and $k$ constant is enhanced by the compressibility of the fluid, since its growth rate

$$\sigma^\sharp = \frac{3}{2}\pi^2 \frac{Pr}{1+Pr}\eta \left(1 + \frac{1}{2}m\theta \frac{Pr}{1+Pr}\right), \tag{3.231}$$

is greater than the Boussinesq one. However, as in the case of critical Rayleigh number, any generalization to other types of anelastic systems should not be carried out without careful consideration of the full equations; as we remarked above, a change in the physical properties of the fluid, such as e.g. a change from constant $k$ to constant $\kappa$ leads to an important change in the physical response of the system to convective driving.

Finally, it is of interest to note, that since the solution slightly above convection threshold differs from the marginal one (3.206), (3.207) and contains corrections $\eta(\hat{v}_{0z} + \theta \hat{v}_{1z})$, the convective flow in the exact form of the marginal solution at $Ra = Ra_{crit}$, given in (3.206–3.217), might not be observable experimentally, since exactly at threshold the amplitude of convection $A$ vanishes; to observe convective flow one needs to exceed the threshold value $Ra_{crit}$ at least slightly, so that $0 < \eta = (Ra - Ra_{crit})/Ra_{crit} \ll 1$, when the corrections $\eta(\hat{v}_{0z} + \theta \hat{v}_{1z})$ already appear.

## 3.6 Fully Developed, Stratified Convection

This section recalls the theory of turbulent stratified convection of Jones et al. (2020), with some amendments to present more detailed clarifications. To make the description of the fully developed, turbulent, stratified convection as clear as possible simplifications need to be introduced. Therefore we assume the equation of state of an ideal gas, uniform dynamic viscosity $\mu$ and uniform thermal conductivity $k$, negligible bulk viscosity $\mu_b$, no internal heating $Q = 0$ and constant gravity $\mathbf{g} = -g\hat{\mathbf{e}}_z$, $g = \text{const}$. The equations of motion are expressed in the following way

$$\frac{\partial \mathbf{u}}{\partial t} + (\mathbf{u}\cdot\nabla)\,\mathbf{u} = -\nabla\frac{p'}{\tilde{\rho}} + \frac{s'}{c_p}g\hat{\mathbf{e}}_z + \frac{\mu}{\tilde{\rho}}\nabla^2\mathbf{u} + \frac{\mu}{3\tilde{\rho}}\nabla\left(\nabla\cdot\mathbf{u}\right) \tag{3.232a}$$

$$\nabla\cdot(\tilde{\rho}\mathbf{u}) = 0, \tag{3.232b}$$

$$\tilde{\rho}\tilde{T}\left[\frac{\partial s'}{\partial t}+\mathbf{u}\cdot\nabla\left(\tilde{s}+s'\right)\right]=\nabla\cdot(k\nabla T)+2\mu\mathbf{G}^s:\mathbf{G}^s-\frac{2}{3}\mu\left(\nabla\cdot\mathbf{u}\right)^2, \tag{3.232c}$$

$$\frac{\rho'}{\tilde{\rho}}=-\frac{T'}{\tilde{T}}+\frac{p'}{\tilde{p}}, \qquad s'=-R\frac{p'}{\tilde{p}}+c_p\frac{T'}{\tilde{T}}, \tag{3.232d}$$

where, as remarked, we have assumed

$$\mu=\text{const}, \quad k=\text{const}, \quad g=\text{const}, \quad \mu_b=0, \quad Q=0, \tag{3.233}$$

$$p=\rho RT, \qquad s=c_v\ln\frac{p}{\rho^\gamma}. \tag{3.234}$$

### 3.6.1 The Hydrostatic Conduction Reference State and the Hydrostatic Adiabatic State

The hydrostatic reference state determined by vertical conductive heat transport and hydrostatic force balance is recalled here (cf. (3.67) and (3.68a–3.68c))

$$\tilde{T}=\tilde{T}_B\left(1-\tilde{\theta}\frac{z}{L}\right), \qquad \tilde{\rho}=\tilde{\rho}_B\left(1-\tilde{\theta}\frac{z}{L}\right)^m, \qquad \tilde{p}=\frac{gL\tilde{\rho}_B}{\tilde{\theta}\,(m+1)}\left(1-\tilde{\theta}\frac{z}{L}\right)^{m+1}, \tag{3.235a}$$

$$\tilde{s}=c_p\frac{m+1-\gamma m}{\gamma}\ln\left(1-\tilde{\theta}\frac{z}{L}\right)+\text{const}, \tag{3.235b}$$

where

$$m=\frac{gL}{R\Delta\tilde{T}}-1, \qquad \tilde{\theta}=\frac{\Delta\tilde{T}}{\tilde{T}_B}, \qquad \Delta\tilde{T}=\tilde{T}_B-\tilde{T}_T. \tag{3.236}$$

For the purpose of this section we make a clear distinction between the top and bottom values of the temperature and density in the static reference state and those of a convective state. The reason for doing so is that two cases will be considered, that of isothermal boundaries, when $\tilde{T}_B=T_B$, $\tilde{T}_T=T_T$ and that of isentropic boundaries, when $\tilde{T}_B\neq T_B$, $\tilde{T}_T\neq T_T$. In the latter case the total temperature at boundaries differs from the reference state boundary values by terms of order $\delta$, i.e. $T'(\mathbf{x}_h, z=0, t)$ at the bottom and $T'(\mathbf{x}_h, z=L, t)$ at the top.

We also recall the hydrostatic adiabatic state

$$T_{ad}=T_{ad\,B}\left(1-\frac{gz}{c_pT_{ad\,B}}\right), \qquad \rho_{ad}=\rho_{ad\,B}\left(1-\frac{gz}{c_pT_{ad\,B}}\right)^{\frac{1}{\gamma-1}}, \tag{3.237a}$$

$$p_{ad}=\rho_{ad\,B}RT_{ad\,B}\left(1-\frac{gz}{c_pT_{ad\,B}}\right)^{\frac{\gamma}{\gamma-1}}, \qquad s_{ad}=\text{const}. \tag{3.237b}$$

Most of the time we will assume, that $T_{ad\,B} = \tilde{T}_B$, whereas $\rho_{ad\,B}$ is related to $\tilde{\rho}_B$ through (3.76) by the condition, that the total fluid mass is contained in the hydrostatic reference state, either conduction state or the adiabatic one. Furthermore we define the ratios of the bottom to top temperature values

$$\tilde{\Gamma} = \frac{\tilde{T}_B}{\tilde{T}_T} = \frac{1}{1-\tilde{\theta}} > 1, \qquad \Gamma_{ad} = \frac{T_{ad\,B}}{T_{ad\,T}} = \frac{1}{1-\frac{gL}{c_p T_{ad\,B}}} > 1. \tag{3.238}$$

On defining an alternative measure of small departure from adiabaticity (cf. (3.69))

$$\epsilon_a = \frac{L}{\tilde{T}_B}\left(\frac{\Delta\tilde{T}}{L} - \frac{g}{c_p}\right) = \delta\frac{\tilde{\Gamma}-1}{\tilde{\Gamma}\ln\tilde{\Gamma}} > 0, \qquad \epsilon_a \ll 1, \tag{3.239}$$

where the subscript $a$ stands for '*anelastic*' (to distinguish this new parameter from the Boussinesq small parameter $\epsilon$ defined in (2.6)), for the case of

$$T_{ad\,B} = \tilde{T}_B \tag{3.240}$$

we get the following relations

$$\tilde{\theta} = \frac{gL}{c_p\tilde{T}_B} + \epsilon_a, \qquad \tilde{\Gamma} = \Gamma_{ad} + \epsilon_a\tilde{\Gamma}\Gamma_{ad} = \Gamma_{ad} + \epsilon_a\Gamma_{ad}^2 + \mathcal{O}\left(\Gamma_{ad}^3\epsilon_a^2\right). \tag{3.241}$$

Finally we define the superadiabatic temperature fluctuation, say $T_S(\mathbf{x}, t)$, as a component of the total temperature in the convective state $T(\mathbf{x}, t)$, thus as a correction to hydrostatic adiabatic state so that

$$T(\mathbf{x}, t) = \tilde{T}(z) + T'(\mathbf{x}, t) = T_{ad}(z) + T_S(\mathbf{x}, t). \tag{3.242}$$

By the use of the definitions of $\tilde{T}(z)$ and $T_{ad}(z)$ in (3.235a) and (3.237a) with $T_{ad\,B} = \tilde{T}_B$, the superadiabatic temperature fluctuation $T_S(\mathbf{x}, t)$ is related to the temperature fluctuation about the conduction reference state $T'(\mathbf{x}, t)$ in the following way

$$T'(\mathbf{x}, t) = T_S(\mathbf{x}, t) + \epsilon_a\frac{\tilde{T}_B}{L}z. \tag{3.243}$$

Therefore the fluctuation $T'(\mathbf{x}, t)$ is simply shifted with respect to the superadiabatic temperature fluctuation by the superadiabatic linear profile of the conduction reference state; of course both fluctuations are of the order $\mathcal{O}(\epsilon_a) = \mathcal{O}(\delta)$ by the second fundamental assumption of the anelastic approximation (3.12).

### 3.6.2 *Relation Between Entropy and Temperature Jumps Across the Boundary Layers*

One of the major differences between compressible and Boussinesq convection is that the temperature fluctuation and the entropy fluctuation are not equivalent in the compressible case, contrary to the Boussinesq case (up to a constant, cf. (2.48)). In the compressible case the bulk of turbulent convection, where the fluid is efficiently mixed by a vigorous flow, is characterized by an almost uniform mean total entropy $\langle s\rangle_h$, as can be anticipated from the energy equation (3.232c) which suggests efficient advection of the mean entropy (cf. mean entropy profiles on Figs. 3.2 and 3.3 sketched for the cases of isothermal and isentropic boundaries).[15] On the other hand the mean temperature, according to the Eq. (3.37) or (3.38) for the considered case of a perfect gas, is advected along with the mean pressure and none of these two quantities alone needs to be homogenized in the bulk (only the sum $-R\left\langle p'\right\rangle_h/\tilde{p}+c_p\left\langle T'\right\rangle_h/\tilde{T}+\tilde{s}=\langle s\rangle_h$, which is the total mean entropy, as is clear from (3.232d)). At the top and bottom of the fluid domain the boundary layers are formed to adjust the bulk top and bottom values of the entropy and temperature to their values at boundaries. However, these boundary layers are not symmetric with respect to the mid plane, contrary to the symmetric Boussinesq case, and typically the jumps of the total mean entropy and mean temperature fluctuation across the top boundary layer, denoted by $(\Delta s)_T$ and $\left(\Delta T'\right)_T$ are considerably larger than the relative jumps across the bottom boundary layer, $(\Delta s)_B$ and $\left(\Delta T'\right)_B$.[16] Note, that in the case of entropy both, the variations of the fluctuation $s'$ and of the reference state entropy $\tilde{s}$ are of the order $\mathcal{O}(\epsilon_a)$, thus in developed convection the jump of $\tilde{s}$ across thin boundary layers is negligible with respect to the jumps of $\left\langle s'\right\rangle_h$ across the layers. On the contrary, for the other thermodynamic variables, such as temperature, pressure and density the reference state variables are $\mathcal{O}(\epsilon_a^{-1})$ times stronger than the fluctuations and thus jumps in the values of $\tilde{T}$, $\tilde{p}$ and $\tilde{\rho}$ across the boundary layers may significantly exceed the jumps of the corresponding fluctuations.

Before we proceed to providing a dynamical picture of developed compressible convection we note that in some cases it is also useful to express the temperature jumps across the thermal boundary layers by the entropy jumps. From the state equations (3.232d) we obtain

[15] In a well-mixed bulk the fluctuations about the horizontal means are expected to be small. Since the impermeability of boundaries together with the continuity equation $\nabla\cdot(\tilde{\rho}\mathbf{u})=0$ imply $\langle u_z\rangle_h=0$, the horizontal average of the stationary energy equation (3.232c) leaves a dominant balance between the mean conduction and mean viscous heating; in turn, by the use of the full equation, this leads to $u_z\partial_z\langle s\rangle_h\approx 0$ in the well-mixed bulk of turbulent convection.

[16] The jumps are positive by definition, hence e.g. the entropy jump across the top boundary layer is defined as $(\Delta s)_T=\langle s\rangle_h(z=L-\delta_{th,T})-\langle s\rangle_h(z=L)$.

$$\frac{\left(\Delta\rho'\right)_i}{\tilde{\rho}_i} = \frac{1}{\gamma-1}\left[\frac{\left(\Delta T'\right)_i}{\tilde{T}_i} - \gamma\frac{(\Delta s)_i}{c_p}\right] + \mathcal{O}\left(\epsilon_a^2\right), \tag{3.244a}$$

$$\frac{\left(\Delta p'\right)_i}{\tilde{p}_i} = \frac{\gamma}{\gamma-1}\left[\frac{\left(\Delta T'\right)_i}{\tilde{T}_i} - \frac{(\Delta s)_i}{c_p}\right] + \mathcal{O}\left(\epsilon_a^2\right), \tag{3.244b}$$

where the subscript $i$ stands either for $T$ or $B$, and similarly as for the entropy and temperature we denote the jumps of the mean density and pressure fluctuations across the top and bottom boundary layers by $\left(\Delta\rho'\right)_T$, $\left(\Delta p'\right)_T$ and $\left(\Delta\rho'\right)_B$, $\left(\Delta p'\right)_B$, respectively. Next we observe, that in the boundary layers the mass conservation constraint $\nabla\cdot(\tilde{\rho}\mathbf{u}) = 0$ allows for the following estimate

$$u_{z,i} \sim \frac{\delta_{\nu,i}}{L}, \tag{3.245}$$

where $\delta_{\nu,i}$ denotes the thickness of either the top ($i = T$) or the bottom ($i = B$) viscous boundary layer and $u_{z,i}$ the vertical velocity in each of the layers. Consequently the $z$-component of the nonlinear inertial term in the Navier-Stokes equation in a boundary layer, i.e. $\mathbf{u}_i\cdot\nabla u_{z,i}$, is small, since in fully developed turbulent convection the boundary layers are thin, thus

$$\delta_{\nu,T} \ll L, \qquad \delta_{\nu,B} \ll L. \tag{3.246a}$$

$$\delta_{th,T} \ll L, \qquad \delta_{th,B} \ll L, \tag{3.246b}$$

where $\delta_{th,T}$ and $\delta_{th,B}$ denote the thicknesses of the top and bottom thermal boundary layers respectively. Therefore the balance on the $z$-component of the Navier-Stokes equation (3.232a) in the boundary layers, occurs predominantly between the pressure gradient and the buoyancy force, suggesting

$$\left(\Delta p'\right)_i \approx gL\tilde{\rho}_i\frac{(\Delta s)_i}{c_p}\frac{\delta_{th,i}}{L}, \tag{3.247}$$

which implies, that the pressure jump across the thermal boundary layers is rather small. Inserting (3.247) into (3.244b) leads to

$$\frac{\left(\Delta T'\right)_i}{\tilde{T}_i} \approx \frac{(\Delta s)_i}{c_p}\left(1+\tilde{\theta}\frac{\delta_{th,i}}{L}\frac{\tilde{T}_B}{\tilde{T}_i}\right) + \mathcal{O}\left(\epsilon_a^2\frac{\delta_{th,i}}{L}\right). \tag{3.248}$$

Typically the term $(\tilde{\theta}\delta_{th,i}\tilde{T}_B)/(L\tilde{T}_i)$ resulting from the pressure jump across the boundary layers is expected to be small, however, it is not necessarily the case always. In fact, as we will see in Sect. 3.6.6, the thicknesses of the thermal boundary layers depend in a complicated way on the density scale height $D_\rho = -\tilde{\rho}/\mathrm{d}_z\tilde{\rho}$, and can increase when the scale height decreases and the layer becomes more strongly stratified. Consequently in strongly stratified cases the boundary layers themselves may contain a few density scale heights, i.e. we may have $D_\rho < \delta_{th,i}$, and thus the boundary layers are not necessarily incompressible. Moreover, it is clear, that especially the top boundary layer is more prone to become compressible as the stratification $\tilde{\theta}$ is increased (the scale heights decreased), as suggested by appearance of the factor $\tilde{\Gamma} = \tilde{T}_B/\tilde{T}_T > 1$ in the pressure-jump correction at the top boundary layer in (3.248) (that is in the second term inside the brackets of that equation); in strongly stratified systems this factor makes the top correction significantly greater than the bottom one.

Nevertheless, for clarity of the presentation it is of interest to consider the simplest case when both the boundary layers are incompressible and then it is allowed to use an approximate relation between the jumps of the mean temperature fluctuation and of the mean total entropy across the boundary layers,

$$\frac{\left(\Delta T'\right)_i}{\tilde{T}_i} \approx \frac{(\Delta s)_i}{c_p}. \tag{3.249}$$

The other case, when the pressure-jump correction $(\tilde{\theta}\delta_{th,i}\tilde{T}_B)/(L\tilde{T}_i)$ in the Eq. (3.248) matters will be referred to as the compressible boundary layer case.

### 3.6.3 "Subadiabatic" Temperature Gradient in the Bulk

An important feature of the turbulent, (statistically) stationary anelastic convection is that the magnitude of the mean temperature gradient in the bulk of convection is weaker than $g/c_p$, i.e. than that of the hydrostatic adiabatic state. This is because the bulk is non-static, that is the mean force balance on the vertical direction in the bulk is non-hydrostatic due to the influence of inertia. In other words the vertical equilibrium of mean forces, obtained from horizontally averaging the $z$-component of the Navier-Stokes equation (3.26a) with the reference state balance $\mathrm{d}_z\tilde{p} = -\tilde{\rho}g$ incorporated back into it, reads

$$\frac{\mathrm{d}}{\mathrm{d}z}\left\langle\tilde{\rho}u_z^2\right\rangle_h + \frac{\mathrm{d}\left\langle p\right\rangle_h}{\mathrm{d}z} = -g\left\langle\rho\right\rangle_h. \tag{3.250}$$

In obtaining the above equation we have used the fact, that $\langle u_z\rangle_h = 0$ (cf. (3.101)). As argued, the mean total entropy is uniformly distributed by vigorous advection and thus almost constant in the bulk of turbulent convection at high Rayleigh number. Therefore we can assume that the bulk is approximately adiabatic, i.e.

$$\langle s\rangle_h = \tilde{s} + \left\langle s'\right\rangle_h = \text{const} \quad \text{in the bulk}. \tag{3.251}$$

Consequently for the bulk of convection we may expect that the adiabatic relation between mean pressure and density

$$\langle p\rangle_h = \text{const}\,\langle\rho\rangle_h^{\gamma}, \tag{3.252}$$

is satisfied. Moreover, the fact that the bulk is efficiently mixed and the fluctuations about the means are assumed negligible implies that approximately we may also expect

$$\langle p\rangle_h = \langle\rho\rangle_h R\,\langle T\rangle_h. \tag{3.253}$$

Elimination of the mean total pressure $\langle p\rangle_h$ and density $\langle\rho\rangle_h$ from the Eqs. (3.250), (3.252) and (3.253) yields

$$\frac{d\,\langle T\rangle_h}{dz} = -\frac{g}{c_p} - \frac{1}{c_p\,\langle\rho\rangle_h}\frac{\mathrm{d}}{\mathrm{d}z}\left\langle\tilde{\rho}u_z^2\right\rangle_h, \tag{3.254}$$

where $\langle\rho\rangle_h$ has been retained in the last term; this term, however, is of the order $\mathcal{O}(\epsilon_a)$ (or equivalently $\mathcal{O}(\delta)$, cf. (3.15)), and therefore keeping the order of accuracy at the level of the magnitude of fluctuations, i.e. at $\mathcal{O}(\epsilon_a)$, the mean total density $\langle\rho\rangle_h$ can be replaced by $\tilde{\rho}$, so that

$$\begin{aligned}\frac{d\,\langle T\rangle_h}{dz} &= -\frac{g}{c_p} - \frac{1}{c_p\tilde{\rho}}\frac{\mathrm{d}}{\mathrm{d}z}\left\langle\tilde{\rho}u_z^2\right\rangle_h + \mathcal{O}\left(\epsilon_a^2\frac{g}{c_p}\right)\\ &= -\frac{g}{c_p} + \frac{\left\langle u_z^2\right\rangle_h}{c_p D_\rho} - \frac{1}{c_p}\frac{\mathrm{d}\left\langle u_z^2\right\rangle_h}{\mathrm{d}z} + \mathcal{O}\left(\epsilon_a^2\frac{g}{c_p}\right),\end{aligned} \tag{3.255}$$

where we have used the density scale height $D_\rho = -\tilde{\rho}/\mathrm{d}_z\tilde{\rho}$. Integration of the latter equality across the bulk, hence from $z = \delta_{th,B}$ to $z = L - \delta_{th,T}$ allows to calculate the jump of the mean total temperature across the bulk

$$\begin{aligned}-(\Delta T)_{bulk} &= \langle T\rangle_h\left(z = L - \delta_{th,T}\right) - \langle T\rangle_h\left(z = \delta_{th,B}\right)\\ &= -\frac{gL}{c_p} + \frac{g}{c_p}\left(\delta_{th,B} + \delta_{th,T}\right) + \int_0^L \frac{\left\langle u_z^2\right\rangle_h}{c_p D_\rho}\mathrm{d}z\\ &\quad + \mathcal{O}\left(\epsilon_a\delta_{\nu,B}^2\frac{g}{c_p L}\right) + \mathcal{O}\left(\epsilon_a\delta_{\nu,T}^2\frac{g}{c_p L}\right) + \mathcal{O}\left(\epsilon_a^2\frac{gL}{c_p}\right).\end{aligned} \tag{3.256}$$

To obtain the above relation we have utilized the observation about the magnitude of the vertical velocity in boundary layers made in (3.245), which implies that at the top and bottom of the bulk we have

$$\left\langle u_z^2\right\rangle_h\left(z = L - \delta_{th,T}\right) = \mathcal{O}\left(\epsilon_a\delta_{\nu,T}^2\right), \quad \left\langle u_z^2\right\rangle_h\left(z = \delta_{th,B}\right) = \mathcal{O}\left(\epsilon_a\delta_{\nu,B}^2\right). \tag{3.257}$$

The term resulting from integration of the hydrostatic adiabatic gradient across the boundary layers, i.e. $g(\delta_{th,B} + \delta_{th,T})/c_p$ must be retained, since it is of the order of non-dimensional thicknesses of the boundary layers, $\delta_{th,B}/L$ and $\delta_{th,T}/L$, which although small still have to be much greater than the small anelastic parameter $\epsilon_a$ whose smallness guarantees validity of the anelastic system of equations. The $\mathcal{O}(\epsilon_a)$ correction to the hydrostatic adiabatic temperature jump across the bulk in (3.256) is also retained, because it is of the same order as the thermodynamic fluctuations; it is denoted by

$$(\Delta T)_{vel} = \int_0^L \frac{\left\langle u_z^2 \right\rangle_h}{c_p D_\rho} \mathrm{d}z = \mathcal{O}(\epsilon_a) > 0, \tag{3.258}$$

and is positive definite, hence convection reduces the magnitude of the mean temperature jump across the bulk with respect to the magnitude of the temperature jump between $z = \delta_{th,B}$ and $z = L - \delta_{th,T}$ in a hydrostatic adiabatic state, $gL/c_p + g(\delta_{th,B} + \delta_{th,T})/c_p$. In other words the non-hydrostatic mean temperature gradient in the adiabatic bulk is weaker than the hydrostatic adiabatic gradient $g/c_p$.

From the Eq. (3.256) we can extract the jump of the mean temperature fluctuation across the bulk

$$\begin{aligned} 0 < \left(\Delta T'\right)_{bulk} =& \left\langle T' \right\rangle_h \left(z = L - \delta_{th,T}\right) - \left\langle T' \right\rangle_h \left(z = \delta_{th,B}\right) \\ =& \epsilon_a \tilde{T}_B + (\Delta T)_{vel} \\ &+ \mathcal{O}\left(\epsilon_a \delta_{th,B} \frac{g}{c_p}\right) + \mathcal{O}\left(\epsilon_a \delta_{th,T} \frac{g}{c_p}\right), \end{aligned} \tag{3.259}$$

where the term describing the differences between the temperature jumps across the boundary layers for the hydrostatic reference and adiabatic profiles

$$\epsilon_a \tilde{T}_B \frac{\delta_{th,B} + \delta_{th,T}}{L} \tag{3.260}$$

has been included in the remainders $\mathcal{O}(\epsilon_a \delta_{th,B} g/c_p)$ and $\mathcal{O}(\epsilon_a \delta_{th,T} g/c_p)$. The jump of the mean temperature fluctuation is positive but defined in the opposite way to the jump of the total mean temperature, that is the bottom value is subtracted from the top value. This is because the mean fluctuation $\left\langle T' \right\rangle_h (z)$ increases in the bulk (cf. Figs. 3.2b and 3.3b).

### 3.6.4 Vertical Profiles of Mean Temperature and Entropy

In this section we schematically sketch the vertical profiles of the total, horizontally averaged entropy and temperature and the horizontally averaged temperature fluctuation. Justification of the important characteristics of the profiles is provided. The reason for depicting separately the total mean temperature and the mean temperature fluctuation is simply clarity, because the total temperature is strongly dominated by the contribution from the reference state, which by assumption is $\mathcal{O}(\epsilon_a^{-1})$ times greater than the fluctuation and hence details of the vertical dependence of the total mean temperature are difficult to present; at the same time, however, the general picture is also instructive.

#### 3.6.4.1 The Case of Isothermal Boundaries

We start by considering the case when the temperature at boundaries is held fixed. The vertical profiles of $\langle s\rangle_h$, $\langle T'\rangle_h$ and $\langle T\rangle_h$ are shown on Fig. 3.2 and the justification of the negative sign of the entropy shifts at both boundaries with respect to the reference state values $\tilde{s}_B$ and $\tilde{s}_T$ and their relative magnitudes is provided below. According to the results of Sect. 3.6.3 the gradient of the mean total temperature in the bulk on Fig. 3.2c is marked weaker than that of a hydrostatic adiabatic state. This means, that the dashed line representing the horizontally shifted adiabatic profile must intersect with the dashed horizontal line $z = L - \delta_{th,T}$ at a lower temperature than that of the mean total temperature profile at the top of the bulk (the point of intersection of the two dashed lines must be to the left of the point of intersection of the bold line $T_{\text{conv.}}$ and the horizontal line $z = L - \delta_{th,T}$). We also note an interesting detail of the profiles. The relation between the temperatures at the top of the bulk, i.e. at $z = L - \delta_{th,T}$ for the hydrostatic adiabatic profile

$$T_{ad}\left(z = L - \delta_{th,T}\right) = T_B - \frac{g}{c_p}\left(L - \delta_{th,T}\right) \tag{3.261}$$

hooked at $T_B$ (continuous line) and the mean total temperature $\langle T\rangle_h\,(z = L - \delta_{th,T})$, which is suggested by the figure to be $T_{ad}(z = L - \delta_{th,T}) < \langle T\rangle_h\,(z = L - \delta_{th,T})$, in fact remains unknown and either one could be greater. In other words the point of intersection of the bold line $T_{\text{conv}}$ with the horizontal dashed line $z = L - \delta_{th,T}$ could as well be to the left of the point of intersection of the profile $T_{ad}(z)$ hooked at $T_B$ with the line $z = L - \delta_{th,T}$. The relation between $T_{ad}(z = L - \delta_{th,T})$ and $\langle T\rangle_h\,(z = L - \delta_{th,T})$ is determined by the relation between $(\Delta T)_T$ and

$$\begin{aligned}
T_{ad}\left(z = L - \delta_{th,T}\right) - T_T =& \Delta T - \frac{g}{c_p}L + \frac{g}{c_p}\delta_{th,T} \\
=& (\Delta T)_T + (\Delta T)_B + (\Delta T)_{bulk} - \frac{g}{c_p}L + \frac{g}{c_p}\delta_{th,T} \\
=& (\Delta T)_T + (\Delta T)_B - (\Delta T)_{vel} - \frac{g}{c_p}\delta_{th,B} \\
=& (\Delta T)_T + \left(\Delta T'\right)_B - (\Delta T)_{vel} + \epsilon_a\frac{T_B}{L}\delta_{th,B}
\end{aligned} \tag{3.262}$$

where we have used (3.256) and (3.258). This comes down to the relation

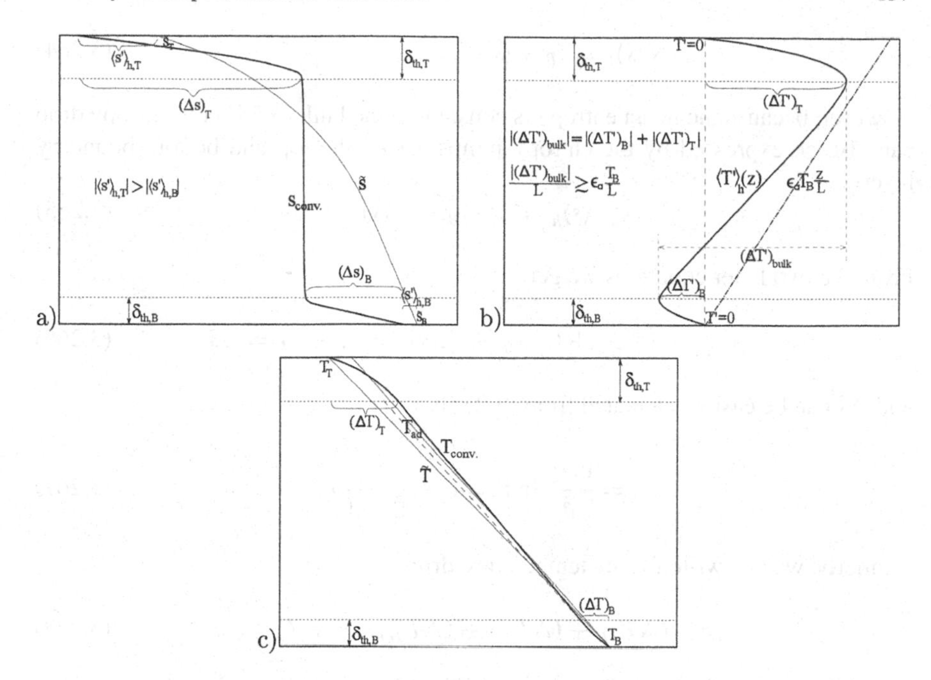


**Fig. 3.2** A schematic picture of vertical profiles of the total entropy $s_{\mathrm{conv}} = \langle s \rangle_h = \tilde{s} + \langle s' \rangle_h$ (**a**), the mean temperature fluctuation $\langle T' \rangle_h$ (**b**), and the total temperature $T_{\mathrm{conv}} = \langle T \rangle_h = \tilde{T} + \langle T' \rangle_h$ (**c**) (marked with bold lines) in developed convection with fixed temperature at boundaries and $Q = 0$. We stress, that the last figure c) is *not in scale*, and therefore may be somewhat misleading, as the superadiabaticity $\delta$, i.e. the departure of the basic profile gradient $\mathrm{d}_z\tilde{T}$ from the adiabatic one had to be made significant for clarity of presentation; in particular the jumps of the mean temperature fluctuation across the boundary layers $(\Delta T')_B$ and $(\Delta T')_T$ are in general $\delta \ll 1$ times smaller than corresponding jumps in the reference temperature profile, i.e. $\tilde{T}(z = 0) - \tilde{T}(\delta_{th,B})$ and $\tilde{T}(z = L - \delta_{th,T}) - \tilde{T}(L)$ respectively. Moreover, the non-dimensional thicknesses of the boundary layers, $\delta_{th,B}/L$ and $\delta_{th,T}/L$ must be much larger than the superadiabaticity $\delta$ for consistency of the anelastic approximation. The dashed line on figure c) represents the same gradient as that of the hydrostatic adiabatic profile $T_{ad} = T_B - gz/c_p$, but is shifted horizontally, so that it is hooked at the point $z = \delta_{th,B}$ and $T = \langle T \rangle_h\,(z = \delta_{th,B})$

$$(\Delta T)_{vel} \overset{?}{\gtrless} \left(\Delta T'\right)_B + \mathcal{O}\left(\epsilon_a \delta_{th,B} \frac{T_B}{L}\right), \tag{3.263}$$

and which of these two quantities is greater is most likely dependent on the Rayleigh number, that is on the strength of driving.

## Negative Entropy Shift at Boundaries

First we express the total mean entropy drop across the entire fluid layer directly from the definition of the total entropy $s = \tilde{s} + s'$

$$\Delta \langle s \rangle_h = \tilde{s}_B + \langle s' \rangle_{h,B} - \tilde{s}_T - \langle s' \rangle_{h,T} \,. \tag{3.264}$$

However, because the mean entropy is constant in the bulk (3.251) the entropy drop can also be expressed by the entropy jumps across the top and bottom boundary layers

$$\Delta \langle s \rangle_h = (\Delta s)_B + (\Delta s)_T \,. \tag{3.265}$$

From the two latter equations we get

$$\langle s' \rangle_{h\,T} - \langle s' \rangle_{h\,B} + (\Delta s)_B + (\Delta s)_T = \tilde{s}_B - \tilde{s}_T = \Delta\tilde{s}, \tag{3.266}$$

and $\Delta\tilde{s}$ can be easily calculated from (3.235b)

$$\Delta\tilde{s} = \frac{c_p \epsilon_a}{\tilde{\theta}} \ln \tilde{\Gamma} = c_p \epsilon_a \frac{\tilde{\Gamma}}{\tilde{\Gamma} - 1} \ln \tilde{\Gamma}. \tag{3.267}$$

Similarly, we can write for the temperature drop

$$(\Delta T)_B + (\Delta T)_T + (\Delta T)_{bulk} = \Delta T, \tag{3.268}$$

where of course $\langle T' \rangle_{h,T} = \langle T' \rangle_{h,T} = 0$ because by assumption of this Sect. 3.6.4.1 the boundaries are held at a constant temperature. Extracting the temperature fluctuation jumps one obtains

$$(\Delta T')_B + (\Delta T')_T = (\Delta T')_{bulk} \,. \tag{3.269}$$

Introducing the ratio of temperature jumps across the boundary layers

$$r_T = \frac{(\Delta T')_T}{(\Delta T')_B} > 1, \tag{3.270}$$

which in stratified developed convection is typically significantly greater than one (cf. Sect. 3.6.6), we can express the jumps of the mean temperature fluctuation by the bulk jump in the following way

$$(\Delta T')_B = \frac{1}{1 + r_T} (\Delta T')_{bulk} \,, \qquad (\Delta T')_T = \frac{r_T}{1 + r_T} (\Delta T')_{bulk} \,. \tag{3.271}$$

Consequently by the use of (3.249) the mean entropy jumps take the form

$$(\Delta s)_B \approx \frac{c_p (\Delta T')_B}{T_B} = \frac{c_p}{T_B} \frac{1}{1 + r_T} (\Delta T')_{bulk} \,, \tag{3.272a}$$

$$(\Delta s)_T \approx \frac{c_p (\Delta T')_T}{T_T} = \frac{c_p}{T_B} \frac{\tilde{\Gamma} r_T}{1 + r_T} (\Delta T')_{bulk} \,; \tag{3.272b}$$

of course here $T_B = \tilde{T}_B$ and $T_T = \tilde{T}_T$ since the boundaries are isothermal. Substitution of the latter expressions into (3.266) yields

$$\left\langle s'\right\rangle_{h,T} - \left\langle s'\right\rangle_{h,B} + \frac{c_p}{T_B}\frac{1+\tilde{\Gamma}r_T}{1+r_T}\left(\Delta T'\right)_{bulk} = c_p\epsilon_a\frac{\tilde{\Gamma}}{\tilde{\Gamma}-1}\ln\tilde{\Gamma}, \tag{3.273}$$

and since $\left(\Delta T'\right)_{bulk} = \epsilon_a T_B + (\Delta T)_{vel}$ (cf. (3.259)) we get finally

$$\left\langle s'\right\rangle_{h,T} - \left\langle s'\right\rangle_{h,B} = c_p\epsilon_a\left(\frac{\tilde{\Gamma}\ln\tilde{\Gamma}}{\tilde{\Gamma}-1} - \frac{1+\tilde{\Gamma}r_T}{1+r_T}\right) - \frac{c_p}{\tilde{T}_B}\frac{1+\tilde{\Gamma}r_T}{1+r_T}(\Delta T)_{vel}. \tag{3.274}$$

We will now show, that the right hand side of the Eq. (3.274) is negative. First we observe, that obviously

$$-\frac{c_p}{T_B}\frac{1+\tilde{\Gamma}r_T}{1+r_T}(\Delta T)_{vel} < 0, \tag{3.275}$$

so that we only need to demonstrate, that

$$\frac{\tilde{\Gamma}\ln\tilde{\Gamma}}{\tilde{\Gamma}-1} - \frac{1+\tilde{\Gamma}r_T}{1+r_T} < 0. \tag{3.276}$$

In order to do this, based on the property $r_T > 1$ let us express the temperature jump ratio $r_T$ by a positive power of $\tilde{\Gamma}$, i.e.,[17]

$$r_T = \tilde{\Gamma}^a, \qquad a \in \mathbb{R}_+. \tag{3.277}$$

With the use of the latter relation the inequality (3.276) may be cast in the following form

$$\frac{\left(\tilde{\Gamma}-1\right)\left(1+\tilde{\Gamma}^{1+a}\right)}{\tilde{\Gamma}+\tilde{\Gamma}^{1+a}} - \ln\tilde{\Gamma} > 0. \tag{3.278}$$

Since at $\tilde{\Gamma} = 1$ the left hand side (l.h.s.) of the inequality (3.278) vanishes it is enough if we prove, that the l.h.s. is a monotonically increasing function of $\tilde{\Gamma} > 1$. A straightforward calculation leads to

---

[17] In that way we imply a scaling law relation between $r_T$ and $\tilde{\Gamma}$, which will be concretised in Sect. 3.6.6.

$$\frac{\mathrm{d}}{\mathrm{d}\tilde{\Gamma}}\left[\frac{\left(\tilde{\Gamma}-1\right)\left(1+\tilde{\Gamma}^{1+a}\right)}{\tilde{\Gamma}+\tilde{\Gamma}^{1+a}}-\ln\tilde{\Gamma}\right]$$
$$=\frac{\tilde{\Gamma}\left(\tilde{\Gamma}-1\right)}{\left(\tilde{\Gamma}+\tilde{\Gamma}^{1+a}\right)^{2}}\left[\tilde{\Gamma}^{2a+1}-1+(1+a)\tilde{\Gamma}^{a}\left(\tilde{\Gamma}-1\right)\right]>0 \tag{3.279}$$

for all $\tilde{\Gamma} > 1$ which together with (3.274) and (3.275) proves, that

$$\left\langle s'\right\rangle_{h,T}-\left\langle s'\right\rangle_{h,B}<0. \tag{3.280}$$

The final step is based on an exactly analogous argumentation as in Sect. 3.3.1.1, which allows to express the values of the mean entropy fluctuation at isothermal boundaries by the values of the mean pressure fluctuation directly from (3.103) and the second relation in (3.232d) as follows

$$\left\langle s'\right\rangle_{h,B}=-\frac{\left\langle p'\right\rangle_{h,B}}{\tilde{\rho}_B T_B},\qquad \left\langle s'\right\rangle_{h,T}=-\frac{\left\langle p'\right\rangle_{h,B}}{\tilde{\rho}_T T_T}=\tilde{\Gamma}^{m+1}\left\langle s'\right\rangle_{h,B}, \tag{3.281}$$

where we have used $\tilde{\rho}_B/\tilde{\rho}_T=\tilde{\Gamma}^m>1$. Therefore it is clear, that $\left\langle s'\right\rangle_{h,T}$ and $\left\langle s'\right\rangle_{h,B}$ are of the same sign and $\left|\left\langle s'\right\rangle_{h,T}\right|>\left|\left\langle s'\right\rangle_{h,B}\right|$, which in light of (3.280) necessarily implies

$$\left\langle s'\right\rangle_{h,T}<0,\qquad \left\langle s'\right\rangle_{h,B}<0. \tag{3.282}$$

#### 3.6.4.2 The Case of Isentropic Boundaries

We now turn to the case when the entropy is fixed at boundaries for which the dynamical description of fully developed stratified convection will be further developed in the next few sections. The vertical profiles of $\langle s\rangle_h$, $\left\langle T'\right\rangle_h$ and $\langle T\rangle_h$ are shown on Fig. 3.3. When the boundaries are isentropic the temperature at the boundaries is shifted with respect to the reference state values $\tilde{T}_B$ and $\tilde{T}_T$ by $\left\langle T'\right\rangle_{h,B}$ and $\left\langle T'\right\rangle_{h,T}$ respectively, but in this case the shift is positive. We elaborate on this below. According to the observation made in Sect. 3.6.3 the gradient of the mean total temperature in the bulk on Fig. 3.3c is marked weaker than that of a hydrostatic adiabatic state. However, we stress again, that also in the current case the relation between the temperatures at the top of the bulk for the total mean $\langle T\rangle_h\,(z=L-\delta_{th,T})$ and the hydrostatic adiabatic profile hooked at $\tilde{T}_B$, i.e. $T_{ad}(z=L-\delta_{th,T})$ remains unknown. On similar grounds as in the case of isothermal boundaries explained at the beginning of Sect. 3.6.4.1 it can be shown, that the relation between $\langle T\rangle_h\,(z=L-\delta_{th,T})$ and $T_{ad}(z=L-\delta_{th,T})$ (with the latter hooked at $\tilde{T}_B$) comes down to the relation between the positive quantities $(\Delta T)_{vel}$ and $(\Delta T')_B-\left\langle T'\right\rangle_{h,B}$, which is most likely non-universal and depends

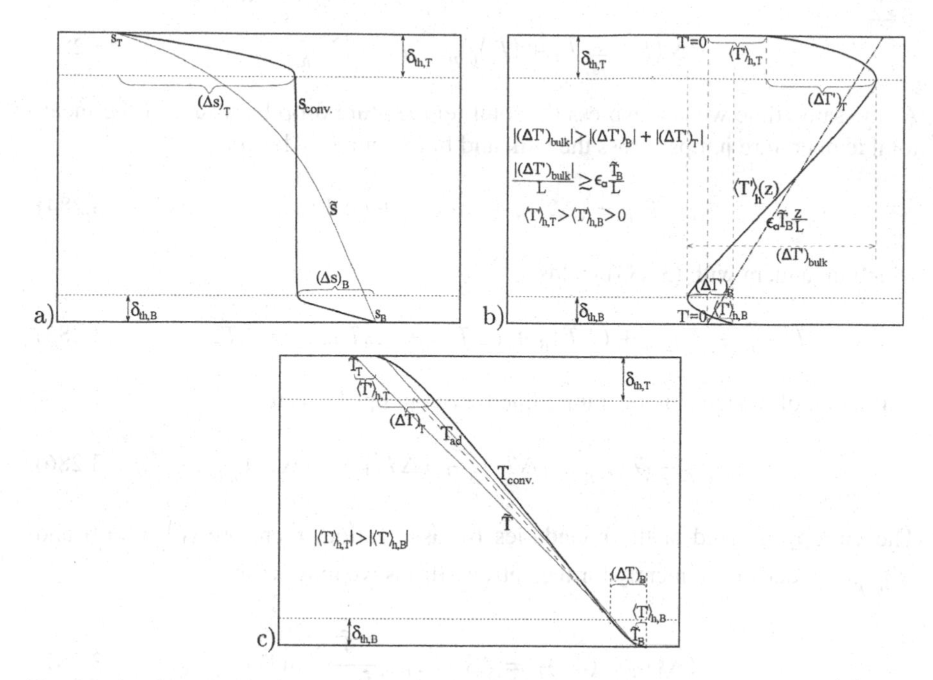

**Fig. 3.3** A schematic picture of vertical profiles of the total entropy $s_{\text{conv}} = \langle s \rangle_h = \tilde{s} + \langle s' \rangle_h$ (**a**), the mean temperature fluctuation $\langle T' \rangle_h$ (**b**) and the total temperature $T_{\text{conv}} = \langle T \rangle_h = \tilde{T} + \langle T' \rangle_h$ (**c**) (marked with bold lines) in developed convection with fixed entropy at boundaries and $Q = 0$. Note, that the last figure (**c**) is *not in scale*, as the superadiabaticity $\delta$, i.e. the departure of the basic profile gradient $d_z\tilde{T}$ from the adiabatic one had to be made significant for clarity of presentation; in particular the jumps of the mean temperature fluctuation across the boundary layers $(\Delta T')_B$ and $(\Delta T')_T$ are in general $\delta \ll 1$ times smaller than corresponding jumps in the reference temperature profile, i.e. $\tilde{T}(z = 0) - \tilde{T}(\delta_{th,B})$ and $\tilde{T}(z = L - \delta_{th,T}) - \tilde{T}(L)$ respectively. Moreover, the non-dimensional thicknesses of the boundary layers, $\delta_{th,B}/L$ and $\delta_{th,T}/L$ must be much larger than the superadiabaticity $\delta$ for consistency of the anelastic approximation. The dashed line on figure (**c**) represents the same gradient as that of the hydrostatic adiabatic profile $T_{ad}$, but is shifted horizontally, so that it is hooked at the point $z = \delta_{th,B}$ and $T = \langle T \rangle_h (z = \delta_{th,B})$

on the Rayleigh number. Moreover, it should be realized, that also the relations between the temperature fluctuation jumps across the boundary layers and the temperature shifts at boundaries, i.e. between $(\Delta T')_B$ and $\langle T' \rangle_{h,B}$ and likewise between $(\Delta T')_T$ and $\langle T' \rangle_{h,T}$, can not be easily established; in particular this implies, that the mean temperature fluctuation profile $\langle T' \rangle_h (z)$ on Fig. 3.3b does not necessarily cross the vertical dashed line indicating $T' = 0$ and it is allowed, that $\langle T' \rangle_h (z) > 0$ for all $z$.

### Positive Temperature Shift at Boundaries

The mean total temperature drop across the entire fluid layer can be expressed directly from the definition of the total temperature $T = \tilde{T} + T'$ in the following way

$$\Delta \langle T \rangle_h = \tilde{T}_B + \langle T' \rangle_{h,B} - \tilde{T}_T - \langle T' \rangle_{h,T}. \tag{3.283}$$

At the same time we can express the total temperature drop by a sum of the mean total temperature jumps across the bulk and both boundary layers

$$\Delta \langle T \rangle_h = (\Delta T)_B + (\Delta T)_T + (\Delta T)_{bulk}, \tag{3.284}$$

which in tandem with (3.283) yields

$$\langle T' \rangle_{h,T} - \langle T' \rangle_{h,B} + (\Delta T)_B + (\Delta T)_T + (\Delta T)_{bulk} = \Delta \tilde{T}. \tag{3.285}$$

Extraction of the mean temperature fluctuation jumps leads to

$$\langle T' \rangle_{h,T} - \langle T' \rangle_{h,B} + (\Delta T')_B + (\Delta T')_T = (\Delta T')_{bulk}. \tag{3.286}$$

The entropy is fixed at the boundaries by assumption, therefore $\langle s' \rangle_{h,B} = 0$ and $\langle s' \rangle_{h,T} = 0$ and consequently for the entropy drops we may write

$$(\Delta s)_B + (\Delta s)_T = \Delta \tilde{s} = c_p \epsilon_a \frac{\tilde{\Gamma}}{\tilde{\Gamma} - 1} \ln \tilde{\Gamma}, \tag{3.287}$$

where $(\Delta s)_{bulk} \approx 0$ is negligibly small due to efficient advection of entropy in the bulk (cf. (3.251)). Next we introduce the ratio of mean total entropy jumps across the boundary layers,

$$r_s = \frac{(\Delta s)_T}{(\Delta s)_B} = \tilde{\Gamma} r_T > 1, \qquad r_T = \frac{(\Delta T')_T}{(\Delta T')_B} > 1, \tag{3.288}$$

which typically in developed stratified convection significantly exceeds unity (cf. Sect. 3.6.6) and which is related by (3.249) to the temperature jump ratio $r_T$ already introduced in (3.270). By the use of (3.249), (3.287) and (3.288) the mean temperature fluctuation jumps across the boundary layers $(\Delta T)_B$ and $(\Delta T)_T$ can be expressed as follows

$$(\Delta T')_B \approx \frac{\tilde{T}_B (\Delta s)_B}{c_p} = \epsilon_a \tilde{T}_B \frac{\tilde{\Gamma} \ln \tilde{\Gamma}}{\left(1 + \tilde{\Gamma} r_T\right)\left(\tilde{\Gamma} - 1\right)}, \tag{3.289a}$$

$$(\Delta T')_T \approx \frac{\tilde{T}_T (\Delta s)_T}{c_p} = \epsilon_a \tilde{T}_B \frac{\tilde{\Gamma} r_T \ln \tilde{\Gamma}}{\left(1 + \tilde{\Gamma} r_T\right)\left(\tilde{\Gamma} - 1\right)}. \tag{3.289b}$$

Substituting the latter expressions into (3.286) and making use of $(\Delta T')_{bulk} = \epsilon_a \tilde{T}_B + (\Delta T)_{vel}$ (cf. (3.259)) we obtain

$$\langle T' \rangle_{h,T} - \langle T' \rangle_{h,B} = \epsilon_a \tilde{T}_B \left[ 1 - \frac{\tilde{\Gamma}\,(1 + r_T) \ln \tilde{\Gamma}}{\left(1 + \tilde{\Gamma} r_T\right)\left(\tilde{\Gamma} - 1\right)} \right] + (\Delta T)_{vel} . \tag{3.290}$$

Since we already know from (3.276)–(3.279) that

$$\frac{\tilde{\Gamma}\,(1 + r_T) \ln \tilde{\Gamma}}{\left(1 + \tilde{\Gamma} r_T\right)\left(\tilde{\Gamma} - 1\right)} < 1, \tag{3.291}$$

and since $(\Delta T)_{vel} > 0$ (cf. (3.258)) it is evident, that

$$\langle T' \rangle_{h,T} - \langle T' \rangle_{h,B} > 0. \tag{3.292}$$

Finally we make use of the observation made below (3.104a, 3.104b), that at the isentropic boundaries $\langle T' \rangle_{h,T}$ and $\langle T' \rangle_{h,B}$ are of the same sign and $\left|\langle T' \rangle_{h,T}\right| > \left|\langle T' \rangle_{h,B}\right|$, which allows to conclude that

$$\langle T' \rangle_{h,T} > 0, \qquad \langle T' \rangle_{h,B} > 0. \tag{3.293}$$

The actual values of $\langle T' \rangle_{h,T}$ and $\langle T' \rangle_{h,B}$ can be easily expressed in terms of the ratio $r_T$ (or alternatively $r_s$), $\tilde{\Gamma}$, $\epsilon_a \tilde{T}_B$ and $(\Delta T)_{vel}$ by the use of (3.290) and $\langle T' \rangle_{h,T} = \tilde{\Gamma}^m \langle T' \rangle_{h,B}$ (cf. (3.104b)).

### 3.6.5 The Nusselt, Rayleigh and Reynolds Numbers

At the top and bottom boundaries all the heat is carried by conduction, thus the horizontally averaged superadiabatic heat flux at the bottom boundary is

$$F_S(z = 0) = -k \left. \frac{\mathrm{d} \langle T_S \rangle_h}{\mathrm{d}z} \right|_{z=0} . \tag{3.294}$$

Of course in a (statistically) stationary turbulent state the superadiabatic flux at the bottom equals that at the top, as is clear from (3.145) and the fact that under current assumptions the superadiabatic flux of the conduction reference state,

$$\tilde{F}_S = -k \frac{\mathrm{d}}{\mathrm{d}z} \left( \tilde{T} - T_{ad} \right) = k \epsilon_a \frac{\tilde{T}_B}{L}, \tag{3.295}$$

is independent of height. As a result

$$F_S(z = 0) = -k \left. \frac{\mathrm{d} \langle T_S \rangle_h}{\mathrm{d}z} \right|_{z=0} = -k \left. \frac{\mathrm{d} \langle T_S \rangle_h}{\mathrm{d}z} \right|_{z=L} = F_S(z = L). \tag{3.296}$$

This allows to define the Nusselt number $Nu$ as the ratio of the mean superadiabatic heat flux either at the top or at the bottom boundary in a stationary convective state divided by the superadiabatic heat flux in the conduction state (3.295), which yields

$$Nu = \frac{F_S(z=0)}{\tilde{F}_S} = \frac{-L\left.\frac{\mathrm{d}\langle T_S\rangle_h}{\mathrm{d}z}\right|_{z=0}}{\epsilon_a \tilde{T}_B} \approx \frac{(\Delta T_S)_B\, L}{\epsilon_a \tilde{T}_B \delta_{th,B}}, \tag{3.297}$$

or equivalently

$$Nu = \frac{F_S(z=L)}{\tilde{F}_S} = \frac{-L\left.\frac{\mathrm{d}\langle T_S\rangle_h}{\mathrm{d}z}\right|_{z=L}}{\epsilon_a \tilde{T}_B} \approx \frac{(\Delta T_S)_T\, L}{\epsilon_a \tilde{T}_B \delta_{th,T}}. \tag{3.298}$$

The so-defined Nusselt number is unity at convection onset and large in fully developed convection. Similarly as it is done in the theory of Boussinesq turbulent convection, in the above definition of the Nusselt number we have approximated the bottom and top values of the superadiabatic temperature gradient with the ratios of the superadiabatic temperature jumps across the boundary layers to the respective thicknesses of boundary layers, so that

$$-\left.\frac{\mathrm{d}\langle T_S\rangle_h}{\mathrm{d}z}\right|_{z=0} \approx \frac{(\Delta T_S)_B}{\delta_{th,B}}, \quad -\left.\frac{\mathrm{d}\langle T_S\rangle_h}{\mathrm{d}z}\right|_{z=L} \approx \frac{(\Delta T_S)_T}{\delta_{th,T}}. \tag{3.299}$$

The Rayleigh number is a measure of the strength of the thermal driving. For the case of isothermal boundaries it was defined in (3.157), however, we now consider the case of isentropic boundaries, when the flow is driven by the entropy jump across the fluid layer, that is $\Delta\tilde{s}$ (cf. (3.267)). Therefore we define the Rayleigh number in the following way

$$Ra = \frac{g\,\Delta\tilde{s}L^3\tilde{\rho}_B^2}{\mu k} \approx \frac{c_p\,\Delta\tilde{s}\,\Delta\tilde{T}L^2\tilde{\rho}_B^2}{\mu k}, \tag{3.300}$$

where in obtaining the second expression we have used $g/c_p = \Delta\tilde{T}/L + \mathcal{O}(\epsilon_a g/c_p)$.

Furthermore, the central idea of the theory of fully developed Boussinesq convection is based on the assumption that the structure of turbulent convective flow is always characterized by the presence of a large-scale convective roll called the *wind of turbulence*. This idea, which in the non-stratified case stems from a vast numerical and experimental evidence, is retained in the case of anelastic convection, however, it must be realized that the significant stratification in the anelastic case breaks the Boussinesq up-down symmetry. Thus we must distinguish between the magnitude of the wind of turbulence near the bottom of the bulk and its magnitude near the top of the bulk, denoted by $U_B$ and $U_T$ respectively, which can now significantly differ (cf. e.g. Jones et al. 2020 for numerical evidence of the 'wind of turbulence' with height-dependent magnitude in stratified convection). With the use of the introduced notation we define the bottom and top Reynolds numbers

$$Re_B = \frac{U_B L \tilde{\rho}_B}{\mu}, \qquad Re_T = \frac{U_T L \tilde{\rho}_T}{\mu}. \tag{3.301}$$

We emphasize, that these Reynolds numbers are based on the large length scale $L$, and such an approach is consistent with the up-to-date available data from numerical simulations of strongly stratified convective flows (cf. e.g. Verhoeven et al. 2015; Jones et al. 2020); these data, however, were obtained for a rather weak driving, as large Rayleigh numbers are currently unachievable for large stratifications, thus it is not clear if they yet correspond to the problem of fully developed, strongly stratified convection. In fact it is possible that at large stratifications both the horizontal and vertical length scales of variation of the large scale convective flow $\langle \mathbf{u}(\mathbf{x}, t)\rangle_t$ (where $\langle\cdot\rangle_t$ denotes a time average) are determined by the scale heights $D_p$, $D_\rho$ and $D_T$; these scale heights vary with height and are significantly smaller at the top, than at the bottom of the fluid domain. It seems rather likely that at larger stratifications the wind of turbulence becomes split into two or more, still large scale convection rolls, whose sizes are determined by the values of the scale heights. However, since currently there is not enough evidence, neither numerical nor experimental and observational for vertical splitting of large scale rolls in strongly stratified convection, it seems more reasonable to postulate the total depth of the layer as the typical length scale of horizontal variation.

#### 3.6.5.1 Thicknesses of Thermal Boundary Layers, $\delta_{th,B}$ and $\delta_{th,T}$

It is possible to express the Nusselt number by the mean temperature fluctuation jumps likewise by the entropy jumps across the boundary layers, which in turn allows to directly relate the thicknesses of the thermal boundary layers to the Nusselt number. By the use of (3.243) we get (cf. also Fig. 3.3b)

$$\begin{aligned}(\Delta T_S)_B &= \langle T_S\rangle_h (z=0) - \langle T_S\rangle_h \left(z=\delta_{th,B}\right)\\ &= \langle T'\rangle_h (z=0) - \langle T'\rangle_h \left(z=\delta_{th,B}\right) + \epsilon_a \tilde{T}_B \frac{\delta_{th,B}}{L}\\ &= \left(\Delta T'\right)_B + \epsilon_a \frac{\delta_{th,B}}{L}\tilde{T}_B,\end{aligned} \tag{3.302a}$$

$$\begin{aligned}(\Delta T_S)_T &= \langle T_S\rangle_h \left(z=L-\delta_{th,T}\right) - \langle T_S\rangle_h (z=L)\\ &= \langle T'\rangle_h \left(z=L-\delta_{th,T}\right) - \langle T'\rangle_h (z=L) + \epsilon_a \tilde{T}_B \frac{\delta_{th,T}}{L}\\ &= \left(\Delta T'\right)_T + \epsilon_a \frac{\delta_{th,T}}{L}\tilde{T}_B,\end{aligned} \tag{3.302b}$$

hence the superadiabatic temperature jumps and the temperature fluctuation jumps are equal at leading order, since the corrections are much smaller. Next we can substitute for $\left(\Delta T'\right)_B$ and $\left(\Delta T'\right)_T$ from (3.249) and using (3.287) and (3.288) obtain

$$(\Delta T_S)_B \approx \frac{\tilde{T}_B}{c_p}\frac{\Delta\tilde{s}}{1+r_s} + \epsilon_a \frac{\delta_{th,B}}{L}\tilde{T}_B, \tag{3.303a}$$

$$(\Delta T_S)_T \approx \frac{\tilde{T}_T}{c_p}\frac{r_s\Delta\tilde{s}}{1+r_s} + \epsilon_a \frac{\delta_{th,T}}{L}\tilde{T}_B. \tag{3.303b}$$

Finally we substitute for $(\Delta T_S)_B$ and $(\Delta T_S)_T$ from the latter equations in the Eqs. (3.297) and (3.298), which with the aid of (3.267) yields

$$Nu \approx \frac{\tilde{\Gamma}\ln\tilde{\Gamma}}{(1+r_s)\left(\tilde{\Gamma}-1\right)}\frac{L}{\delta_{th,B}} + 1, \tag{3.304}$$

$$Nu \approx \frac{r_s\ln\tilde{\Gamma}}{(1+r_s)\left(\tilde{\Gamma}-1\right)}\frac{L}{\delta_{th,T}} + 1. \tag{3.305}$$

This leads to the following expressions for thicknesses of thermal boundary layers

$$\frac{\delta_{th,B}}{L} \approx \frac{\tilde{\Gamma}\ln\tilde{\Gamma}}{(1+r_s)\left(\tilde{\Gamma}-1\right)}(Nu+1)^{-1} \approx \frac{\tilde{\Gamma}\ln\tilde{\Gamma}}{(1+r_s)\left(\tilde{\Gamma}-1\right)}Nu^{-1}, \tag{3.306a}$$

$$\frac{\delta_{th,T}}{L} \approx \frac{r_s\ln\tilde{\Gamma}}{(1+r_s)\left(\tilde{\Gamma}-1\right)}(Nu+1)^{-1} \approx \frac{r_s\ln\tilde{\Gamma}}{(1+r_s)\left(\tilde{\Gamma}-1\right)}Nu^{-1}, \tag{3.306b}$$

where the last approximations concerning neglection of unity with respect to the Nusselt number were made based on the fact, that in fully developed convection the Nusselt number is large, $Nu \gg 1$ (they are of course equivalent to neglection of the small corrections in (3.303a, 3.303b)). The ratio of the thicknesses of the thermal boundary layers is now easily expressed by $r_s$

$$r_\delta \stackrel{\text{def.}}{=} \frac{\delta_{th,T}}{\delta_{th,B}} \approx \tilde{\Gamma}^{-1} r_s = r_T. \tag{3.307}$$

#### 3.6.5.2 Thicknesses of the Viscous Boundary Layers $\delta_{\nu,B}$ and $\delta_{\nu,T}$

The viscous boundary layers are assumed laminar, as the intuition developed from the Boussinesq theory suggests, that the Rayleigh numbers necessary for the boundary layers to become turbulent are huge. We therefore concentrate on the case, when the Rayleigh number is large enough for convection to be already fully turbulent, but at the same time does not exceed a critical much larger value above which the boundary layers become turbulent. Consequently the thicknesses of the viscous boundary layers are defined according to the standard laminar Blasius theory

$$\frac{\delta_{\nu,B}}{L} = Re_B^{-1/2}, \qquad \frac{\delta_{\nu,T}}{L} = Re_T^{-1/2}. \tag{3.308}$$

### 3.6.6 *Estimates of the Mean Superadiabatic Heat Flux in a Fully Developed State*

The first formula for the mean superadiabatic heat flux (3.150) for the analysed case of an ideal gas with isentropic boundaries allows to write

$$\begin{aligned}
F_S(z=0) = &-k\frac{\mathrm{d}}{\mathrm{d}z}\left(\tilde{T} + \langle T'\rangle_h - T_{ad}\right)\Big|_{z=0} \\
= &-k\frac{\mathrm{d}}{\mathrm{d}z}\left(\tilde{T} + \langle T'\rangle_h - T_{ad}\right) + \tilde{\rho}\tilde{T}\left\langle u_z s'\right\rangle_h - \int_0^z \tilde{\rho}\frac{\mathrm{d}\tilde{T}}{\mathrm{d}z}\left\langle u_z s'\right\rangle_h \mathrm{d}z \\
&-\mu\int_0^z \langle q\rangle_h \,\mathrm{d}z - 2\mu\left[\frac{1}{2}\frac{\mathrm{d}\left\langle u_z^2\right\rangle_h}{\mathrm{d}z} - \frac{m\Delta\tilde{T}}{L\tilde{T}}\left\langle u_z^2\right\rangle_h\right],
\end{aligned} \tag{3.309}$$

whereas the second formula for the mean superadiabatic heat flux (3.152) implies

$$\begin{aligned}
F_S(z=0) = &-k\frac{\mathrm{d}}{\mathrm{d}z}\left(\tilde{T} + \langle T'\rangle_h - T_{ad}\right)\Big|_{z=0} \\
= &-k\frac{\tilde{T}_B}{\tilde{T}}\frac{\mathrm{d}}{\mathrm{d}z}\left(\tilde{T} + \langle T'\rangle_h - T_{ad}\right) + \tilde{\rho}\tilde{T}_B\left\langle u_z s'\right\rangle_h - \mu\int_0^z \frac{\tilde{T}_B}{\tilde{T}}\langle q\rangle_h \,\mathrm{d}z \\
&-\epsilon_a k\frac{\tilde{T}_B}{L}\left(\frac{\tilde{T}_B}{\tilde{T}} - 1\right) + k\tilde{T}_B\frac{\Delta\tilde{T}}{L}\left(\frac{\langle T'\rangle_h}{\tilde{T}^2} - \frac{\langle T'\rangle_{h,B}}{\tilde{T}_B^2}\right) \\
&-2k\tilde{T}_B\left(\frac{\Delta\tilde{T}}{L}\right)^2\int_0^z \frac{\langle T'\rangle_h}{\tilde{T}^3}\mathrm{d}z - 2\mu\left[\frac{1}{2}\frac{\tilde{T}_B}{\tilde{T}}\frac{\mathrm{d}\left\langle u_z^2\right\rangle_h}{\mathrm{d}z}\right. \\
&\left.-\left(m+\frac{1}{2}\right)\frac{\tilde{T}_B\,\Delta\tilde{T}}{\tilde{T}^2 L}\left\langle u_z^2\right\rangle_h + (m+1)\,\tilde{T}_B\left(\frac{\Delta\tilde{T}}{L}\right)^2\int_0^z \frac{\left\langle u_z^2\right\rangle_h}{\tilde{T}^3}\mathrm{d}z\right],
\end{aligned} \tag{3.310}$$

where we have introduced

$$q = \nabla \mathbf{u} : \nabla \mathbf{u} + \frac{1}{3} (\nabla \cdot \mathbf{u})^2 . \tag{3.311}$$

Before we proceed let us first demonstrate that in fact a lot of terms in (3.309) and (3.310) are negligible in comparison to the mean superadiabatic heat flux $F_S (z = 0) = F_S(z = L)$ entering the system at the bottom, or leaving at the top; the latter two are equal in a stationary state according to the Eq. (3.145) with subtracted constant adiabatic gradient $-g/c_P$ from both sides or by (3.309) taken at $z = L$. We start with the sum of three temperature terms in (3.310)

$$\Sigma_T \overset{\text{def.}}{=} -\epsilon_a k \frac{\tilde{T}_B}{L} \left( \frac{\tilde{T}_B}{\tilde{T}} - 1 \right) + k\tilde{T}_B \frac{\Delta \tilde{T}}{L} \left( \frac{\langle T' \rangle_h}{\tilde{T}^2} - \frac{\langle T' \rangle_{h,B}}{\tilde{T}_B^2} \right)$$

$$- 2k\tilde{T}_B \left( \frac{\Delta \tilde{T}}{L} \right)^2 \int_0^z \frac{\langle T' \rangle_h}{\tilde{T}^3} \mathrm{d}z, \tag{3.312}$$

which operatively was denoted by $\Sigma_T$. Examination of the mean temperature fluctuation profile (Fig. 3.3b) allows to estimate the integral from above

$$-2k\tilde{T}_B \left( \frac{\Delta \tilde{T}}{L} \right)^2 \int_0^z \frac{\langle T' \rangle_h}{\tilde{T}^3} \mathrm{d}z \leq -2k\tilde{T}_B \left( \frac{\Delta \tilde{T}}{L} \right)^2 \left[ \langle T' \rangle_{h,B} - \left( \Delta T' \right)_B \right] \int_0^z \frac{\mathrm{d}z}{\tilde{T}^3}$$

$$= k \frac{\tilde{\theta}}{L} \left[ \left( \Delta T' \right)_B - \langle T' \rangle_{h,B} \right] \left( \frac{\tilde{T}_B^2}{\tilde{T}^2} - 1 \right). \tag{3.313}$$

The second term in the sum (3.312) can be bounded from above by substituting the maximal value of the mean temperature fluctuation, which according to the vertical profile of $\langle T' \rangle_h (z)$ sketched on Fig. 3.3b is estimated at $\left( \Delta T' \right)_T + \langle T' \rangle_{h,T}$. Consequently the sum $\Sigma_T$ satisfies

$$\Sigma_T \leq -\epsilon_a k \frac{\tilde{T}_B}{L} \left( \frac{\tilde{T}_B}{\tilde{T}} - 1 \right) + k \frac{\tilde{\theta}}{L} \frac{\tilde{T}_B^2}{\tilde{T}^2} \left[ \left( \Delta T' \right)_T + \langle T' \rangle_{hT} - \langle T' \rangle_{h,B} \right]$$

$$+ k \frac{\tilde{\theta}}{L} \left( \Delta T' \right)_B \left( \frac{\tilde{T}_B^2}{\tilde{T}^2} - 1 \right), \tag{3.314}$$

and with the aid of (3.286) and (3.259) we can write

$$\Sigma_T \leq -\epsilon_a k \frac{\tilde{T}_B}{L} \left( \frac{\tilde{T}_B}{\tilde{T}} - 1 \right) + k \frac{\tilde{\theta}}{L} \frac{\tilde{T}_B^2}{\tilde{T}^2} \left( \Delta T' \right)_{bulk} - k \frac{\tilde{\theta}}{L} \left( \Delta T' \right)_B$$

$$= \epsilon_a k \frac{\tilde{T}_B}{L} \left( \tilde{\theta} \frac{\tilde{T}_B^2}{\tilde{T}^2} - \frac{\tilde{T}_B}{\tilde{T}} + 1 \right) + k \frac{\tilde{\theta}}{L} \frac{\tilde{T}_B^2}{\tilde{T}^2} (\Delta T)_{vel} - k\tilde{\theta} \frac{\left( \Delta T' \right)_B}{\delta_{th,B}} \frac{\delta_{th,B}}{L}. \tag{3.315}$$

It follows from the definition (3.258), that

$$(\Delta T)_{vel} = \int_0^L \frac{\left\langle u_z^2 \right\rangle_h}{c_p D_\rho} \mathrm{d}z \leq \frac{m\left(\tilde{\Gamma} - 1\right)}{c_p} \left\langle u_z^2 \right\rangle, \tag{3.316}$$

which in turn allows to write

$$\begin{aligned}\Sigma_T \leq & \frac{\epsilon_a k \frac{\tilde{T}_B}{L}\left(\tilde{\Gamma}^2 - \tilde{\Gamma} + 1\right)}{k\frac{(\Delta T_S)_B}{\delta_{th,B}}} k\frac{(\Delta T_S)_B}{\delta_{th,B}} + \frac{mk}{c_p L}\tilde{\Gamma}\left(\tilde{\Gamma} - 1\right)^2 \left\langle u_z^2 \right\rangle \\ & - \frac{\tilde{\Gamma} - 1}{\tilde{\Gamma}} k\frac{(\Delta T_S)_B}{\delta_{th,B}} \frac{\delta_{th,B}}{L}\end{aligned} \tag{3.317}$$

and finally

$$\begin{aligned}\Sigma_T \leq & \frac{mk}{c_p L}\tilde{\Gamma}\left(\tilde{\Gamma} - 1\right)^2 \left\langle u_z^2 \right\rangle + \left[\left(\tilde{\Gamma}^2 - \tilde{\Gamma} + 1\right) Nu^{-1} - \frac{\tilde{\Gamma} - 1}{\tilde{\Gamma}} \frac{\delta_{th,B}}{L}\right] F_S(z=0) \\ = & \frac{mk}{c_p L}\tilde{\Gamma}\left(\tilde{\Gamma} - 1\right)^2 \left\langle u_z^2 \right\rangle + \mathcal{O}\left(\frac{\delta_{th,B}}{L} F_S(z=0)\right),\end{aligned} \tag{3.318}$$

where we have used (3.297) and (3.306a) to write

$$\mathcal{O}\left(Nu^{-1} F_S(z=0)\right) = \mathcal{O}\left(\frac{\delta_{th,B}}{L} F_S(z=0)\right). \tag{3.319}$$

It remains to prove, that the velocity term in (3.318),

$$\frac{mk}{c_p L}\tilde{\Gamma}\left(\tilde{\Gamma} - 1\right)^2 \left\langle u_z^2 \right\rangle, \tag{3.320}$$

is also negligibly small compared to the superadiabatic flux $F_S(z=0)$, which we demonstrate along with negligibility of the viscous terms in (3.309),

$$-2\mu \left[\frac{1}{2}\frac{\mathrm{d}\left\langle u_z^2 \right\rangle_h}{\mathrm{d}z} - \frac{m\Delta\tilde{T}}{L\tilde{T}} \left\langle u_z^2 \right\rangle_h\right] \ll F_S(z=0), \tag{3.321}$$

and in (3.310)

$$-2\mu \left[\frac{1}{2}\frac{\tilde{T}_B}{\tilde{T}}\frac{\mathrm{d}\left\langle u_z^2 \right\rangle_h}{\mathrm{d}z} - \left(m + \frac{1}{2}\right)\frac{\tilde{T}_B \Delta\tilde{T}}{\tilde{T}^2 L} \left\langle u_z^2 \right\rangle_h\right.$$

$$+ (m+1)\,\tilde{T}_B \left(\frac{\Delta \tilde{T}}{L}\right)^2 \int_0^z \frac{\left\langle u_z^2 \right\rangle_h}{\tilde{T}^3} \mathrm{d}z \Bigg] \ll F_S(z=0). \tag{3.322}$$

The terms are either of order $\sim \mu \mathrm{d}\left\langle u_z^2 \right\rangle_h / \mathrm{d}z$, or $\sim \mu \left\langle u_z^2 \right\rangle_h / L$, or $\sim k \left\langle u_z^2 \right\rangle / c_p L$, hence it is enough if we demonstrate that the squared vertical velocity averaged over the horizontal plane and its first '$z$'-derivative multiplied by the dissipative coefficients $\mu$ or $k/c_p$ are negligible compared to the viscous dissipation terms

$$Q_\nu(z) \overset{\text{def.}}{=} \mu \int_0^z \langle q \rangle_h \, \mathrm{d}z \quad \text{and} \quad Q_{\nu/T}(z) \overset{\text{def.}}{=} \mu \int_0^z \frac{\tilde{T}_B}{\tilde{T}} \langle q \rangle_h \, \mathrm{d}z. \tag{3.323}$$

First of all it is important to realize that in the viscous boundary layers (of thicknesses $\delta_{\nu,B}$ and $\delta_{\nu,T}$) and hence also at the bottom and at the top of the bulk, i.e. in the vicinity of $z = \delta_{\nu,B}$ and $z = L - \delta_{\nu,T}$, the mass conservation constraint $\nabla \cdot (\tilde{\rho}\mathbf{u}) = 0$ implies that the vertical velocity $u_z$ must be very small, of the order of $\delta_{\nu,B}/L$ and $\delta_{\nu,T}/L$ at the bottom and top respectively (cf. (3.245)). The viscous dissipation is either dominated by the contributions from viscous boundary layers or from the bulk, therefore in the former case it is straightforward to see that the terms in (3.321), likewise the terms in (3.320) and (3.322) are $\mathcal{O}(\delta_{\nu,i}) = \mathcal{O}(Re_i^{-1/2})$ times smaller than

$$Q_\nu\left(z \gtrsim \delta_{\nu,B}\right),\ Q_\nu\left(z \lesssim L - \delta_{\nu,T}\right); \qquad Q_{\nu/T}\left(z \gtrsim \delta_{\nu,B}\right),\ Q_{\nu/T}\left(z \lesssim L - \delta_{\nu,T}\right), \tag{3.324}$$

respectively, because of smallness of $\left\langle u_z^2 \right\rangle_h$ at the bottom and top of the bulk (see (3.323) for definitions of $Q_\nu(z)$ and $Q_{\nu/T}(z)$). More precisely all the aforementioned terms can be estimated as follows

$$\frac{\mu \left\langle u_z^2 \right\rangle_h}{L} \sim \frac{\mu}{L} \frac{\delta_{\nu,i}^2}{L^2} U_i^2 = \frac{\mu^3}{\tilde{\rho}_i^2 L^3} Re_i, \tag{3.325a}$$

$$\mu \frac{\mathrm{d}\left\langle u_z^2 \right\rangle_h}{\mathrm{d}z} \lesssim \frac{\mu}{L} \frac{\delta_{\nu,i}}{L} U_i^2 = \frac{\mu^3}{\tilde{\rho}_i^2 L^3} Re_i^{3/2}, \tag{3.325b}$$

$$\begin{aligned}
\mu \tilde{T}_B \left(\frac{\Delta \tilde{T}}{L}\right)^2 \int_0^z \frac{\left\langle u_z^2 \right\rangle_h}{\tilde{T}^3} \mathrm{d}z &\le \frac{\mu}{L} \tilde{\Gamma}^3 \tilde{\theta}^2 \frac{1}{L} \int_0^z \left\langle u_z^2 \right\rangle_h \mathrm{d}z \\
&\sim \frac{\mu}{L} \tilde{\Gamma}^3 \tilde{\theta}^2 \frac{\delta_{\nu,i}^3}{L^3} U_i^2 \\
&\sim \frac{\mu^3}{\tilde{\rho}_i^2 L^3} \tilde{\Gamma} \left(\tilde{\Gamma} - 1\right)^2 Re_i^{1/2},
\end{aligned} \tag{3.325c}$$

$$\frac{k\left\langle u_z^2\right\rangle}{c_p L} = Pr^{-1}\frac{\mu}{L}\left\langle u_z^2\right\rangle \sim Pr^{-1}\frac{\mu}{L}\left(\frac{\delta_{\nu,B}^3}{L^3}U_B^2 + \frac{\delta_{\nu,T}^3}{L^3}U_T^2\right)$$

$$\sim \frac{\mu^3}{\tilde{\rho}_B^2 L^3}\tilde{\Gamma}\left(\tilde{\Gamma}-1\right)^2 Pr^{-1}\left(Re_B^{1/2} + \frac{\tilde{\rho}_B^2}{\tilde{\rho}_T^2}Re_T^{1/2}\right), \quad (3.325d)$$

where

$$Pr = \frac{c_p \mu}{k}, \quad (3.326)$$

is the Prandtl number, $U_i$ and $\tilde{\rho}_i$ are the maximal horizontally averaged velocity and reference density either at the top or the bottom of the bulk, whichever leads to a larger estimate and $Re_i$ is the Reynolds number based on them.

The second case, when the viscous dissipation takes place predominantly in the bulk is a little bit more subtle. The viscous dissipation terms $Q_\nu(L)$ and $Q_{\nu/T}(L)$ can be estimated in a similar way as for the Boussinesq convection (cf. Grossmann and Lohse 2000), i.e. by the use of the fact, that in such a case the dissipative effects are expected to balance the inertial effects in the bulk,

$$Q_\nu(L) \sim Q_{\nu/T}(L) \sim \tilde{\rho}_i U_i^3 = \frac{\mu^3}{\tilde{\rho}_i^2 L^3}Re_i^3. \quad (3.327)$$

The same estimate can, in fact, be obtained by introducing the Kolmogorov cascade picture and thus taking the Kolmogorov scale for velocity $u_K = U_i Re_i^{-1/4}$ and the dynamical length scale $l_K = L Re_i^{-3/4}$ to estimate dissipation, i.e.

$$Q_\nu(L) \sim Q_{\nu/T}(L) \sim \mu L \frac{u_K^2}{l_K^2} = \frac{\mu^3}{\tilde{\rho}_i^2 L^3}Re_i^3. \quad (3.328)$$

This idea provides also estimates for

$$\frac{\mu\left\langle u_z^2\right\rangle_h}{L} \sim \mu\tilde{T}_B\left(\frac{\Delta\tilde{T}}{L}\right)^2\int_0^z \frac{\left\langle u_z^2\right\rangle_h}{\tilde{T}^3}\mathrm{d}z \sim \mu\frac{u_K^2}{L} = \frac{\mu^3}{\tilde{\rho}_i^2 L^3}Re_i^{3/2}, \quad (3.329a)$$

$$\mu\frac{\mathrm{d}\left\langle u_z^2\right\rangle_h}{\mathrm{d}z} \sim \mu\frac{u_K^2}{l_K} = \frac{\mu^3}{\tilde{\rho}_i^2 L^3}Re_i^{9/4}, \quad (3.329b)$$

$$\frac{k\left\langle u_z^2\right\rangle}{c_p L} \sim \frac{k}{c_p}\frac{u_K^2}{L} = \frac{\mu^3}{\tilde{\rho}_i^2 L^3}Pr^{-1}Re_i^{3/2}, \quad (3.329c)$$

in the bulk. Finally by the use of (3.300), (3.297) and (3.267) one obtains

$$\begin{aligned}\frac{\mu^3}{\tilde{\rho}_i^2 L^3} &\approx \frac{\mu^3}{\tilde{\rho}_i^2 L^2 k\epsilon_a \tilde{T}_B} \frac{k\epsilon_a \tilde{T}_B \delta_{th,B}}{Lk\left(\Delta T_S\right)_B} \frac{k\left(\Delta T_S\right)_B}{\delta_{th,B}} \\ &\approx \left(\frac{\tilde{\rho}_B}{\tilde{\rho}_i}\right)^2 \frac{c_p^2 \mu^2}{k^2} \frac{\mu k}{c_p^2 \tilde{\rho}_B^2 L^2 \epsilon_a \Delta \tilde{T}} \tilde{\theta} N u^{-1} F_S(z=0) \\ &\approx \left(\frac{\tilde{\rho}_B}{\tilde{\rho}_i}\right)^2 \ln \tilde{\Gamma} Pr^2 Ra^{-1} Nu^{-1} F_S(z=0). \end{aligned} \tag{3.330}$$

It follows, that all the terms in (3.321), (3.322) and (3.320) are always much smaller than the dissipative terms $Q_\nu(z)$ and $Q_{\nu/T}(z)$ in (3.323). Therefore the Eqs. (3.309) and (3.310) can be written in a simpler, approximate form

$$\begin{aligned} F_S(z=0) \approx &- k\frac{\mathrm{d}}{\mathrm{d}z}\left(\tilde{T} + \left\langle T'\right\rangle_h - T_{ad}\right) + \tilde{\rho}\tilde{T}\left\langle u_z s'\right\rangle_h \\ &+ \frac{\Delta \tilde{T}}{L}\int_0^z \tilde{\rho}\left\langle u_z s'\right\rangle_h \mathrm{d}z - \mu\int_0^z \langle q\rangle_h \,\mathrm{d}z, \end{aligned} \tag{3.331a}$$

$$F_S(z=0) \approx -k\frac{\tilde{T}_B}{\tilde{T}}\frac{\mathrm{d}}{\mathrm{d}z}\left(\tilde{T} + \left\langle T'\right\rangle_h - T_{ad}\right) + \tilde{\rho}\tilde{T}_B\left\langle u_z s'\right\rangle_h - \mu\int_0^z \frac{\tilde{T}_B}{\tilde{T}}\langle q\rangle_h \,\mathrm{d}z. \tag{3.331b}$$

up to

$$\mathcal{O}\left(\frac{\delta_{th,B}}{L} F_S(z=0)\right) + \mathcal{O}\left(Ra^{-1} Nu^{-1} Re_i^{9/4} F_S(z=0)\right) \tag{3.332}$$

in the case of viscous dissipation dominated by the bulk contribution or up to

$$\mathcal{O}\left(\frac{\delta_{th,B}}{L} F_S(z=0)\right) + \mathcal{O}\left(Ra^{-1} Nu^{-1} Re_i^{3/2} F_S(z=0)\right) \tag{3.333}$$

in the case when viscous dissipation takes place predominantly in the boundary layers. It will be confirmed later in Sect. 3.6.8, that the rests

$$\mathcal{O}\left(Ra^{-1} Nu^{-1} Re_i^{9/4} F_S(z=0)\right) \quad \text{and} \quad \mathcal{O}\left(Ra^{-1} Nu^{-1} Re_i^{3/2} F_S(z=0)\right) \tag{3.334}$$

are indeed negligibly small compared to the superadiabatic flux $F_S(z=0)$. Finally we take the second relation, i.e. (3.331b) at $z=L$, which by the use of $F_S\,(z=0) = F_S(z=L)$ leads to

$$F_S\,(z=0)\left(\frac{1}{\tilde{T}_T} - \frac{1}{\tilde{T}_B}\right) = \mu\int_0^L \frac{1}{\tilde{T}}\langle q\rangle_h \,\mathrm{d}z = \frac{1}{\tilde{T}_B} Q_{\nu/T}(L). \tag{3.335}$$

### 3.6.7 Estimates of the Ratios $r_s$, $r_\delta$, $r_T$ and $r_U$

First we write down the leading order balance between inertia and diffusion for the thermal and viscous boundary layers. In the simplest case, when the viscous boundary layers are nested in the thermal ones, $\delta_{th,T} > \delta_{\nu,T}$ and $\delta_{th,B} > \delta_{\nu,B}$ ($Pr \lesssim 1$), the inertia-diffusion balance takes the form

$$\frac{\tilde{\rho}_B U_B}{L} \approx \frac{k}{c_p \delta_{th,B}^2}, \qquad \frac{\tilde{\rho}_T U_T}{L} \approx \frac{k}{c_p \delta_{th,T}^2}, \tag{3.336}$$

in the thermal layers, where (3.249) has been used, and

$$\frac{\tilde{\rho}_B U_B}{L} \approx \frac{\mu}{\delta_{\nu,B}^2}, \qquad \frac{\tilde{\rho}_T U_T}{L} \approx \frac{\mu}{\delta_{\nu,T}^2}, \tag{3.337}$$

in the viscous layers. On dividing Eqs. (3.336) by Eqs. (3.337) respectively we get

$$\frac{\delta_{th,T}}{\delta_{\nu,T}} \approx Pr^{-1/2}, \qquad \frac{\delta_{th,B}}{\delta_{\nu,B}} \approx Pr^{-1/2}, \tag{3.338}$$

thus

$$\frac{\delta_{th,T}}{L} \approx Re_T^{-1/2} Pr^{-1/2}, \qquad \frac{\delta_{th,B}}{L} \approx Re_B^{-1/2} Pr^{-1/2}. \tag{3.339}$$

and hence also

$$r_\delta = \frac{\delta_{th,T}}{\delta_{th,B}} = \frac{\delta_{\nu,T}}{\delta_{\nu,B}}. \tag{3.340}$$

In the case of thicker viscous layers $\delta_{th,T} < \delta_{\nu,T}$ and $\delta_{th,B} < \delta_{\nu,B}$ ($Pr \gtrsim 1$) the velocity scale in the thermal layers must be weakened with respect to the thermal wind velocity by a factor $\delta_{th,i}/\delta_{\nu,i}$ (cf. Sect. 2.4 and Fig. 2.4), which implies the inertia-conduction balance in thermal layers in the form

$$\frac{\tilde{\rho}_B U_B}{L}\frac{\delta_{th,B}}{\delta_{\nu,B}} \approx \frac{k}{c_p \delta_{th,B}^2}, \qquad \frac{\tilde{\rho}_B U_T}{L}\frac{\delta_{th,T}}{\delta_{\nu,T}} \approx \frac{k}{c_p \delta_{th,T}^2}. \tag{3.341}$$

The dominant balance in the viscous boundary layers remains the same (3.337), therefore on dividing equations (3.341) by Eqs. (3.337) respectively we get

$$\frac{\delta_{th,T}}{\delta_{\nu,T}} \approx Pr^{-1/3}, \qquad \frac{\delta_{th,B}}{\delta_{\nu,B}} \approx Pr^{-1/3}, \tag{3.342}$$

thus

$$\frac{\delta_{th,T}}{L} \approx Re_T^{-1/2} Pr^{-1/3}, \qquad \frac{\delta_{th,B}}{L} \approx Re_B^{-1/2} Pr^{-1/3}. \tag{3.343}$$

This clearly implies, that the ratios of the top to bottom thicknesses of thermal boundary layers and top to bottom thicknesses of viscous boundary layers are the same, cf. (3.340), no matter the nesting between the thermal and viscous boundary layers. Moreover from the Eqs. (3.339) and (3.343), supplied by the definitions of the Reynolds numbers in (3.301) one obtains for both the cases, i.e. case 1: $\delta_{th,T} > \delta_{\nu,T}$, $\delta_{th,B} > \delta_{\nu,B}$ ($Pr \lesssim 1$) and case 2: $\delta_{th,T} < \delta_{\nu,T}$, $\delta_{th,B} < \delta_{\nu,B}$ ($Pr \gtrsim 1$) the following relation[18]

$$r_\delta = \frac{\delta_{th,T}}{\delta_{th,B}} = \left(\frac{\tilde{\rho}_B U_B}{\tilde{\rho}_T U_T}\right)^{1/2} = \left(\frac{\tilde{\rho}_B}{\tilde{\rho}_T}\right)^{1/2} \frac{1}{r_U^{1/2}} = \left(\frac{\tilde{\Gamma}^m}{r_U}\right)^{1/2}. \tag{3.344}$$

Gathering now the Eqs. (3.307) and (3.344) yields

$$r_\delta = r_T, \quad r_s = \tilde{\Gamma} r_T, \quad r_U r_\delta^2 = \tilde{\Gamma}^m, \tag{3.345}$$

so that expressing things by $r_U$ we get

$$r_\delta = r_T = \frac{\tilde{\Gamma}^{m/2}}{r_U^{1/2}}, \quad r_s = \frac{\tilde{\Gamma}^{m/2+1}}{r_U^{1/2}}. \tag{3.346}$$

The next step is evaluate somehow the velocity ratio $r_U$. It should be made clear, that this step is the most speculative one in the analysis of fully developed convection presented here. Nevertheless, to obtain an estimate of $r_U$ it seems reasonable to consider an analogue of the 'Deardorff' balance between mean inertia and mean buoyancy (cf. Deardorff 1970). We therefore consider the stationary Navier-Stokes equation multiplied by $\mathbf{u}$ and horizontally averaged and assume, that the dominant terms in the resulting equation are the inertial and buoyancy terms

$$\frac{1}{2\tilde{\rho}} \frac{\partial}{\partial z} \left(\tilde{\rho} \left\langle u_z u^2 \right\rangle_h\right) \approx \frac{g}{c_p} \left\langle u_z s' \right\rangle_h, \tag{3.347}$$

[18]We emphasize, that to estimate inertia in the boundary layers

$$\tilde{\rho}_i \left(\mathbf{u}_h \cdot \nabla + u_z \frac{\partial}{\partial z}\right) \mathbf{u}_h \approx \frac{\tilde{\rho}_i U_i^2}{L},$$

the layer thickness $L$ was assumed as the horizontal length scale of variation of velocity. This is suggested by results of numerical simulations (cf. Verhoeven et al. 2015, Jones et al. 2020) and the reason for it may be, that although the dominant vertical length scales in the bulk do scale with the pressure scale height, the boundary layer wind of turbulence is selected by the longest horizontal length scale over which the flow is coherent. With this approach the thicknesses of the viscous boundary layers are simply given by (3.308), but as remarked below (3.301) it is possible that at large stratifications $\tilde{\Gamma} \gg 1$, the scale heights determine both the vertical and horizontal length scales of variation of the wind of turbulence; in such a case the inertial term in the boundary layers can be estimated by $\tilde{\rho}_i U_i^2 / D_\rho$, hence also the definitions of the boundary layer thicknesses involve the scale heights.

which may also be rewritten in the form

$$-\frac{1}{2D_\rho}\left\langle u_z u^2\right\rangle_h + \frac{1}{2}\left\langle\frac{\partial u_z}{\partial z}u^2\right\rangle_h + \frac{1}{2}\left\langle u_z\frac{\partial u^2}{\partial z}\right\rangle_h \approx \frac{g}{c_p}\left\langle u_z s'\right\rangle_h . \tag{3.348}$$

Assuming that the *vertical* scale of variation of velocity outside the boundary layers is determined by the density scale heights at top and bottom we get

$$\frac{c_p}{g}\frac{U_T^3}{D_{\rho,T}} \approx \left[\left\langle u_z s'\right\rangle_h\right]_T, \qquad \frac{c_p}{g}\frac{U_B^3}{D_{\rho,B}} \approx \left[\left\langle u_z s'\right\rangle_h\right]_B . \tag{3.349}$$

It is postulated, that in turbulent convection the vertical velocity, which is small in the boundary layers and in their vicinity in the bulk, is quickly amplified by strong buoyancy which becomes important away from the boundary layers within upwelling large-scale convective currents (convection cells of the wind of turbulence). The vertical velocity is effectively assumed to become comparable with the horizontal one within the distances $D_{\rho,B}$ and $D_{\rho,T}$ away from the bottom and top boundaries respectively. Nevertheless we will still assume in this case, that the magnitudes of velocities can be approximated by $U_B$ and $U_T$ near the bottom and top respectively, which allows to estimate the means $\left[\left\langle u_z\partial_z u^2\right\rangle_h\right]_i \sim \left[\left\langle u_z u^2\right\rangle_h\right]_i / D_\rho \sim U_i^3/D_\rho$, where $i = B$ or $T$ for the bottom and top balance respectively.

At this stage one needs to consider separately the different cases determined by whether the dominant contributions to viscous and thermal dissipation come from the bulk or boundary layers. For the sake of simplicity and due to lack of sufficient experimental and numerical data, we will consider only the two, perhaps simplest cases, when the thermal dissipation takes place predominantly in the boundary layers, but the viscous dissipation can be dominant either also in the boundary layers or in the bulk. It is clear from Fig. 2.5, that at least for the Boussinesq convection the two aforementioned regimes are the first to appear as the Rayleigh number increases and exceeds a critical value for fully developed convection, whereas other regimes appear for even much higher values of the Rayleigh number.

#### 3.6.7.1 Viscous and Thermal Dissipation Predominantly in the Boundary Layers

Since the dissipation in the bulk is negligible, estimates of the superadiabatic heat flux at the top and bottom of the bulk, therefore just above the bottom boundary layer and below the top one, according to (3.331b) involve at the leading order only advection. This is because conduction is negligible, and the viscous dissipation integral at both locations, $z = \delta_{th,B}$ and $z = L - \delta_{th,T}$ (and in fact in the entire bulk) is dominated by the contribution from the bottom boundary layer, which is approximately $-\mu\int_0^{\delta_{th,B}}\langle q\rangle_h\,\mathrm{d}z$. This allows to write

$$\tilde{\rho}_B\left[\left\langle u_z s'\right\rangle_h\right]_B \approx \tilde{\rho}_T\left[\left\langle u_z s'\right\rangle_h\right]_T, \tag{3.350}$$

so that the latter together with (3.349)[19] produce

$$\frac{\left[\left\langle u_z s'\right\rangle_h\right]_T}{\left[\left\langle u_z s'\right\rangle_h\right]_B} \approx \frac{\tilde{\rho}_B}{\tilde{\rho}_T} = \tilde{\Gamma}^m \approx \frac{U_T^3}{U_B^3}\frac{D_{\rho,B}}{D_{\rho,T}}, \tag{3.351}$$

or equivalently

$$r_U = \tilde{\Gamma}^{(m-1)/3}. \tag{3.352}$$

It follows from (3.346), that

$$r_\delta = r_T = \tilde{\Gamma}^{(2m+1)/6}, \quad r_s = \tilde{\Gamma}^{(2m+7)/6}. \tag{3.353}$$

#### 3.6.7.2 Viscous Dissipation Predominantly in the Bulk, Thermal Dissipation Dominated by Contributions from Boundary Layers

Now the dominant contributions to the mean superadiabatic heat flux at the top and bottom of the bulk, by the use of the first heat production formula (3.331a) are

$$F_S(z=0) \approx \tilde{\rho}_T\tilde{T}_T\left[\left\langle u_z s'\right\rangle_h\right]_T \approx \tilde{\rho}_B\tilde{T}_B\left[\left\langle u_z s'\right\rangle_h\right]_B, \tag{3.354}$$

since at the bottom the work of the buoyancy force and viscous dissipation integral are negligible and at the top, according to the global balance $g\left\langle\tilde{\rho}u_z s'\right\rangle/c_p = \mu\left\langle q\right\rangle$ in (3.141) they are approximately equal and thus cancel out. Consequently

$$\frac{\left[\left\langle u_z s'\right\rangle_h\right]_T}{\left[\left\langle u_z s'\right\rangle_h\right]_B} \approx \frac{\tilde{\rho}_B\tilde{T}_B}{\tilde{\rho}_T\tilde{T}_T} = \tilde{\Gamma}^{m+1} \approx \frac{U_T^3}{U_B^3}\frac{D_{\rho,B}}{D_{\rho,T}}, \tag{3.355}$$

or equivalently

$$r_U = \tilde{\Gamma}^{m/3}. \tag{3.356}$$

Calculating the other ratios from (3.346) we get

$$r_\delta = r_T = \tilde{\Gamma}^{m/3}, \quad r_s = \tilde{\Gamma}^{m/3+1}. \tag{3.357}$$

[19] Once again, we stress, that the relation (3.349) is the weakest point of the presented analysis, despite the fact, that it leads to a rather satisfactory agreement with results of numerical simulations of Jones et al. (2020).

### 3.6.8 Scaling Laws for Fully Developed Stratified Convection with Isentropic Boundaries

We are now ready to derive the scaling laws for the Nusselt and Reynolds numbers versus the driving force measured by the Rayleigh number. We start by observing, that the relation between the Nusselt number and the Reynolds number is now easily obtained. In the first case, when thermal boundary layers are thicker than the viscous layers, $\delta_{th,T} > \delta_{\nu,T}$ and $\delta_{th,B} > \delta_{\nu,B}$ ($Pr \lesssim 1$), from (3.306a) and (3.339) we immediately get

$$Nu = \frac{\tilde{\Gamma}\ln\tilde{\Gamma}}{(1+r_s)\left(\tilde{\Gamma}-1\right)} Re_B^{1/2} Pr^{1/2}. \tag{3.358}$$

For the second case of thermal layers nested in the viscous ones, $\delta_{th,T} < \delta_{\nu,T}$ and $\delta_{th,B} < \delta_{\nu,B}$ ($Pr \gtrsim 1$), by (3.306a) and (3.343) we immediately get

$$Nu = \frac{\tilde{\Gamma}\ln\tilde{\Gamma}}{(1+r_s)\left(\tilde{\Gamma}-1\right)} Re_B^{1/2} Pr^{1/3}. \tag{3.359}$$

These results are due to the fact, that the thermal dissipation in all the cases considered here is dominated by the contributions from boundary layers and is independent of whether the viscous dissipation takes place predominantly in the bulk or in the boundary layers. Let us now turn to these cases separately.

#### 3.6.8.1 Viscous and Thermal Dissipation Predominantly in the Boundary Layers

We take the relation (3.335) and estimate the term $Q_{\nu/T}(L)$ with a sum of the dominant contributions from boundary layers

$$Q_{\nu/T}(L) \approx \mu\tilde{T}_B\left(\frac{U_B^2}{\delta_{\nu,B}\tilde{T}_B} + \frac{U_T^2}{\delta_{\nu,T}\tilde{T}_T}\right) = \frac{\mu U_B^2}{\delta_{\nu,B}}\left(1+\tilde{\Gamma}\frac{r_U^2}{r_\delta}\right), \tag{3.360}$$

which in light of (3.300) and

$$F_S(z=0) \approx k\frac{\left(\Delta T'\right)_B}{\delta_{th,B}} \approx k\frac{\tilde{T}_B}{c_p(1+r_s)\,\delta_{th,B}}\frac{\Delta\tilde{s}}{}, \tag{3.361}$$

(cf. (3.299), (3.302a) and (3.303a)) allows to write down

$$\frac{\tilde{\Gamma}}{(1+r_s)} Ra Pr^{-2} \approx \frac{\delta_{th,B}}{\delta_{\nu,B}} Re_B^2\left(1+\tilde{\Gamma}\frac{r_U^2}{r_\delta}\right). \tag{3.362}$$

The latter, by the use of (3.352) and (3.353) is equivalent to

$$Re_B^2 \approx \frac{\delta_{\nu,B}}{\delta_{th,B}} \frac{\tilde{\Gamma}}{\left(1+\tilde{\Gamma}^{(2m+1)/6}\right)\left(1+\tilde{\Gamma}^{(2m+7)/6}\right)} Ra Pr^{-2}. \tag{3.363}$$

**Case 1: Thermal Layers Thicker Than Viscous Layers, $\delta_{th,T} > \delta_{\nu,T}$ and $\delta_{th,B} > \delta_{\nu,B}$ $(Pr \lesssim 1)$**

From (3.358), (3.363) and (3.338) we easily get the scaling laws

$$\begin{aligned} Nu &\approx \frac{\tilde{\Gamma}^{5/4} \ln \tilde{\Gamma}}{\left(1+\tilde{\Gamma}^{(2m+7)/6}\right)^{5/4}\left(1+\tilde{\Gamma}^{(2m+1)/6}\right)^{1/4}\left(\tilde{\Gamma}-1\right)} Pr^{1/8} Ra^{1/4} \\ &\xrightarrow{\tilde{\Gamma}\gg 1} \frac{\ln \tilde{\Gamma}}{\tilde{\Gamma}^{(2m+5)/4}} Pr^{1/8} Ra^{1/4}, \end{aligned} \tag{3.364}$$

$$\begin{aligned} Re_B = \tilde{\Gamma}^{(2m+1)/3} Re_T &\approx \left[\frac{\tilde{\Gamma}}{\left(1+\tilde{\Gamma}^{(2m+1)/6}\right)\left(1+\tilde{\Gamma}^{(2m+7)/6}\right)}\right]^{1/2} Pr^{-3/4} Ra^{1/2} \\ &\xrightarrow{\tilde{\Gamma}\gg 1} \tilde{\Gamma}^{-(2m+1)/6} Pr^{-3/4} Ra^{1/2}, \end{aligned} \tag{3.365}$$

supplied by (3.306a, 3.306b), which now take the form

$$\frac{\delta_{th,B}}{L} \approx \frac{\tilde{\Gamma} \ln \tilde{\Gamma}}{(1+r_s)\left(\tilde{\Gamma}-1\right)} Nu^{-1} \xrightarrow{\tilde{\Gamma}\gg 1} \tilde{\Gamma}^{(2m+1)/12} Pr^{-1/8} Ra^{-1/4}, \tag{3.366a}$$

$$\frac{\delta_{th,T}}{\delta_{th,B}} \approx \frac{r_s}{\tilde{\Gamma}} \approx \tilde{\Gamma}^{(2m+1)/6} > 1. \tag{3.366b}$$

For completeness we recall (3.358) in the form

$$Nu = \frac{\tilde{\Gamma} \ln \tilde{\Gamma}}{\left(1+\tilde{\Gamma}^{(2m+7)/6}\right)\left(\tilde{\Gamma}-1\right)} Re_B^{1/2} Pr^{1/2}. \tag{3.367}$$

The large $\tilde{\Gamma}$ limit requires care, because we have assumed incompressibility of the boundary layers, in other words by assumption the density stratification cannot exceed a critical large value, at which the density scale height becomes comparable and less than the thicknesses of the boundary layers; of course the smallest value of $D_\rho(z)$ is achieved at the top, where also the boundary layer is thicker than at the bottom, therefore the strongest constraint on validity of the results of this section is obtained at the top. More precisely from (3.248) it is clear, that the term $(\tilde{\theta}\delta_{th,i}\tilde{T}_B)/(L\tilde{T}_i)$ must be small and the most restrictive constraint is, indeed, obtained at the top, $\left(\tilde{\Gamma}-1\right)\delta_{th,T}/L \ll 1$, so that the large $\tilde{\Gamma}$ limit in the above formulae has an upper bound, i.e. it corresponds to

$$1 \ll \tilde{\Gamma} \ll Pr^{1/(4m+10)} Ra^{1/(2m+5)}, \tag{3.368}$$

which guarantees, that the boundary layers remain incompressible.

It remains to be verified, whether the remainders (3.325a–3.325d), which were assumed negligible are indeed small. The expression (3.325b) with the largest power of the Reynolds number is most likely to be the largest of the remainders, therefore consistency requires (cf. 3.330)

$$\left(\frac{\tilde{\rho}_B}{\tilde{\rho}_i}\right)^2 \ln\tilde{\Gamma}\, Pr^2 Ra^{-1} Nu^{-1} Re_i^{3/2} \ll 1, \tag{3.369}$$

which is largest at the top, that is for $i = T$, as long as $m > 1/2$ or equivalently $\gamma < 3$; since $\gamma > 3$ for a fluid is highly unlikely, we proceed with the assumption $m > 1/2$. It follows from (3.369) taken at the top and the scaling laws (3.364), (3.365), that the stratification parameter $\tilde{\Gamma}$ must also satisfy $\tilde{\Gamma} \ll Ra^{1/(2m+1)} Pr^{-3/(4m+2)}$, which is already satisfied by (3.368) since

$$\left(\frac{Ra}{Pr}\right)^{4/(2m+5)(2m+1)} > 1, \tag{3.370}$$

in fully developed convection.

**Case 2: Viscous Layers Thicker Than Thermal Layers, $\delta_{th,T} < \delta_{\nu,T}$ and $\delta_{th,B} < \delta_{\nu,B}$ ($Pr \gtrsim 1$)**

In this case we use the Eqs. (3.359), (3.363) and (3.342) to obtain the scaling laws

$$Nu \approx \frac{\tilde{\Gamma}^{5/4}\ln\tilde{\Gamma}}{\left(1+\tilde{\Gamma}^{(2m+7)/6}\right)^{5/4}\left(1+\tilde{\Gamma}^{(2m+1)/6}\right)^{1/4}\left(\tilde{\Gamma}-1\right)} Pr^{-1/12} Ra^{1/4} \xrightarrow{\tilde{\Gamma}\gg 1} \frac{\ln\tilde{\Gamma}}{\tilde{\Gamma}^{(2m+5)/4}} Pr^{-1/12} Ra^{1/4}, \tag{3.371}$$

$$Re_B = \tilde{\Gamma}^{(2m+1)/3} Re_T \approx \left[\frac{\tilde{\Gamma}}{\left(1+\tilde{\Gamma}^{(2m+1)/6}\right)\left(1+\tilde{\Gamma}^{(2m+7)/6}\right)}\right]^{1/2} Pr^{-5/6} Ra^{1/2} \xrightarrow{\tilde{\Gamma}\gg 1} \tilde{\Gamma}^{-(2m+1)/6} Pr^{-5/6} Ra^{1/2}, \tag{3.372}$$

and from (3.306a, 3.306b) it follows, that

$$\frac{\delta_{th,B}}{L} \approx \frac{\tilde{\Gamma}\ln\tilde{\Gamma}}{(1+r_s)\left(\tilde{\Gamma}-1\right)} Nu^{-1} \xrightarrow{\tilde{\Gamma}\gg 1} \tilde{\Gamma}^{(2m+1)/12} Pr^{1/12} Ra^{-1/4}, \tag{3.373a}$$

$$\frac{\delta_{th,T}}{\delta_{th,B}} \approx \frac{r_s}{\tilde{\Gamma}} \approx \tilde{\Gamma}^{(2m+1)/6} > 1. \tag{3.373b}$$

For completeness we recall (3.359)

$$Nu = \frac{\tilde{\Gamma}\ln\tilde{\Gamma}}{\left(1+\tilde{\Gamma}^{(2m+7)/6}\right)\left(\tilde{\Gamma}-1\right)} Re_B^{1/2} Pr^{1/3}. \tag{3.374}$$

The large $\tilde{\Gamma}$ limit can only be taken up to

$$\tilde{\Gamma} \ll Pr^{-1/(6m+15)} Ra^{1/(2m+5)}, \tag{3.375}$$

so that the boundary layers remain incompressible. Consistency with (3.369) requires $\tilde{\Gamma} \ll Ra^{1/(2m+1)} Pr^{-5/(6m+3)}$, which is satisfied by (3.375) in fully developed convection at high Rayleigh numbers and moderately high Prandtl numbers,

$$\left(\frac{Ra}{Pr^{2m/3+2}}\right)^{4/(2m+5)(2m+1)} > 1. \tag{3.376}$$

#### 3.6.8.2 Viscous Dissipation Predominantly in the Bulk, Thermal Dissipation Dominated by Contributions from Boundary Layers

Again, we start with the relation (3.335). First we make the previous estimate of $Q_{\nu/T}(L)$, provided in (3.327), somewhat more precise. The viscous integral $Q_\nu(L)$ can be estimated from the Navier-Stokes equation as follows

$$Q_\nu(L) = \mu\int_0^L \langle q\rangle_h \,\mathrm{d}z \approx \mu\int_{\delta_{th,B}}^{L-\delta_{th,T}} \langle q\rangle_h \,\mathrm{d}z \approx \tilde{\rho}_B U_B^3 = \frac{\mu^3}{\tilde{\rho}_B^2 L^3} Re_B^3, \tag{3.377}$$

since the viscous dissipation is dominant in the bulk, where it is expected to balance the nonlinear inertial term. Moreover, since $r_U^3 = \tilde{\Gamma}^m$ in the current case, the maximal estimate of $Q_\nu(L)$ is obtained either by taking $\tilde{\rho}_B U_B^3$ or equivalently $\tilde{\rho}_T U_T^3 \approx \tilde{\rho}_B U_B^3$. Next we observe, that

$$Q_{\nu/T}(L) \lesssim \tilde{\Gamma} Q_\nu(L), \tag{3.378}$$

so that by (3.335), (3.377), (3.378) and (3.330) we can finally write down the following rough estimate

$$\tilde{\Gamma}\ln\tilde{\Gamma}Pr^2Ra^{-1}Nu^{-1}Re_B^3 \approx \tilde{\Gamma} - 1, \tag{3.379}$$

or equivalently

$$RaNuPr^{-2} \approx \frac{\tilde{\Gamma}\ln\tilde{\Gamma}}{\tilde{\Gamma} - 1}Re_B^3. \tag{3.380}$$

**Case 1: Thermal Layers Thicker Than Viscous Layers, $\delta_{th,T} > \delta_{\nu,T}$ and $\delta_{th,B} > \delta_{\nu,B}$ $(Pr \lesssim 1)$**

Using (3.358), (3.380) and (3.357) one obtains the following scaling laws

$$\begin{aligned} Nu &\approx \frac{\tilde{\Gamma}\ln\tilde{\Gamma}}{\left(1+\tilde{\Gamma}^{m/3+1}\right)^{6/5}\left(\tilde{\Gamma}-1\right)}Pr^{1/5}Ra^{1/5} \\ &\xrightarrow{\tilde{\Gamma}\gg 1} \frac{\ln\tilde{\Gamma}}{\tilde{\Gamma}^{(2m+6)/5}}Pr^{1/5}Ra^{1/5}, \end{aligned} \tag{3.381}$$

$$\begin{aligned} Re_B = \tilde{\Gamma}^{2m/3}Re_T &\approx \frac{1}{\left(1+\tilde{\Gamma}^{m/3+1}\right)^{2/5}}Pr^{-3/5}Ra^{2/5} \\ &\xrightarrow{\tilde{\Gamma}\gg 1} \tilde{\Gamma}^{-(2m+6)/15}Pr^{-3/5}Ra^{2/5}, \end{aligned} \tag{3.382}$$

and consequently (3.306a, 3.306b) implies

$$\frac{\delta_{th,B}}{L} \approx \frac{\tilde{\Gamma}\ln\tilde{\Gamma}}{(1+r_s)\left(\tilde{\Gamma}-1\right)}Nu^{-1} \xrightarrow{\tilde{\Gamma}\gg 1} \tilde{\Gamma}^{(m+3)/15}Pr^{-1/5}Ra^{-1/5}, \tag{3.383a}$$

$$\frac{\delta_{th,T}}{\delta_{th,B}} \approx \frac{r_s}{\tilde{\Gamma}} \approx \tilde{\Gamma}^{m/3} > 1. \tag{3.383b}$$

For completeness (3.358) is recalled in an explicit form

$$Nu = \frac{\tilde{\Gamma}\ln\tilde{\Gamma}}{\left(1+\tilde{\Gamma}^{m/3+1}\right)\left(\tilde{\Gamma}-1\right)}Re_B^{1/2}Pr^{1/2}. \tag{3.384}$$

Due to the assumption of incompressibility of the boundary layers (cf. (3.248) which implies $\tilde{\theta}\tilde{\Gamma}\delta_{th,T}/L \ll 1$) the large $\tilde{\Gamma}$ limit can only be taken up to

$$\tilde{\Gamma} \ll (Pr\,Ra)^{1/(2m+6)}, \tag{3.385}$$

so that the boundary layers remain incompressible.

On the other hand neglection of the terms (3.329a–3.329c) in the process of derivation of the scaling laws requires verification, whether these terms are indeed small in comparison with the total superadiabatic heat flux $F_S(z=0)$. The expression (3.329b) with the largest power of the Reynolds number is the largest, therefore consistency requires (cf. 3.330)

$$\left(\frac{\tilde{\rho}_B}{\tilde{\rho}_i}\right)^2 \ln\tilde{\Gamma}\, Pr^2 Ra^{-1} Nu^{-1} Re_i^{9/4} \ll 1, \tag{3.386}$$

and the strongest restriction is obtained at the top, that is for $i = T$. It follows, that the stratification parameter $\tilde{\Gamma}$ must also satisfy $\tilde{\Gamma} \ll Ra^{1/(2m+1)} Pr^{-3/(4m+2)}$, which is already satisfied by (3.368) since

$$\left(\frac{Ra}{Pr^{m+2}}\right)^{5/(2m+6)(2m+1)} > 1, \tag{3.387}$$

in fully developed convection.

**Case 2: Viscous Layers Thicker Than Thermal Layers, $\delta_{th,T} < \delta_{\nu,T}$ and $\delta_{th,B} < \delta_{\nu,B}$ $(Pr \gtrsim 1)$**

From (3.359), (3.380) and (3.357) we get the scaling laws

$$\begin{aligned} Nu &\approx \frac{\tilde{\Gamma}\ln\tilde{\Gamma}}{\left(1+\tilde{\Gamma}^{m/3+1}\right)^{6/5}\left(\tilde{\Gamma}-1\right)} Ra^{1/5} \\ &\xrightarrow{\tilde{\Gamma}\gg 1} \frac{\ln\tilde{\Gamma}}{\tilde{\Gamma}^{(2m+6)/5}} Ra^{1/5}, \end{aligned} \tag{3.388}$$

$$\begin{aligned} Re_B = \tilde{\Gamma}^{2m/3} Re_T &\approx \frac{1}{\left(1+\tilde{\Gamma}^{m/3+1}\right)^{2/5}} Pr^{-2/3} Ra^{2/5} \\ &\xrightarrow{\tilde{\Gamma}\gg 1} \tilde{\Gamma}^{-(2m+6)/15} Pr^{-2/3} Ra^{2/5}, \end{aligned} \tag{3.389}$$

and by the use of (3.306a, 3.306b)

$$\frac{\delta_{th,B}}{L} \approx \frac{\tilde{\Gamma}\ln\tilde{\Gamma}}{(1+r_s)\left(\tilde{\Gamma}-1\right)} Nu^{-1} \xrightarrow{\tilde{\Gamma}\gg 1} \tilde{\Gamma}^{(m+3)/15} Ra^{-1/5}, \tag{3.390a}$$

$$\frac{\delta_{th,T}}{\delta_{th,B}} \approx \frac{r_s}{\tilde{\Gamma}} \approx \tilde{\Gamma}^{m/3} > 1. \tag{3.390b}$$

For completeness we provide (3.359) in the form

$$Nu = \frac{\tilde{\Gamma} \ln \tilde{\Gamma}}{\left(1 + \tilde{\Gamma}^{m/3+1}\right)\left(\tilde{\Gamma} - 1\right)} Re_B^{1/2} Pr^{1/3}. \tag{3.391}$$

The large $\tilde{\Gamma}$ limit can only be taken up to

$$\tilde{\Gamma} \ll Ra^{1/(2m+6)}, \tag{3.392}$$

so that the boundary layers remain incompressible. Consistency with (3.386) requires $\tilde{\Gamma} \ll Ra^{1/(2m+1)} Pr^{-5/(6m+3)}$, which is satisfied by (3.392) in fully developed convection at high Rayleigh numbers and moderately high Prandtl numbers,

$$\left(\frac{Ra^{1/(2m+6)}}{Pr^{1/3}}\right)^{5/(2m+1)} > 1. \tag{3.393}$$

Finally we observe, that in principle direct quantitative comparisons of the results obtained in this section taken at $\tilde{\theta} = 0$ with the Boussinesq case could be used to provide the prefactors in the final scaling laws.

### *3.6.9 Discussion*

The influence of stratification on the dynamics of fully developed convection is, indeed, substantial. In a stationary state the heat flux entering at the bottom is equal to the heat flux leaving the system at the top as in the Boussinesq case, but the work done by the buoyancy force and the viscous heating are no longer negligible and are of the same order as the total heat flux in the system. Therefore the total heat flux passing through every plane $z = \text{const.}$ within the fluid domain is no longer the same. In the case when most of the viscous dissipation takes place in the boundary layers the heat flux entering at the bottom is increased by the viscous heating in the bottom boundary layer, and then the work done by the buoyancy force reduces the flux in the bulk so that it falls below the value at $z = 0$ and then it is again boosted by the viscous heating in the top boundary layer to reach the same value at the top boundary as at the bottom boundary (cf. (3.148) and (3.141)).

Furthermore, we note that the top boundary layer is always thicker than the bottom one and that the entropy jump across the boundary layer is always greater at the top, that is $r_\delta > 1$ and $r_s > 1$. Moreover, the thicknesses of the boundary layers generally increase with the stratification parameter $\tilde{\Gamma}$, in other words the thermal boundary layers thicken as the density scale height decreases. This means, that for strong enough stratifications the boundary layers become compressible and since the top boundary layer is thicker, it is also much more prone to such a transition.

Next we may observe, that since typically $r_U$ also exceeds unity (always when $m > 1$), the convective velocities are also larger close to the top of the flow domain than in the bottom region. The latter fact, together with $\delta_{th,T} > \delta_{th,B}$ mean that the top boundary layer is also more prone to instability, i.e. is more likely to become turbulent at high Rayleigh numbers.

The validity restrictions for the presented approach, that is the upper bounds in (3.368), (3.375), (3.385) and (3.392), which result from the assumed incompressibility of the boundary layers indicate, that for moderately high Rayleigh numbers, about $10^7$ and $10^8$, even relatively weak stratifications, with $\tilde{\Gamma} \approx 10$ lead to compressible boundary layers (at least the top one), thus fall out of the regime of validity for the presented theory. Smaller values of the polytropic index $m$ allow for larger stratifications at which $\delta_{th,T} \ll D_\rho(L)$ is still satisfied and the theory remains valid, but generally speaking the large stratification limit with incompressible boundary layers is expected for larger values of $Ra \gtrsim 10^{10}$.

## 3.7 Validity of the Approximation and Summary

The anelastic approximation is designed for description of systems with significant stratification

$$L \sim D_\rho,\ D_T,\ D_p, \tag{3.394}$$

such as e.g. planetary atmospheres or planetary and stellar interiors. The shear and bulk dynamical viscosities $\mu(x, y, z)$ and $\mu_b(x, y, z)$, the specific heats $c_v(x, y, z)$ and $c_p(x, y, z)$ are allowed any spatial variation, whereas the thermal conduction $k(z)$ and the gravitational acceleration $\mathbf{g} = -g(z)\hat{\mathbf{e}}_z$ are typically assumed only depth-dependent. If the boundary conditions are assumed time-independent the hydrostatic reference state satisfies the following equations

$$\frac{\mathrm{d}\tilde{p}}{\mathrm{d}z} = -\tilde{\rho}g, \quad \frac{\mathrm{d}}{\mathrm{d}z}\left(k\frac{\mathrm{d}\tilde{T}}{\mathrm{d}z}\right) = -\tilde{Q}, \quad \tilde{\rho} = \rho(\tilde{p}, \tilde{T}), \quad \tilde{s} = s(\tilde{p}, \tilde{T}). \tag{3.395}$$

The actual form of the reference state depends on the functions $g(z)$ and $k(z)$, which have to be known beforehand. It must be emphasized, that a comparison between two anelastic systems, which is often necessary when two different numerical codes are expected to produce the same results, must be done with great care and inclusion of the form of the reference state. We elaborate on this issue later.

The *first fundamental assumption* leading to the anelastic approximation is that the convective system is only slightly superadiabatic, that is

$$0 < \delta \equiv \left\langle \frac{L}{\tilde{T}}\Delta_S \right\rangle = -\left\langle \frac{L}{\tilde{T}}\left(\frac{\mathrm{d}\tilde{T}}{\mathrm{d}z} + \frac{g\tilde{\alpha}\tilde{T}}{\tilde{c}_p}\right)\right\rangle = -\left\langle \frac{L}{\tilde{c}_p}\frac{\mathrm{d}\tilde{s}}{\mathrm{d}z}\right\rangle \ll 1. \tag{3.396}$$

*Secondly*, one must require that the weak superadiabaticity which drives convection implies weak fluctuations of thermodynamic variables

$$\left|\frac{\rho'}{\tilde{\rho}}\right| \sim \left|\frac{T'}{\tilde{T}}\right| \sim \left|\frac{p'}{\tilde{p}}\right| \sim \left|\frac{s'}{\tilde{s}}\right| \sim \left|\frac{\psi'}{\tilde{\psi}}\right| \sim \mathcal{O}(\delta) \ll 1. \tag{3.397}$$

Furthermore, the convective velocity and time scales are

$$\mathscr{U} \sim \delta^{1/2}\sqrt{\bar{g}L}, \qquad \mathscr{T} \sim \delta^{-1/2}\sqrt{\frac{L}{\bar{g}}}, \tag{3.398}$$

and hence consistency of the approximation requires, that the scales of viscosity and thermal conductivity also have to be small,

$$\mu_b/\bar{\rho} \lesssim \nu \sim \delta^{1/2}\sqrt{\bar{g}L}L, \qquad k \sim \delta^{1/2}\bar{\rho}\bar{c}_p\sqrt{\bar{g}L}L. \tag{3.399}$$

The *third assumption* involves the equation of state, that is to say we require, that the derivative $\partial p/\partial \rho$ is of the same order of magnitude in terms of the small parameter $\delta$ as $p/\rho$, which is typically satisfied by fluids and thus does not impose a strong restriction on the system. The latter assumption, together with (3.398) imply small Mach number $Ma^2 = \mathscr{U}^2/\left\langle\left(\frac{\partial p}{\partial \rho}\right)_s\right\rangle = \mathcal{O}(\delta) \ll 1$.

The full system of anelastic equations reads

$$\tilde{\rho}\left[\frac{\partial \mathbf{u}}{\partial t} + (\mathbf{u}\cdot\nabla)\mathbf{u}\right] = -\nabla p' + \rho'\mathbf{g} - \tilde{\rho}\nabla\psi' + \mu\nabla^2\mathbf{u} + \left(\frac{\mu}{3} + \mu_b\right)\nabla(\nabla\cdot\mathbf{u})$$
$$+ 2\nabla\mu\cdot\mathbf{G}^s + \nabla\left(\mu_b - \frac{2}{3}\mu\right)\nabla\cdot\mathbf{u}, \tag{3.400a}$$

$$\nabla\cdot(\tilde{\rho}\mathbf{u}) = 0, \qquad \nabla^2\psi' = 4\pi G\rho', \tag{3.400b}$$

$$\tilde{\rho}\tilde{T}\left(\frac{\partial s'}{\partial t} + \mathbf{u}\cdot\nabla s'\right) - \tilde{\rho}\tilde{c}_p u_z \Delta_S = \nabla\cdot\left(k\nabla T'\right) + 2\mu\mathbf{G}^s : \mathbf{G}^s$$
$$+ \left(\mu_b - \frac{2}{3}\mu\right)(\nabla\cdot\mathbf{u})^2 + Q', \tag{3.400c}$$

$$\frac{\rho'}{\tilde{\rho}} = -\tilde{\alpha}T' + \tilde{\beta}p', \qquad s' = -\tilde{\alpha}\frac{p'}{\tilde{\rho}} + \tilde{c}_p\frac{T'}{\tilde{T}}. \tag{3.400d}$$

The entropy equation may be replaced by the temperature equation, for which the most general form is

$$
\tilde{\rho}\tilde{c}_p\left(\frac{\partial T'}{\partial t}+\mathbf{u}\cdot\nabla T'\right)-\tilde{\alpha}\tilde{T}\left(\frac{\partial p'}{\partial t}+\mathbf{u}\cdot\nabla p'\right)+\rho c_p u_z\frac{\mathrm{d}\tilde{T}}{\mathrm{d}z}-\alpha T u_z\frac{\mathrm{d}\tilde{p}}{\mathrm{d}z}
$$
$$
=\nabla\cdot\left(k\nabla T'\right)+2\mu\mathbf{G}^s:\mathbf{G}^s+\left(\mu_b-\frac{2}{3}\mu\right)(\nabla\cdot\mathbf{u})^2+Q'. \tag{3.401}
$$

We also recall here the results of Sect. 3.1.2. In the case of uniform gravity the conservation of mass implies, that in order for the thermodynamic fluctuations to be correctly resolved, the jump of the mean pressure fluctuation across the depth of the fluid layer must vanish at all times, i.e. $\langle p'\rangle_h\,(z=L)-\langle p'\rangle_h\,(z=0)=0$. This constitutes a boundary condition, which must be imposed on the pressure field.

The system of anelastic equations can be expressed solely in terms of the velocity field and the pressure and entropy fluctuations, which is known as the "entropy formulation". Under the assumptions that the fluid satisfies the ideal gas equation, the volume cooling can be modelled by $\tilde{Q}=\kappa g\mathrm{d}_z\tilde{p}$, the gravity $g=\mathrm{const}$, thermal diffusivity $\kappa=\mathrm{const}$ and specific heat $\tilde{c}_p=\mathrm{const}$ are uniform, the entropy formulation takes the form

$$
\frac{\partial\mathbf{u}}{\partial t}+(\mathbf{u}\cdot\nabla)\,\mathbf{u}=-\nabla\left(\frac{p'}{\tilde{\rho}}\right)+\frac{gs'}{c_p}\hat{\mathbf{e}}_z+\frac{\mu}{\tilde{\rho}}\nabla^2\mathbf{u}+\left(\frac{\mu}{3\tilde{\rho}}+\frac{\mu_b}{\tilde{\rho}}\right)\nabla\left(\nabla\cdot\mathbf{u}\right)
$$
$$
+\frac{2}{\tilde{\rho}}\nabla\mu\cdot\mathbf{G}^s+\frac{1}{\tilde{\rho}}\nabla\left(\mu_b-\frac{2}{3}\mu\right)\nabla\cdot\mathbf{u}, \tag{3.402a}
$$

$$
\nabla\cdot(\tilde{\rho}\mathbf{u})=0, \tag{3.402b}
$$

$$
\tilde{\rho}\tilde{T}\left(\frac{\partial s'}{\partial t}+\mathbf{u}\cdot\nabla s'\right)-\tilde{\rho}c_p u_z\Delta_S=\nabla\cdot\left(\kappa\tilde{\rho}\tilde{T}\nabla s'\right)+\mathcal{J}+Q', \tag{3.402c}
$$

where $\mathcal{J}$ is given in (3.107). In such a way the pressure fluctuation is entirely eliminated from the energy balance and it appears only in the momentum equation. The pressure problem is easily removable by taking a curl of the momentum balance. It follows, that the boundary conditions for the pressure which physically cannot be controlled at the boundaries are no longer necessary, if one is searching for the velocity and entropy fields only (however, imposition of explicit boundary conditions on the entropy, not temperature is necessary). Such a two-variable approach gives an advantage from the point of view of effectiveness of numerical simulations of anelastic convection under the aforementioned assumptions.

The energetic properties of the anelastic systems are presented in Sect. 3.4. We recall here *the second formula for superadiabatic heat flux*

$$
\begin{aligned}
F_S(z=0) =& -k\frac{\mathrm{d}}{\mathrm{d}z}\left(\tilde{T}+\langle T'\rangle_h - T_{ad}\right)\Big|_{z=0}\\
=& -k\frac{T_B}{\tilde{T}}\frac{\mathrm{d}}{\mathrm{d}z}\left(\tilde{T}+\langle T'\rangle_h - T_{ad}\right)\\
&-\left[k\frac{\mathrm{d}\left(\tilde{T}-T_{ad}\right)}{\mathrm{d}z}\Bigg|_{z=0} - \frac{T_B}{\tilde{T}}k\frac{\mathrm{d}\left(\tilde{T}-T_{ad}\right)}{\mathrm{d}z}\right]\\
&-T_B k\frac{\mathrm{d}\tilde{T}}{\mathrm{d}z}\left(\frac{\langle T'\rangle_h}{\tilde{T}^2} - \frac{\langle T'\rangle_h\big|_{z=0}}{T_B^2}\right) - 2T_B k\frac{\mathrm{d}\tilde{T}}{\mathrm{d}z}\int_0^z \frac{\langle T'\rangle_h}{\tilde{T}^3}\frac{\mathrm{d}\tilde{T}}{\mathrm{d}z}\mathrm{d}z\\
&+\tilde{\rho}T_B\langle u_z s'\rangle_h\\
&-2T_B\int_0^z\left\langle\frac{\mu}{\tilde{T}}\mathbf{G}^s:\mathbf{G}^s\right\rangle_h \mathrm{d}z + T_B\int_0^z\left\langle\frac{2\mu-3\mu_b}{3\tilde{T}}(\nabla\cdot\mathbf{u})^2\right\rangle_h \mathrm{d}z.
\end{aligned}
\tag{3.403}
$$

which is one of two provided general formulae for the superadiabatic heat flux, satisfied by any anelastic system. If for simplicity one assumes, that the fluid satisfies the equation of state of an ideal gas, isothermal boundaries, uniform fluid properties $k=$ const., $\mu=$ const., $\mu_b=$ const., $c_p=$ const., no radiation $Q=0$ and uniform gravity $g=$ const. and $\mathbf{g}'=0$, the latter formula implies

$$
\begin{aligned}
\frac{k^2}{\rho_B^2 c_p^2 L^4} Ra\,(Nu-1)(\Gamma-1) =& \; 2\Delta T\left\langle\frac{1}{\tilde{T}}\mathbf{G}^s:\mathbf{G}^s\right\rangle\\
&-\Delta T\left(\frac{2}{3}-\frac{\mu_b}{\mu}\right)\left\langle\frac{(\nabla\cdot\mathbf{u})^2}{\tilde{T}}\right\rangle\\
&+2c_p Pr^{-1} L\left(\frac{\Delta T}{L}\right)^3\left\langle\frac{T'}{\tilde{T}^3}\right\rangle,
\end{aligned}
\tag{3.404}
$$

where $Pr=\mu c_p/k$, $\Gamma=T_B/T_T$ and the Nusselt and Rayleigh numbers are defined as follows

$$
Nu = \frac{F_S(z=0)}{k\Delta_S} = \frac{-k\frac{\mathrm{d}}{\mathrm{d}z}\left(\tilde{T}+\langle T'\rangle_h - T_{ad}\right)\Big|_{z=0}}{k\Delta_S},
\tag{3.405}
$$

$$
Ra = \frac{g\,\Delta_S\,L^4\rho_B^2 c_p}{T_B\mu k} = \frac{g\Delta T\,\Delta\tilde{s}\,L^3\rho_B^2}{\ln\Gamma\, T_B\mu k}.
\tag{3.406}
$$

In the above $k\Delta_S = k(\Delta T/L - g/c_p)$ is the superadiabatic conductive heat flux in the hydrostatic basic state.

On the other hand *the first formula for the total heat flux*

$$\begin{aligned} F_{total}(z=0) =& -k\frac{\mathrm{d}}{\mathrm{d}z}\left(\tilde{T}+\langle T'\rangle_h\right)\Big|_{z=0} \\ =& -k\frac{\mathrm{d}}{\mathrm{d}z}\left(\tilde{T}+\langle T'\rangle_h\right) \\ &+\tilde{\rho}\tilde{T}\left\langle u_z s'\right\rangle_h - \int_0^z \tilde{\rho}\frac{\mathrm{d}\tilde{T}}{\mathrm{d}z}\left\langle u_z s'\right\rangle_h \mathrm{d}z \\ &-2\int_0^z \left\langle \mu\mathbf{G}^s:\mathbf{G}^s\right\rangle_h \mathrm{d}z + \int_0^z \left\langle\left(\frac{2}{3}\mu-\mu_b\right)(\nabla\cdot\mathbf{u})^2\right\rangle_h \mathrm{d}z, \end{aligned} \tag{3.407}$$

comes useful when the heat flux is kept fixed at the boundaries, which under the same assumptions of constant $k$, $\mu$, $\mu_b$, $c_p$, $Q$, $g$ and $\mathbf{g}'=0$, implies

$$\begin{aligned} \frac{k^2}{\rho_B^2 c_p^2 L^4} Ra\left(Nu_Q - \frac{k\frac{\Delta\langle T'\rangle_h}{2L}}{k\Delta_S}\right)\frac{\Gamma}{\Gamma-1} =& \left\langle\left(\frac{\tilde{T}}{\Delta T}-1\right)\mathbf{G}^s:\mathbf{G}^s\right\rangle \\ &-\left(\frac{1}{3}-\frac{\mu_b}{2\mu}\right)\left\langle\left(\frac{\tilde{T}}{\Delta T}-1\right)(\nabla\cdot\mathbf{u})^2\right\rangle. \end{aligned} \tag{3.408}$$

where the Rayleigh number is still defined as in (3.406) but a new definition of the Nusselt number is required

$$Nu_Q = \frac{\left\langle\tilde{\rho}\tilde{T}u_z s'\right\rangle}{k\Delta_S}, \tag{3.409}$$

and $\Delta\left\langle T'\right\rangle_h = \left\langle T'\right\rangle_h\big|_{z=L} - \left\langle T'\right\rangle_h\big|_{z=0}$.

It is important to note, that contrary to the Boussinesq case, it is clear from (3.407), that

$$\begin{aligned} F_{total}(z) =& F_{total}(z=0) - \int_0^z \frac{\tilde{\alpha}\tilde{T}g\tilde{\rho}}{\tilde{c}_p}\left\langle u_z s'\right\rangle_h \mathrm{d}z \\ &+2\int_0^z \left\langle\mu\mathbf{G}^s:\mathbf{G}^s\right\rangle_h \mathrm{d}z - \int_0^z\left\langle\left(\frac{2}{3}\mu-\mu_b\right)(\nabla\cdot\mathbf{u})^2\right\rangle_h \mathrm{d}z, \end{aligned} \tag{3.410}$$

therefore anelastic systems are *not* characterized by constant heat flux at every $z$, but the heat flux is strongly influenced by the viscous heating and the work of the buoyancy force.

Linear stability of an anelastic ideal gas at constant $\nu$, $k$, $\mathbf{g}$ and $c_p$, $\mathbf{g}'=0$, $Q=0$, with isothermal, stress-free and impermeable boundaries is characterized by the critical Rayleigh number

$$Ra_{crit} \approx \frac{27}{4}\pi^4 \left[1 + \frac{1}{2}\theta\,(m-1)\right], \tag{3.411}$$

and the growth rate

$$\sigma = \frac{\kappa_B}{L^2}\frac{3}{2}\pi^2 \frac{Pr}{1+Pr}\eta \left(1 + \frac{1}{2}m\theta \frac{Pr}{1+Pr}\right), \tag{3.412}$$

where $Ra = g\Delta_S L^4/T_B\kappa_B\nu$ and $\eta = (Ra - Ra_{crit})/Ra_{crit}$; both the growth rate and the Rayleigh number are greater than in the Boussinesq, that is non-stratified (very weakly compressible) case. The total heat per unit mass accumulated in the fluid layer (between top and bottom boundaries) in the marginal state, $c_p\kappa_B\nu T_B Ra_{crit}/gL^3$, is also greater in the stratified case, than in the Boussinesq one, but it is known, that the assumptions regarding the depth dependence of transport coefficients, in particular the thermal diffusivity, can strongly influence the latter results concerning the threshold of convection.

### 3.7.1 Comparison of Anelastic Systems and Relation to the Adiabatic Reference State Formulation

There are different forms of anelastic equations used in the literature, and the differences result from various possible assumptions regarding the fluid properties $k$, $\mu$, $\mu_b$, $c_p$, $c_v$, the form of the heating source term $Q$ and whether or not the gravity is assumed uniform or influenced by the density fluctuations. Moreover the different formulations can result from two most common possibilities for the choice of the reference state, that is either a weakly superadiabatic hydrostatic reference state, satisfying the boundary conditions can be chosen, as e.g. in (3.68a–3.68c) or the adiabatic state can serve as the reference one (see e.g. (3.73a)), but then the boundary conditions on convective fluctuations are non-uniform. However, the reference state can in fact be chosen in an arbitrary way, even as a time-dependent state, but in the latter case its time evolution must be included in the set of equations.

It is often necessary to compare the results of different anelastic formulations, e.g. in computing, when testing numerical codes against some benchmarking solutions. The choice of the reference state does not matter in the sense, that two formulations with different reference state can obviously still lead to the same results. What matters, is first of all, that the physical assumptions are the same, thus in particular that the fluid properties $k$, $\mu$, $\mu_b$, $c_p$, $c_v$, the heating source $Q$ and gravity have the same spatial (and possibly temporal) dependence, the equation of state for the fluid is the same, and the boundary conditions are of the same type. The latter implies, that the departure from adiabatic state, which drives the convective flow and physically can be realized by heating up the bottom boundary and cooling the upper one to generate superadiabatic temperature gradients, must also be the same, if two anelastic systems are expected to produce the same results. In other words one needs to take care so that the excess in temperature jump across the layer over the temperature jump in the adiabatic state relative to the bottom temperature $T_B$

$$\frac{\Delta T}{T_B} - \int_0^L \frac{g\tilde{\alpha}\tilde{T}}{\tilde{c}_p T_B} \mathrm{d}z, \tag{3.413}$$

(equal to $\Delta T/T_B - gL/c_p T_B$ for a perfect gas when $g$ and $c_p$ are uniform), is the same in both compared anelastic formulations. The thermodynamic fluctuations about the conduction and adiabatic reference states by the use of (3.77) are related through

$$T_S(\mathbf{x}, t) = T'(\mathbf{x}, t) + \tilde{T}(z) - T_{ad}(z), \tag{3.414a}$$

$$\rho_S(\mathbf{x}, t) = \rho'(\mathbf{x}, t) + \tilde{\rho}(z) - \rho_{ad}(z), \tag{3.414b}$$

$$p_S(\mathbf{x}, t) = p'(\mathbf{x}, t) + \tilde{p}(z) - p_{ad}(z), \tag{3.414c}$$

$$s_S(\mathbf{x}, t) = s'(\mathbf{x}, t) + \tilde{s}(z) + \text{const}, \tag{3.414d}$$

where the subscript $S$ denotes the superadiabatic fluctuation about the adiabatic state. When the total mass of the fluid is the same in both formulations direct correspondence is achieved, see Sect. 3.2 and Eqs. (3.75a, 3.75b). It is a matter of choice whether or not the total mass is contained in the reference state, but the former option is certainly useful and more clear. In such a case the bottom densities $\rho_B$ and $\rho_{ad\,B}$ in the conduction and adiabatic reference states are related through

$$\int_0^L \tilde{\rho} \mathrm{d}z = \int_0^L \rho_{ad} \mathrm{d}z, \tag{3.415}$$

(cf. (3.76) for an explicit relation between $\rho_B$ and $\rho_{ad\,B}$ in the case of a perfect gas with uniform material properties). The relation (3.415) involves corrections of the order $\mathcal{O}(\delta\rho_B)$, which are important in the transformations (3.414b, 3.414c) for the density and pressure fluctuations, as the bottom values of density $\rho_B$ and $\rho_{ad\,B}$ have to explicitly appear in the expressions for $\tilde{\rho}(z)$, $\rho_{ad}(z)$, $\tilde{p}(z)$ and $p_{ad}(z)$. For example in the case when all the fluid properties and gravity are uniform, and the fluid is described by the equation of state of a perfect gas, by the use of (3.71b), (3.73a) and (3.76) the density fluctuation transformation between the two anelastic formulations takes the form (3.79) with the ratio $\rho_B/\rho_{ad\,B}$ given by (3.80), cf. Sect. 3.2.

The anelastic numerical codes are often constructed based on a non-dimensional form of the dynamical equations. But here again the comparison requires extra care, since the non-dimensional variables are often defined with the use of different scales, such as the density scale, temperature scale, etc. E.g., it is vital to compare Rayleigh numbers, which utilize the same scale definitions, e.g. scales defined by the bottom values of density, temperature, etc. Nevertheless, comparison and full correspondence between two anelastic formulations, with conduction and adiabatic reference state is possible, and in fact not very difficult. Let us provide an example of a non-dimensional form of the dynamical equations under the anelastic approximation, where for simplicity we assume that the viscosities are uniform, $\mu = \text{const}$ and $\mu_b = \text{const}$; this assumption is by no means necessary, and it is used only to make the Navier-Stokes equation somewhat simpler, as it allows to remove some viscous dissipation terms. We introduce the following non-dimensional variables (which can, of course be chosen differently)

$$\mathbf{x} = L\mathbf{x}^\sharp, \quad t = \frac{L^2}{\kappa_B} t^\sharp, \quad \mathbf{u} = \frac{\kappa_B}{L}\mathbf{u}^\sharp, \tag{3.416a}$$

$$\rho = \rho_B \rho^\sharp, \quad T = T_B T^\sharp, \quad p = \frac{\rho_B \kappa_B^2}{L^2} p^\sharp, \quad s = c_{p\,B} s^\sharp, \quad \psi = g_B L \psi^\sharp, \tag{3.416b}$$

$$\alpha = \frac{\alpha^\sharp}{T_B}, \quad \beta = \frac{L^2}{\rho_B \kappa_B^2} \beta^\sharp, \tag{3.416c}$$

$$\mathbf{g} = g_B \mathbf{g}^\sharp, \quad Q = Q_B Q^\sharp, \quad c_p = c_{p\,B} c_p^\sharp, \quad k = k_B k^\sharp. \tag{3.416d}$$

It follows, that the non-dimensional dynamical equations take the form

$$\rho_r^\sharp \left[\frac{\partial \mathbf{u}^\sharp}{\partial t^\sharp} + \left(\mathbf{u}^\sharp \cdot \nabla^\sharp\right)\mathbf{u}^\sharp\right] = -\nabla p'^\sharp + Ra Pr \rho'^\sharp \mathbf{g}_r^\sharp - Ra Pr \rho_r^\sharp \nabla \psi'^\sharp + Pr \nabla^{\sharp 2}\mathbf{u}^\sharp + Pr\left(\frac{1}{3} + \frac{\mu_b}{\mu}\right)\nabla^\sharp\left(\nabla^\sharp \cdot \mathbf{u}^\sharp\right), \tag{3.417a}$$

$$\nabla^\sharp \cdot \left(\rho_r^\sharp \mathbf{u}^\sharp\right) = 0, \qquad \nabla^{\sharp 2}\psi'^\sharp = \frac{4\pi G \rho_B L}{g_B}\rho'^\sharp, \tag{3.417b}$$

$$\rho_r^\sharp T_r^\sharp \left(\frac{\partial s'^\sharp}{\partial t^\sharp} + \mathbf{u}^\sharp \cdot \nabla^\sharp s'^\sharp\right) - \rho_r^\sharp c_{p\,r}^\sharp u_z^\sharp \frac{\Delta_S}{\langle \Delta_S \rangle} = \nabla^\sharp \cdot \left(k^\sharp \nabla T'^\sharp\right) + \frac{2 g_B L}{c_{p\,B} T_B Ra}\left[\mathbf{G}^s : \mathbf{G}^s + \left(\frac{\mu_b}{\mu} - \frac{2}{3}\right)\left(\nabla^\sharp \cdot \mathbf{u}^\sharp\right)^2\right] + \frac{Q_B L}{\kappa_B \rho_B c_{p\,B} \Delta_S} Q'^\sharp. \tag{3.417c}$$

$$\frac{\rho'^\sharp}{\rho_r^\sharp} = -\alpha_r^\sharp T'^\sharp + \beta_r^\sharp p'^\sharp, \qquad s'^\sharp = -\frac{\nu_B \kappa_B}{g_B L^3} \frac{g_B L}{c_{p\,B} T_B Pr} \alpha_r^\sharp \frac{p'^\sharp}{\rho_r^\sharp} + c_{p\,r}^\sharp \frac{T'^\sharp}{T_r^\sharp}, \tag{3.417d}$$

where the subscript $r$ denotes a reference state variable, either the conduction reference state ($\rho_r = \tilde{\rho}$, $T_r = \tilde{T}$, etc.) or the adiabatic reference state ($\rho_r = \rho_{ad}$, $T_r = T_{ad}$, etc.), depending on the formulation used; the term $-\rho_r^\sharp c_{p\,r}^\sharp u_z^\sharp \Delta_S / \langle \Delta_S \rangle$ in the energy equation (3.417c) is absent in the case $r = ad$, i.e. when the reference state is adiabatic. In the above non-dimensional equations the Rayleigh number, which measures the relative strength of buoyancy with respect to dissipation (and provides a useful measure of departure from the critical state), and the Prandtl number (which both must have the same values for the two compared formulations) are defined as follows

$$Ra = \frac{g_B L^3}{\nu_B \kappa_B}\left(\frac{\Delta T}{T_B} - \frac{L}{T_B}\left\langle\frac{g\tilde{\alpha}\tilde{T}}{\tilde{c}_p}\right\rangle\right), \qquad Pr = \frac{\nu_B}{\kappa_B} = \frac{\mu c_{p\,B}}{k_B}. \tag{3.418}$$

The expression in the brackets in the Rayleigh number definition is the aforementioned superadiabatic excess in the temperature jump across the layer, which as remarked must be the same in the both considered formulations of the anelastic approximation. It follows that the non-dimensional parameter $g_B L^3/\nu_B \kappa_B$ must also be the same in two anelastic, physically equivalent systems. Finally the non-dimensional measure of stratification (compressibility) $\theta = \Delta T/T_B$ (or equivalently $\Gamma = T_B/T_T$ or $g_B L/c_{p\,B} T_B$ instead) must coincide as well. Eventually an exemplary set of parameters which need to be compared between two different anelastic formulations in order to expect from them the same physical results (i.e. describing the same physical system) can be written as follows

$$\frac{\Delta T}{T_B} - \frac{L}{T_B}\left\langle\frac{g\tilde{\alpha}\tilde{T}}{\tilde{c}_p}\right\rangle, \quad \frac{g_B L^3}{\nu_B \kappa_B}, \quad \theta = \frac{\Delta T}{T_B}, \quad Pr = \frac{\nu_B}{\kappa_B}, \tag{3.419a}$$

$$\frac{\mu_b}{\mu}, \quad \tilde{\alpha}_B T_B, \quad \frac{Q_B L}{\kappa_B \rho_B c_{p\,B} \Delta_S}, \quad \frac{G \rho_B L}{g_B}, \tag{3.419b}$$

where, as mentioned $\left\langle g\tilde{\alpha}\tilde{T}/\tilde{c}_p\right\rangle = g/c_p$ in the simplest case of an ideal gas and constant $g$ and $c_p$. Of course the parameter $\theta$ could be replaced by another measure of stratification, e.g. $\Gamma = T_B/T_T$ or the non-dimensional measure of the adiabatic gradient $g_B L/c_{p\,B} T_B$. In addition $\tilde{\alpha}$, $\tilde{T}$ and $\tilde{c}_p$ could be replaced by their profiles (values) in the adiabatic state; in any case the conduction reference state is determined by the boundary conditions, which must be the same in both formulations, since they are responsible for the driving and consequently $\tilde{\alpha}$, $\tilde{T}$ and $\tilde{c}_p$ and therefore all the above parameters in (3.419a, 3.419b) can be evaluated no matter the formulation. The three bottom-row parameters in (3.419b) are irrelevant in the case of a perfect gas with negligible bulk viscosity, no heating source, $Q = 0$ and uniform gravity. Note, that the scales of thermodynamic variables do not need to be defined as their values at the bottom, and e.g. the mid-plane values are often used instead.

### 3.7.2 Anelastic Versus Boussinesq Approximations

The transformation of the anelastic system of equations under the Boussinesq limit $\mathrm{d}\tilde{\rho}/\mathrm{d}z \ll \bar{\rho}/L$, $\mathrm{d}\tilde{T}/\mathrm{d}z \ll \bar{T}/L$ and $\mathrm{d}\tilde{p}/\mathrm{d}z \ll \bar{p}/L$ has been thoroughly described in Sect. 3.1.3. Since $\theta = \Delta T/T_B = \mathcal{O}(\epsilon)$, where $\epsilon = \Delta\tilde{\tilde{\rho}}/\bar{\rho}$, the Boussinesq limit simply corresponds to taking the formal limit $\theta \to 0$ (or equivalently $\Gamma \to 1$) in the anelastic equations, but one must bare in mind, that this also implies $\delta \lesssim \mathcal{O}(\theta)$, $c_p = \mathcal{O}(\theta^{-1} gL/\tilde{T})$ likewise $\mathscr{U} = \mathcal{O}(\theta^{1/2}\sqrt{gL})$ and $\mathscr{T} = \mathcal{O}(\theta^{-1/2}\sqrt{L/g})$. The thermodynamic variables then scale as in (2.28a–2.28d), where $\epsilon$ may be replaced by $\theta$, that is $T'/\tilde{T} = \mathcal{O}(\theta)$, $\rho'/\tilde{\rho} = \mathcal{O}(\theta)$, but the mean pressure is boosted so that $p'/\tilde{p} = \mathcal{O}(\theta T'/\tilde{T}) = \mathcal{O}(\theta^2)$; moreover $s'/c_p = \mathcal{O}(T'/\tilde{T}) = \mathcal{O}(\theta)$ likewise $\tilde{\tilde{s}}/c_p = \mathcal{O}(\theta)$, but $\tilde{s}/c_p = \mathcal{O}(1)$. Note, that the scalings for the thermodynamic variables are fully consistent with the reference state solutions for perfect gas given in (3.68a–3.68c) and with (3.69).

The significant difference between the hydrostatic reference states in the anelastic and Boussinesq approximations can be seen on Fig. 3.4. The characteristic feature of Boussinesq systems is that the mean temperature, density and pressure are large compared to the static state variations, whereas it is not the case in the anelastic systems. Thus the latter are stratified and the density and temperature stratification can be strong, which means, that the convective flow is compressible $\nabla \cdot \mathbf{u} \neq 0$, contrary to the Boussinesq case. Moreover, the leading-order hydrostatic balance within the Boussinesq approximation, $\mathrm{d}_z\tilde{\tilde{p}} \approx -\bar{\rho}g$, requires relatively strong hydrostatic pressure variations, $\tilde{\tilde{p}} \sim \bar{\rho}gL$, and this aided by the fundamental assumption $|\tilde{\tilde{p}}|/\bar{p} \sim \epsilon \ll 1$ implies that the mean pressure has to be extremely high, of the order $\bar{p} \sim \epsilon^{-1}\bar{\rho}gL$. A particular consequence is that within the Boussinesq approximation the entropy fluctuation is equivalent to the temperature fluctuation up to a constant factor, $s' \approx \bar{c}_p T'/\bar{T}$; the latter coincidence of $s'$ and $T'$ is *not* satisfied under the anelastic approximation and in the anelastic formulation the mean pressure is of the same order as pressure in the static reference state. It is also important to realize, that Boussinesq convection is not necessarily weakly superadiabatic and the ratio of the temperature gradient in the static reference state to the adiabatic gradient, $\left|c_p \mathrm{d}_z\tilde{\tilde{T}}/g\alpha T\right|$, can be much greater than unity for Boussinesq systems. A schematic Fig. 3.5 depicts regions of validity of the anelastic and Boussinesq approximations on the $\delta\epsilon$ plane; the limit marked AB obtained for $\delta \ll 1$ and $\epsilon \ll 1$ is where the two approximations merge.

Finally, a very important difference between the two approximations is that in Boussinesq convection the thermal energy strongly dominates over the kinetic energy (even the part of the thermal energy associated with temperature fluctuation only, is still $\epsilon^{-1}$ times greater than the kinetic energy). Consequently the viscous heating in the energy balance is negligible and thus in the absence of radiation the only process contributing to the heat exchange between a fluid parcel and its surroundings within the Boussinesq approximation is the molecular heat conduction,

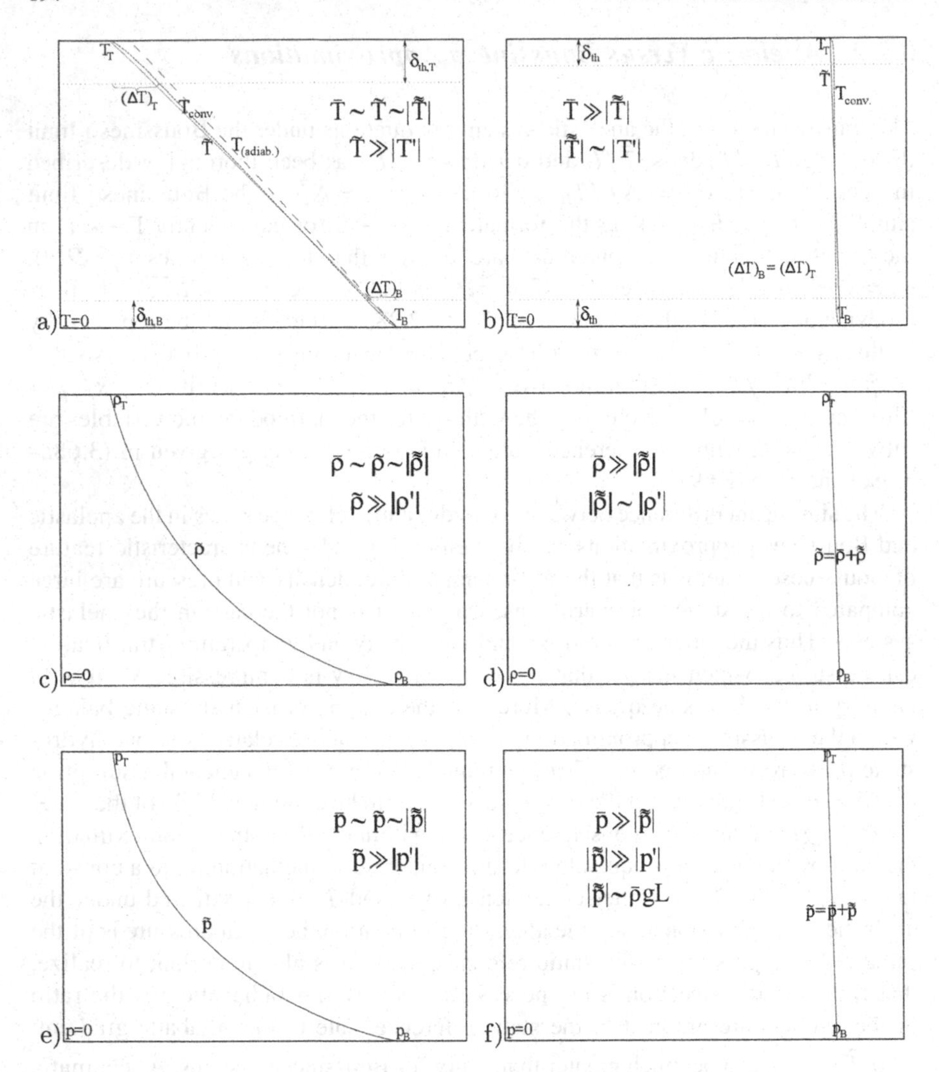


**Fig. 3.4** A schematic comparison of physical situations described by anelastic (left column) and Boussinesq (right column) approximations. The Boussinesq approximation is characterized by the fact, that the total variations of temperature, density and pressure are very small compared to their means, whereas in the anelastic approximation only the fluctuations induced by the convective flow are small. Moreover, within the Boussinesq approximation the vertical profiles of temperature, density and pressure in the hydrostatic state are approximately linear (2.29a–2.29c) and the mean pressure is extremely high, $\bar{p} \sim \epsilon^{-1}\bar{\rho} g L$

$$\frac{\text{đ}q_{Bq}}{\mathrm{d}t} = T\frac{\mathrm{d}s}{\mathrm{d}t} = \frac{1}{\rho}\nabla \cdot (k\nabla T), \tag{3.420}$$

where $Bq$ stands for *Boussinesq*. On the contrary, in anelastic convection the kinetic energy is comparable with the thermal one and thus the viscous heating strongly

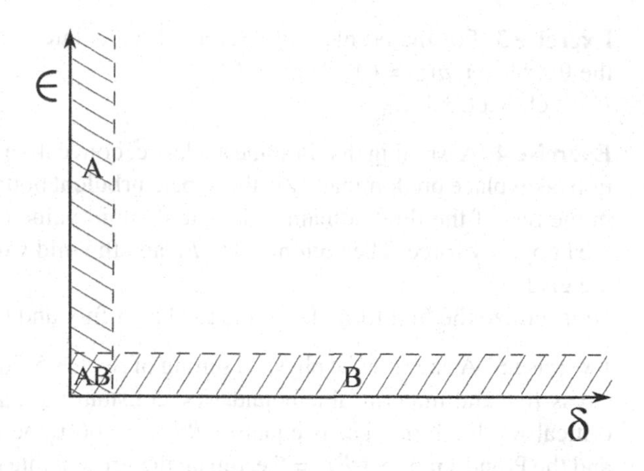


**Fig. 3.5** Schematic comparison of the anelastic ($\delta \ll 1$, cf. (3.11); marked A) and Boussinesq ($\epsilon \ll 1$, cf. (2.6); marked B) approximations on the $\delta\epsilon$ plane. The small region where $\delta \ll 1$ and $\epsilon \ll 1$ marked AB is where both the approximations are valid

contributes to the total heat flux in the system. Therefore in the absence of radiation the contributions to heat exchanged by a fluid parcel with its surroundings come from both, the molecular conduction and viscous friction,

$$\frac{\text{đ}q_{Ac}}{\mathrm{d}t} = T\frac{\mathrm{d}s}{\mathrm{d}t} = \frac{1}{\rho}\nabla \cdot (k\nabla T) + 2\frac{\mu}{\rho}\mathbf{G}^s : \mathbf{G}^s + \left(\frac{\mu_b}{\rho} - \frac{2\mu}{3\rho}\right)(\nabla \cdot \mathbf{u})^2, \quad (3.421)$$

where $Ac$ stands for *Anelastic*. Furthermore, the work done by the buoyancy force is also comparable with the thermal energy. Therefore the effects of viscous heating and of the work done by buoyancy have a substantial influence on the total, horizontally averaged convective heat flux, which varies with height, contrary to Boussinesq systems, where the mean heat flux is uniform.

## Review Exercises

**Exercise 1** Solve for a hydrostatic reference state in a perfect gas under presence of uniform heat sink $Q = -Q_0$, $Q_0 = \text{const} > 0$. Assume $g = \text{const}$, $k = \text{const}$ and that the bottom and top temperatures $T_B$ and $T_T$ are held constant; consider the values of $\rho_B$, $T_B$ and $T_T$ as given. Determine the small anelastic parameter $\delta$.
*Hint*: use Eqs. (3.67) with the second one modified by the presence of a heat sink $\mathrm{d}^2\tilde{T}/\mathrm{d}z^2 = Q_0/k$.

**Exercise 2** For the previous problem determine the explicit transformations between the superadiabatic variables $T_S = T - T_{ad}$, $\rho_S = \rho - \rho_{ad}$ and the fluctuations about the calculated reference state $T'$, $\rho'$.
*Hint*: cf. (3.78) and (3.79).

**Exercise 3** For the problem of Exercise 1, calculate the total pressure jump across the fluid layer $p(z = 0) - p(z = L)$.
*Hint*: cf. Sect. 3.1.2.

**Exercise 4** Assuming that in some anelastic, convective system the viscous dissipation takes place predominantly in the upper, turbulent boundary layer and is negligible in the rest of the fluid domain, calculate the mid value (at $z = L/2$) of the work of the buoyancy force. The bottom value $F_B$ and the mid value $F_M$ of the total heat flux are given.
*Hint*: utilize the first formula for the total heat flux and (3.141).

**Exercise 5** Assuming the physical setting of Sect. 3.5 (ideal gas, $Q = 0$, isothermal, stress-free and impermeable boundaries, constant $\nu$, $k$, $\mathbf{g}$ and $c_p$) and given that the critical Rayleigh number is equal to $Ra_{crit} = 660$, the polytropic index $m = 1.49$ and the Prandtl number $Pr = 5$ estimate the growth rate of convection at $Ra = 670$.
*Answer*: $\sigma \approx 0.189 L^2 c_p \rho_B / k$, ($\theta \approx 0.015$).

## References

Anufriev, A.P., C.A. Jones, and A.M. Soward. 2005. The Boussinesq and anelastic liquid approximations for convection in the Earth's core. *Physics of the Earth and Planetary Interiors* 152: 163–190.

Bannon, P.R. 1996. On the anelastic approximation for a compressible atmosphere. *Journal of the Atmospheric Sciences* 53 (23): 3618–3628.

Braginsky, S.I., and P.H. Roberts. 1995. Equations governing convection in Earth's core and the geodynamo. *Geophysical and Astrophysical Fluid Dynamics* 79: 1–97.

Calkins, M.A., K. Julien, and P. Marti. 2015. The breakdown of the anelastic approximation in rotating compressible convection: implications for astrophysical systems. *Proceedings of the Royal Society A* 471: 20140689.

Deardorff, J.W. 1970. Convective velocity and temperature scales for the unstable planetary boundary layer and for Rayleigh convection. *Journal of the Atmospheric Sciences* 27: 1211–1213.

Drew, S.J., C.A. Jones, and K. Zhang. 1995. Onset of convection in a rapidly rotating compressible fluid spherical shell. *Geophysical and Astrophysical Fluid Dynamics* 80: 241–254.

Durran, D.R. 1989. Improving the anelastic approximation. *Journal of the Atmospheric Sciences* 46 (11): 1453–1461.

Durran, D.R. 2008. A physically motivated approach for filtering acoustic waves from the equations governing compressible stratified flow. *Journal of Fluid Mechanics* 601: 365–379.

Feireisl, E., and A. Novotný. 2017. *Singular limits in thermodynamics of viscous fluids*. Cham, Birkhäuser: Springer International Publishing AG.

Gilman, P.A., and G.A. Glatzmaier. 1981. Compressible convection in a rotating spherical shell I. Anelastic equations. *The Astrophysical Journal Supplement Series* 45: 335–349.

Glatzmaier, G.A., and P.A. Gilman. 1981a. Compressible convection in a rotating spherical shell II. A linear anelastic model. *The Astrophysical Journal Supplement Series* 45: 351–380.

Glatzmaier, G.A., and P.A. Gilman. 1981b. Compressible convection in a rotating spherical shell IV. Effects of viscosity, conductivity, boundary conditions, and zone depth. *The Astrophysical Journal Supplement Series* 47: 103–115.

Goody, R.M. 1956. The influence of radiative transfer on cellular convection. *Journal of Fluid Mechanics* 1 (4): 424–435 (Note: Goody, R.M. 1956. Corrigendum. *Journal of Fluid Mechanics* 1 (6): 670).

Goody, R.M., and Y.L. Yung. 1989. *Atmospheric radiation: theoretical basis*. New York: Oxford University Press.

Gough, D.O. 1969. The anelastic approximation for thermal convection. *Journal of the Atmospheric Sciences* 26: 448–456.

Grossmann, S., and D. Lohse. 2000. Scaling in thermal convection: a unifying theory. *Journal of Fluid Mechanics* 407: 27–56.

Jones, C.A., K.A. Mizerski, and M. Kessar. 2020. Fully developed anelastic convection with no-slip boundaries. *Journal of Fluid Mechanics* (in preparation).

Kamari, A., A.H. Mohammadi, A. Bahadori, and S. Zendehboudi. 2014. Prediction of air specific heat ratios at elevated pressures using a novel modeling approach. *Chemical Engineering and Technology* 37 (12): 2047–2055.

Klein, R. 2009. Asymptotics, structure, and integration of sound-proof atmospheric flow equations. *Theoretical and Computational Fluid Dynamics* 23: 161–195.

Klein, R., and O. Pauluis. 2012. Thermodynamic consistency of a pseudoincompressible approximation for general equations of state. *Journal of the Atmospheric Sciences* 69: 961–968.

Klein, R., U. Achatz, D. Bresch, O.M. Knio, and P.K. Smolarkiewicz. 2010. Regime of validity of soundproof atmospheric flow models. *Journal of the Atmospheric Sciences* 67: 3226–3237.

Korn, G.A., and T.M. Korn. 1961. *Mathematical handbook for scientists and engineers*. Mineola, New York: Dover Publications.

Lantz, S., and Y. Fan. 1999. Anelastic magnetohydrodynamic equations for modeling solar and stellar convection zones. *The Astrophysical Journal Supplement Series* 121: 247–264.

Lilly, D.K. 1996. A comparison of incompressible, anelastic and Boussinesq dynamics. *Atmospheric Research* 40: 143–151.

Lipps, F.B., and R.S. Hemler. 1982. A scale analysis of deep moist convection and some related numerical calculations. *Journal of the Atmospheric Sciences* 39: 2192–2210.

Mizerski, K.A. 2017. Rigorous entropy formulation of the anelastic liquid equations in an ideal gas. *Journal of Fluid Mechanics* 833: 677–686.

Mizerski, K.A., and S.M. Tobias. 2011. The effect of stratification and compressibility on anelastic convection in a rotating plane layer. *Geophysical and Astrophysical Fluid Dynamics* 105 (6): 566–585.

Ogura, Y., and N. Phillips. 1962. Scale analysis of deep and shallow convection in the atmosphere. *Journal of the Atmospheric Sciences* 19: 173–179.

Perry, R.H., D.W. Green, and J.O. Maloney. 1997. *Perry's chemical engineers' handbook*. New York: McGraw-Hill.

Phillips, W.D. 1998. Nobel lecture: laser cooling and trapping of neutral atoms. *Reviews of Modern Physics* 70 (3): 721–741.

Soward, A.M. 1991. The Earth's dynamo. *Geophysical and Astrophysical Fluid Dynamics* 62: 191–209.

Tobias, S.M., and N.O. Weiss. 2007a. The solar dynamo and the tachocline. In *The solar tachocline*, ed. D.W. Hughes, R. Rosner, and N.O. Weiss. Cambridge: Cambridge University Press.

Tobias, S.M., and N.O. Weiss. 2007b. Stellar dynamos. In *Mathematical aspects of natural dynamos*, ed. E. Dormy, and A.M. Soward. Boca Raton: CRC Press.

Vasil, G.M., D. Lecoanet, B.P. Brown, T.S. Wood, and E.G. Zweibel. 2013. Energy conservation and gravity waves in sound-proof treatments of stellar interiors. II. Lagrangian constrained analysis. *The Astrophysical Journal* 773: 169 (23pp.).

Verhoeven, J., and G.A. Glatzmaier. 2018. Validity of sound-proof approaches in rapidly-rotating compressible convection: marginal stability versus turbulence. *Geophysical and Astrophysical Fluid Dynamics* 112 (1): 36–61.

Verhoeven, J., T. Wiesehöfer, and S. Stellmach. 2015. Anelastic versus fully compressible turbulent Rayleigh-Bénard convection. *The Astrophysical Journal* 805: 62.

Wilhelmson, R., and Y. Ogura. 1972. The pressure perturbation and the numerical modeling of a cloud. *Journal of the Atmospheric Sciences* 29: 1295–1307.

# Chapter 4
# Inclusion of Compositional Effects

Thermal convection considered up to now is not the only type of buoyancy-driven flows. A very common type of convective flows occurring in natural systems is the compositional convection in binary alloys, composed of light and heavy constituents, when the buoyancy is generated via local increase in the concentration of the light constituent. This type of driving is e.g. an important energy source for convective motions in the Earth's liquid core, where at the inner-outer core boundary the process of iron solidification leads to excess in the concentration of the light constituents in the core alloy, composed mainly of iron.

We now turn to the fundamental ideas regarding buoyancy generation in liquids via thermal and compositional effects. A comprehensive derivation of the full set of equations describing the dynamics of convection driven by the compositional and thermal effects can also be found e.g. in Landau and Lifshitz (1987) or Braginsky and Roberts (1995); the latter is done in the Earth's core context. The case, when driving is solely thermal has been explained in Sect. 1.5.1. Let us start by introducing the symbol

$$\xi = \frac{\rho_l}{\rho}, \tag{4.1}$$

to denote the mass fraction of the light component in a binary alloy, where $\rho_l$ is the mass of the light constituent in a unitary volume and $\rho$ is the total mass of that volume; in other words $\rho$ is the density field of the entire liquid alloy, $\rho = \rho_l + \rho_h$, where subscript $h$ denotes the heavy constituent. In the following we assume that there are *no chemical reactions* between the light and heavy components of the alloy. In a static state, i.e. when no motions are present the thermodynamic fields $\rho$, $T$, $p$, $s$ and $\xi$ are stationary but height-dependent. Stability of a fluid layer, with temperature and mass fraction gradients, can be studied by considering two infinitesimally spaced horizontal (perpendicular to gravity) fluid layers situated at heights $z_0$ and $z_0 + \mathrm{d}z$.



K. A. Mizerski, *Foundations of Convection with Density Stratification*, GeoPlanet: Earth and Planetary Sciences, https://doi.org/10.1007/978-3-030-63054-6_4

A very similar type of reasoning as in Sect. 1.5.1 can be put forward now. A fluid volume of unitary mass $V = 1/\rho$ taken from the level $z_0$ and placed slightly higher at $z_0 + \mathrm{d}z$ experiences a thermodynamic transformation, since its temperature, pressure and chemical potential,[1] denoted here by $\mu_c$, will start adjusting to the environment at the higher level. If the fluid volume after the transformation becomes denser than the surroundings, the gravity will act to put it back at the original level $z_0$ and then the situation is stable; in the opposite case the buoyancy is non-zero and the system looses stability.

Let us take the pressure $p(z)$, the entropy $s(z)$ and the mass fraction $\xi(z)$ as the system parameters, then the fluid volume, which initially is

$$V\left(p(z_0), s(z_0), \xi(z_0)\right), \tag{4.2}$$

after the shift and the thermodynamic transformation which adjusts the pressure to the value $p(z_0 + \mathrm{d}z)$ of the surrounding fluid at the higher level changes to

$$V\left(p(z_0 + \mathrm{d}z), s(z_0) + \mathrm{D}s, \xi(z_0) + \mathrm{D}\xi\right), \tag{4.3}$$

where $s(z_0) + \mathrm{D}s$ and $\xi(z_0) + \mathrm{D}\xi$ denote the entropy and mass fraction after the thermodynamic process ($\mathrm{D}s$ is the entropy change and $\mathrm{D}\xi$ is the mass fraction change in the process). We can now simplify things slightly, by considering the most constrained case of adiabatic transformation of the fluid volume at constant $s$ and $\xi$, thus with $\mathrm{D}s = 0$ and $\mathrm{D}\xi = 0$; in that way we will obtain a sufficient and strongest stability restriction. The hydrostatic state is stable as long as the fluid parcel after the transformation is denser than the surroundings at the new level $z_0 + \mathrm{d}z$, i.e. when

---

[1] Because we have used $\mu$ to denote the shear dynamic viscosity throughout the previous chapters, the chemical potential will be denoted here by $\mu_c$; moreover, to avoid any confusion we will only use the symbols $\nu$ and $\nu_b$ to denote kinematic shear and bulk viscosities, whereas the dynamic ones will be simply denoted by $\tilde{\rho}\nu$ and $\tilde{\rho}\nu_b$.

The introduction of one chemical potential $\mu_c$ for a binary mixture composed of light and heavy constituents follows from writing down the differential of the internal energy

$$\mathrm{d}\mathcal{E} = T\mathrm{d}S - p\mathrm{d}V + \mu^{(l)}\mathrm{d}N^{(l)} + \mu^{(h)}\mathrm{d}N^{(h)},$$

where $\mathcal{E}$ and $S$ are the actual internal energy and entropy of the alloy (as opposed to the mass densities $\varepsilon$ and $s$), $V$ is the volume, $(\mu^{(l)}, N^{(l)})$ and $(\mu^{(h)}, N^{(h)})$ are the pairs of chemical potentials and the numbers of particles of the light and heavy constituents respectively; division by the total mass $M = m_l N^{(l)} + m_h N^{(h)}$, where $m_l$ and $m_h$ denote the molecular masses for both constituents, allows to transform the above into the differential for the internal energy mass density, which takes the form

$$\mathrm{d}\varepsilon = T\mathrm{d}s - p\mathrm{d}\left(\frac{1}{\rho}\right) + \mu_c \mathrm{d}\xi,$$

where

$$\mu_c = \frac{\mu^{(l)}}{m_l} - \frac{\mu^{(h)}}{m_h}, \quad \xi = \frac{m_l N^{(l)}}{M} = \frac{\rho_l}{\rho}.$$

$$V\left(p(z_0+\mathrm{d}z), s(z_0), \xi(z_0)\right) < V\left(p(z_0+\mathrm{d}z), s(z_0+\mathrm{d}z), \xi(z_0+\mathrm{d}z)\right). \tag{4.4}$$

Expanding the right hand side of the latter inequality in the entropy and mass fraction about the values at the level $z_0$ we get

$$0 < \left(\frac{\partial V}{\partial s}\right)_{p,\xi} \mathrm{d}s + \left(\frac{\partial V}{\partial \xi}\right)_{p,s} \mathrm{d}\xi, \tag{4.5}$$

where $\mathrm{d}s = s(z_0+\mathrm{d}z) - s(z_0)$ and $\mathrm{d}\xi = \xi(z_0+\mathrm{d}z) - \xi(z_0)$. Furthermore, the coefficient of thermal expansion $\alpha$ (1.39) likewise the coefficient of isentropic compositional expansion

$$\chi = \frac{1}{V}\left(\frac{\partial V}{\partial \xi}\right)_{p,s} = -\frac{1}{\rho}\left(\frac{\partial \rho}{\partial \xi}\right)_{p,s}, \tag{4.6}$$

are both positive for all standard fluids/binary alloys, therefore

$$\left(\frac{\partial V}{\partial s}\right)_{p,\xi} = \frac{\left(\frac{\partial V}{\partial T}\right)_{p,\xi}}{\left(\frac{\partial s}{\partial T}\right)_{p,\xi}} = \frac{\alpha T}{\rho c_{p,\xi}} > 0, \tag{4.7}$$

$$\left(\frac{\partial V}{\partial \xi}\right)_{p,s} = \frac{\chi}{\rho} > 0. \tag{4.8}$$

which further implies, that the stability condition (4.5) can be expressed in the following way

$$0 < \frac{\alpha T}{\rho c_{p,\xi}} \frac{\mathrm{d}s}{\mathrm{d}z} + \frac{\chi}{\rho} \frac{\mathrm{d}\xi}{\mathrm{d}z}. \tag{4.9}$$

This means that an adiabatic well-mixed state characterized by uniform entropy $\mathrm{d}_z s = 0$ and mass fraction $\mathrm{d}_z \xi = 0$ corresponds to instability threshold in the absence of dissipative effects; in particular if both the entropy and mass fraction gradients become negative in the absence of diffusion, the system becomes unstable. Note however, that in fact the stability problem is governed by a sum of the two gradients and the system may become unstable when only one, say the mass fraction gradient, is negative and overcomes the effect of positive entropy gradient.

It can be seen now, that anelastic convection driven by both mechanisms—thermal and compositional, has to be based on the requirement, that the convective state departures only weakly from the well-mixed, adiabatic state and therefore both the gradients $\mathrm{d}_z s$ and $\mathrm{d}_z \xi$ must be small. Next let us obtain the expressions for vertical gradients of the intensive parameters $T$, $p$ and $\mu_c$ in the well-mixed, adiabatic, hydrostatic state. The pressure gradient is easily obtained from the hydrostatic momentum balance, $\mathrm{d}_z p = -\rho g$. This allows to express the entropy gradient in terms of the temperature, pressure and mass fraction gradients in the following way,

$$0 = \frac{\mathrm{d}s}{\mathrm{d}z} = \left(\frac{\partial s}{\partial T}\right)_{p,\xi}\frac{\mathrm{d}T}{\mathrm{d}z} + \left(\frac{\partial s}{\partial p}\right)_{T,\xi}\frac{\mathrm{d}p}{\mathrm{d}z} + \left(\frac{\partial s}{\partial \xi}\right)_{T,p}\frac{\mathrm{d}\xi}{\mathrm{d}z} = \frac{c_{p,\xi}}{T}\frac{\mathrm{d}T}{\mathrm{d}z} + \alpha g, \tag{4.10}$$

where we have used the Maxwell identity $(\partial s/\partial p)_T = -(\partial V/\partial T)_p$ and $\mathrm{d}_z\xi = 0$ for the well-mixed state. From the latter we can infer, that the standard formula for the adiabatic gradient still applies, that is the adiabatic, well-mixed state is characterized by

$$\frac{\mathrm{d}T}{\mathrm{d}z} = -\frac{\alpha T g}{c_{p,\xi}}. \tag{4.11}$$

Expressing now the mass fraction gradient in terms of the temperature, pressure and chemical potential gradients

$$\begin{aligned} 0 = \frac{\mathrm{d}\xi}{\mathrm{d}z} &= \left(\frac{\partial \xi}{\partial T}\right)_{p,\mu_c}\frac{\mathrm{d}T}{\mathrm{d}z} + \left(\frac{\partial \xi}{\partial p}\right)_{T,\mu_c}\frac{\mathrm{d}p}{\mathrm{d}z} + \left(\frac{\partial \xi}{\partial \mu_c}\right)_{p,T}\frac{\mathrm{d}\mu_c}{\mathrm{d}z} \\ &= -\left(\frac{\partial \xi}{\partial T}\right)_{p,\mu_c}\frac{\alpha T g}{c_{p,\xi}} - \left(\frac{\partial \xi}{\partial p}\right)_{T,\mu_c}\rho g + \left(\frac{\partial \xi}{\partial \mu_c}\right)_{p,T}\frac{\mathrm{d}\mu_c}{\mathrm{d}z}, \end{aligned} \tag{4.12}$$

leads to

$$\frac{\mathrm{d}\mu_c}{\mathrm{d}z} = \Upsilon g\left[\left(\frac{\partial \xi}{\partial p}\right)_{T,\mu_c}\rho + \left(\frac{\partial \xi}{\partial T}\right)_{p,\mu_c}\frac{\alpha T}{c_{p,\xi}}\right], \tag{4.13}$$

where we have introduced

$$\Upsilon = \left(\frac{\partial \mu_c}{\partial \xi}\right)_{p,T}. \tag{4.14}$$

However, utilizing the implicit function theorem we may write

$$\left(\frac{\partial \xi}{\partial p}\right)_{T,\mu_c} = -\frac{\left(\frac{\partial \mu_c}{\partial p}\right)_{T,\xi}}{\left(\frac{\partial \mu_c}{\partial \xi}\right)_{p,T}} = -\frac{\chi_T}{\rho\Upsilon}, \tag{4.15}$$

$$\left(\frac{\partial \xi}{\partial T}\right)_{p,\mu_c} = -\frac{\left(\frac{\partial \mu_c}{\partial T}\right)_{p,\xi}}{\left(\frac{\partial \mu_c}{\partial \xi}\right)_{p,T}} = \frac{h_{p,T}}{T\Upsilon}, \tag{4.16}$$

where

$$\chi_T = -\frac{1}{\rho}\left(\frac{\partial \rho}{\partial \xi}\right)_{p,T} = \rho\left(\frac{\partial \mu_c}{\partial p}\right)_{T,\xi}, \tag{4.17}$$

is the coefficient of isothermal compositional expansion (the last equality is simply a Maxwell relation) and the coefficient

$$h_{p,T} = T\left(\frac{\partial s}{\partial \xi}\right)_{p,T} = -T\left(\frac{\partial \mu_c}{\partial T}\right)_{p,\xi}, \tag{4.18}$$

(again, the last equality is a Maxwell relation) is a thermodynamic property of the fluid, which is related (but not directly ![2]) to the amount of heat delivered to the system when the mass fraction of the light constituent is increased at constant temperature and pressure (sometimes termed "the heat of reaction").

Equation (4.13) can now be rewritten in the form

$$\frac{\mathrm{d}\mu_c}{\mathrm{d}z} = -g\left[\chi_T - \frac{\alpha h_{p,T}}{c_{p,\xi}}\right]. \tag{4.19}$$

Furthermore, the coefficients of isentropic and isothermal compositional expansions are related through the following equation

$$\begin{aligned}\chi = &-\frac{1}{\rho}\left(\frac{\partial \rho}{\partial \xi}\right)_{p,s} = -\frac{1}{\rho}\left[\left(\frac{\partial \rho}{\partial \xi}\right)_{p,T} + \left(\frac{\partial \rho}{\partial T}\right)_{p,\xi}\left(\frac{\partial T}{\partial \xi}\right)_{p,s}\right] \\ &= \chi_T - \frac{\alpha h_{p,T}}{c_{p,\xi}},\end{aligned} \tag{4.20}$$

where we have utilized the implicit function theorem to obtain $(\partial T/\partial \xi)_{p,s} = -h_{p,T}/c_{p,\xi}$, which allows to express the chemical potential gradient in the well-mixed, adiabatic and hydrostatic state in the most compact form

$$\frac{\mathrm{d}\mu_c}{\mathrm{d}z} = -\chi g. \tag{4.21}$$

The full expressions for the entropy and mass fraction gradient in (4.10) and in (4.12) yield,

$$\frac{\mathrm{d}s}{\mathrm{d}z} = \frac{c_{p,\xi}}{T}\left(\frac{\mathrm{d}T}{\mathrm{d}z} + \frac{\alpha T g}{c_{p,\xi}}\right) + \frac{h_{p,T}}{T}\frac{\mathrm{d}\xi}{\mathrm{d}z}, \tag{4.22a}$$

$$\begin{aligned}\frac{\mathrm{d}\xi}{\mathrm{d}z} &= \frac{h_{p,T}}{T\Upsilon}\frac{\mathrm{d}T}{\mathrm{d}z} + \frac{\chi_T g}{\Upsilon} + \frac{1}{\Upsilon}\frac{\mathrm{d}\mu_c}{\mathrm{d}z} \\ &= \frac{h_{p,T}}{T\Upsilon}\left(\frac{\mathrm{d}T}{\mathrm{d}z} + \frac{\alpha T g}{c_{p,\xi}}\right) + \frac{1}{\Upsilon}\left(\chi g + \frac{\mathrm{d}\mu_c}{\mathrm{d}z}\right),\end{aligned} \tag{4.22b}$$

hence the stability condition (4.9), can now be cast in the form

---

[2] The relation is not direct, since as we will show in Sect. 4.1 in a binary alloy composed of the light and heavy constituents the total heat delivered to the unitary mass in an infinitesimal process can not be expressed solely by $T\mathrm{d}s$, and there are contributions from the process of chemical potential equilibration.

$$0 < \frac{\alpha T}{\rho c_{p,\xi}} \frac{\mathrm{d}s}{\mathrm{d}z} + \frac{\chi}{\rho} \frac{\mathrm{d}\xi}{\mathrm{d}z} = \frac{\alpha}{\rho}\left(\frac{\mathrm{d}T}{\mathrm{d}z} + \frac{\alpha T g}{c_{p,\xi}}\right) + \frac{\chi_T}{\rho} \frac{\mathrm{d}\xi}{\mathrm{d}z}$$
$$= \left(\frac{\alpha}{\rho} + \frac{\chi_T h_{p,T}}{\rho T \Upsilon}\right)\left(\frac{\mathrm{d}T}{\mathrm{d}z} + \frac{\alpha T g}{c_{p,\xi}}\right) + \frac{\chi_T}{\rho \Upsilon}\left(\frac{\mathrm{d}\mu_c}{\mathrm{d}z} + \chi g\right). \tag{4.23}$$

We recall, that this condition does not involve the effect of heat exchange and material diffusion between a fluid parcel and surroundings, thus providing a strongest stability restriction, which is typically weakened by the presence of dissipation. Since $\alpha$ and $\chi$ are positive for standard binary alloys we can easily formulate the following sufficient (but *not necessary*) conditions for stability

$$\frac{\mathrm{d}s}{\mathrm{d}z} > 0, \quad \text{and} \quad \frac{\mathrm{d}\xi}{\mathrm{d}z} > 0. \tag{4.24}$$

Additionally, another sufficient (but *not necessary*) set of stability conditions can be formulated, based on $\chi_T > 0$,

$$-\frac{\mathrm{d}T}{\mathrm{d}z} < \frac{g\alpha T}{c_{p,\xi}}, \quad \text{and} \quad \frac{\mathrm{d}\xi}{\mathrm{d}z} > 0, \tag{4.25}$$

but we emphasize, that it is not directly equivalent to the former one (4.24), since the former can be transformed to

$$-\frac{\mathrm{d}T}{\mathrm{d}z} < \frac{g\alpha T}{c_{p,\xi}} + \frac{\chi_T - \chi}{\alpha} \frac{\mathrm{d}\xi}{\mathrm{d}z}, \quad \text{and} \quad \frac{\mathrm{d}\xi}{\mathrm{d}z} > 0, \tag{4.26}$$

and whether it is a stronger or a weaker restriction on the temperature gradient than (4.25) depends on the sign of the "heat of reaction" $h_{p,T} = c_{p,\xi}(\chi_T - \chi)/\alpha$. Nevertheless, we can conclude that convection does not develop when the negative temperature gradient is below the adiabatic one and the mass fraction gradient is positive, i.e. exceeds that of the "well-mixed" uniform profile. Thus the buoyancy forces may only start to appear when the temperature gradient exceeds that of the adiabatic profile and/or the mass fraction gradient exceeds zero (as explained above, convection can develop if only one of the gradients exceeds the threshold value strongly enough, to overcome the stabilizing effect of the other one).

The last line of (4.23) expresses the stability condition in terms of the temperature and chemical potential gradients. It is of interest to observe, that imposing simultaneously $-\mathrm{d}_z T < g\alpha T/c_{p,\xi}$ and $-\mathrm{d}_z \mu_c < \chi g$ does not guarantee stability, because the sign of the coefficient $h_{p,T}$ is not specified. Therefore the stability condition (4.23) in terms of temperature and chemical potential gradients is simply expressed in the following way

$$-\frac{\mathrm{d}\mu_c}{\mathrm{d}z} < \chi g + \left(\frac{\alpha \Upsilon}{\chi_T} + \frac{h_{p,T}}{T}\right)\left(\frac{\mathrm{d}T}{\mathrm{d}z} + \frac{\alpha T g}{c_{p,\xi}}\right), \tag{4.27}$$

where we have utilized the fact, that $\Upsilon > 0$ (cf. Landau and Lifshitz 1980, p. 288, Eq. (96.7), chapter on "Thermodynamic inequalities for solutions", where this property of binary alloys is derived directly from the minimal work principle).

However, in natural systems convection is typically driven by only weak departure from the adiabatic, well-mixed state. In such a case, which corresponds to the first fundamental assumption of the anelastic approximation, all the following quantities are small

$$-\frac{L}{c_{p,\xi}}\frac{\mathrm{d}s}{\mathrm{d}z} \ll 1, \quad -L\frac{\mathrm{d}\xi}{\mathrm{d}z} \ll 1, \quad -\alpha L\left(\frac{\mathrm{d}T}{\mathrm{d}z} + \frac{g\alpha T}{c_{p,\xi}}\right) \ll 1, \tag{4.28}$$

and as long as $\Upsilon$ remains an order unity quantity, also

$$-\frac{\chi L}{\Upsilon}\left(\frac{\mathrm{d}\mu_c}{\mathrm{d}z} + \chi g\right) \ll 1 \tag{4.29}$$

must be small (which is not true in weak solutions defined by $\xi \ll 1$ when $\Upsilon \gg c_{p,\xi}T$ is large, cf. Sect. 4.5).

We end this section by listing a full set of thermodynamic properties of binary alloys which are used in this chapter, for an easy reference

$$h_{p,T} = T\left(\frac{\partial s}{\partial \xi}\right)_{p,T} = -T\left(\frac{\partial \mu_c}{\partial T}\right)_{p,\xi}, \tag{4.30a}$$

$$\chi_T = -\frac{1}{\rho}\left(\frac{\partial \rho}{\partial \xi}\right)_{p,T} = \rho\left(\frac{\partial \mu_c}{\partial p}\right)_{T,\xi} > 0, \tag{4.30b}$$

$$\chi = -\frac{1}{\rho}\left(\frac{\partial \rho}{\partial \xi}\right)_{p,s} = \chi_T - \frac{\alpha h_{p,T}}{c_{p,\xi}} > 0, \tag{4.30c}$$

$$\Upsilon = \left(\frac{\partial \mu_c}{\partial \xi}\right)_{p,T} > 0, \quad c_{p,\xi} = T\left(\frac{\partial s}{\partial T}\right)_{p,\xi} > 0, \quad c_{v,\xi} = T\left(\frac{\partial s}{\partial T}\right)_{\rho,\xi} > 0, \tag{4.30d}$$

$$\alpha = -\frac{1}{\rho}\left(\frac{\partial \rho}{\partial T}\right)_{p,\xi} > 0, \quad \beta = \frac{1}{\rho}\left(\frac{\partial \rho}{\partial p}\right)_{T,\xi} > 0. \tag{4.30e}$$

## 4.1 Compositional and Heat Fluxes

It is essential to derive the formulae for the compositional flux and the heat flux; in the case of convection driven by both, compositional and thermal effects the expression for the heat flux is significantly modified with respect to that for a single-component fluid. We start by introducing the notation for the total molecular flux of entropy and the total molecular compositional flux (without the contribution from the effect of advection by the flow), which will be denoted by $\mathbf{j}_{s,\mathrm{mol}}$ and $\mathbf{j}_{\xi,\mathrm{mol}}$ respectively. The total heat flux will be denoted by $\mathbf{j}_q$. Therefore the general forms of the entropy and mass fraction equations are

$$\rho\left(\frac{\partial s}{\partial t}+\mathbf{u}\cdot\nabla s\right)+\nabla\cdot\mathbf{j}_{s,\mathrm{mol}}=\sigma_s, \tag{4.31}$$

$$\rho\left(\frac{\partial \xi}{\partial t}+\mathbf{u}\cdot\nabla \xi\right)+\nabla\cdot\mathbf{j}_{\xi,\mathrm{mol}}=0, \tag{4.32}$$

where the mass conservation law $\partial_t\rho+\nabla\cdot(\rho\mathbf{u})=0$ has been used,[3] $\sigma_s$ are the volume entropy sources and there are no volume sources of the light constituent. In order to obtain the formulae for the fluxes $\mathbf{j}_{s,\mathrm{mol}}$ and $\mathbf{j}_{\xi,\mathrm{mol}}$, we must first identify the "thermodynamic forces",[4] that is the causative thermodynamic stimuli of the fluxes, which take the form of gradients of intensive thermodynamic variables, such as e.g. the temperature or the chemical potential. To that end, we need to find an explicit expression for the entropy production, which in general terms is known to take the bilinear form of a sum of products of the fluxes and thermodynamic forces associated with irreversible processes, $\sigma_s=\sum_j\mathbf{j}_j\cdot\mathbf{X}_j$, where the symbol $\mathbf{X}$ is commonly used to denote the thermodynamic forces (cf. Landau and Lifshitz 1987, Glansdorff and Prigogine 1971, de Groot and Mazur 1984).

The total energy density per unit mass in a fluid volume $V$ is a sum of the kinetic energy density, the potential energy resulting from the presence of conservative body forces, $\mathbf{F}=-\nabla\psi$, which we will assume stationary and the internal energy,

$$e=\frac{1}{2}\mathbf{u}^2+\psi+\varepsilon. \tag{4.33}$$

The evolution of the total energy is described by the following law (cf. (1.4))

$$\rho\left(\frac{\partial e}{\partial t}+\mathbf{u}\cdot\nabla e\right)+\nabla\cdot\mathbf{j}_{e,\mathrm{mol}}=Q, \tag{4.34}$$

---

[3]The law of conservation of mass has the standard form for a continuous medium, with the density $\rho=\rho_l+\rho_h$ and the flow velocity of the alloy being the centre of mass velocity $\mathbf{u}=\xi\mathbf{u}_l+(1-\xi)\mathbf{u}_h$ for a fluid element, where $\mathbf{u}_l$ and $\mathbf{u}_h$ denote the velocities of motion of each of the two constituents of the alloy. Therefore the law of mass conservation for the alloy, $\partial_t\rho+\nabla\cdot(\rho\mathbf{u})=0$, results in a straightforward way from summing up the equations describing the mass balance for each of the constituents, $\partial_t\rho_l+\nabla\cdot(\rho_l\mathbf{u}_l)=0$, and $\partial_t\rho_h+\nabla\cdot(\rho_h\mathbf{u}_h)=0$, i.e.

$$\partial_t(\rho_l+\rho_h)+\nabla\cdot[\rho(\xi\mathbf{u}_l+(1-\xi)\mathbf{u}_h)]=0.$$

[4]Also called "affinities".

where $\rho e\mathbf{u}$ is the advective energy flux, $\mathbf{j}_{e,\mathrm{mol}}$ is the flux of energy from molecular mechanical, thermal and compositional effects and $Q$ is the energy source (absorbed heat per unit mass per unit volume), here unspecified, which e.g. may describe the effects of radioactive heating, thermal radiation, etc. From Sect. 1.4 we know, that the molecular energy flux is composed of two factors, that is the heat transferred between a fluid volume and the rest of the fluid and the total work done on the volume by the stresses,

$$\mathbf{j}_{e,\mathrm{mol}} = -\boldsymbol{\tau} \cdot \mathbf{u} + \mathbf{j}_q, \tag{4.35}$$

where $\mathbf{j}_q$ denotes the total molecular heat flux. From the expression for the total differential of the internal energy per unit mass

$$\mathrm{d}\varepsilon = T\mathrm{d}s + \frac{p}{\rho^2}\mathrm{d}\rho + \mu_c \mathrm{d}\xi, \tag{4.36}$$

supplied by the mass conservation equation $\partial_t \rho + \mathbf{u} \cdot \nabla\rho = -\rho\nabla \cdot \mathbf{u}$ and the mass fraction balance (4.32) one obtains

$$\rho\left(\frac{\partial\varepsilon}{\partial t} + \mathbf{u} \cdot \nabla\varepsilon\right) = \rho T\left(\frac{\partial s}{\partial t} + \mathbf{u} \cdot \nabla s\right) - p\nabla \cdot \mathbf{u} - \mu_c \nabla \cdot \mathbf{j}_{\xi,\mathrm{mol}}. \tag{4.37}$$

The variation of the kinetic and potential energies is described by the same equations as in Sect. 1.4, thus

$$\rho\left(\frac{\partial}{\partial t} + \mathbf{u} \cdot \nabla\right)\frac{1}{2}\mathbf{u}^2 = \nabla \cdot (\boldsymbol{\tau} \cdot \mathbf{u}) - \boldsymbol{\tau} : \mathbf{G}^s + \rho\mathbf{u} \cdot \mathbf{F}, \tag{4.38}$$

$$\rho\left(\frac{\partial}{\partial t} + \mathbf{u} \cdot \nabla\right)\psi = -\rho\mathbf{u} \cdot \mathbf{F}, \tag{4.39}$$

where

$$\boldsymbol{\tau} : \mathbf{G}^s = -p\nabla \cdot \mathbf{u} + 2\rho\nu\mathbf{G}^s : \mathbf{G}^s + \left(\rho\nu_b - \frac{2}{3}\rho\nu\right)(\nabla \cdot \mathbf{u})^2. \tag{4.40}$$

Substitution of (4.33), (4.37), (4.38) and (4.39) into the total energy balance (4.34) leads to

$$\rho T\left(\frac{\partial s}{\partial t} + \mathbf{u} \cdot \nabla s\right) = -\nabla \cdot \left(\mathbf{j}_{e,\mathrm{mol}} + \boldsymbol{\tau} \cdot \mathbf{u} - \mu_c \mathbf{j}_{\xi,\mathrm{mol}}\right) + \boldsymbol{\tau}_\nu : \mathbf{G}^s - \mathbf{j}_{\xi,\mathrm{mol}} \cdot \nabla\mu_c + Q, \tag{4.41}$$

where the expression

$$\boldsymbol{\tau}_\nu : \mathbf{G}^s = 2\rho\nu\mathbf{G}^s : \mathbf{G}^s + \left(\rho\nu_b - \frac{2}{3}\rho\nu\right)(\nabla \cdot \mathbf{u})^2, \tag{4.42}$$

involves now only the dissipative part of the stress tensor, $\boldsymbol{\tau}_\nu$. Comparison of the Eq. (4.41) with the entropy balance (4.31) multiplied by the temperature $T$, which reads

$$\rho T\left(\frac{\partial s}{\partial t}+\mathbf{u}\cdot\nabla s\right)=-\nabla\cdot\left(T\mathbf{j}_{s,\mathrm{mol}}\right)+\mathbf{j}_{s,\mathrm{mol}}\cdot\nabla T+T\sigma_s, \tag{4.43}$$

allows to express the molecular flux of the total energy and the entropy sources in terms of the entropy and mass fraction fluxes

$$\mathbf{j}_{e,\mathrm{mol}}=-\boldsymbol{\tau}\cdot\mathbf{u}+\mu_c\mathbf{j}_{\xi,\mathrm{mol}}+T\mathbf{j}_{s,\mathrm{mol}}, \tag{4.44}$$

$$T\sigma_s=\boldsymbol{\tau}_\nu:\mathbf{G}^s-\mathbf{j}_{\xi,\mathrm{mol}}\cdot\nabla\mu_c-\mathbf{j}_{s,\mathrm{mol}}\cdot\nabla T+Q. \tag{4.45}$$

Additionally, by the use of (4.35) and (4.44) we get an expression for the total molecular heat flux

$$\mathbf{j}_q=\mu_c\mathbf{j}_{\xi,\mathrm{mol}}+T\mathbf{j}_{s,\mathrm{mol}}. \tag{4.46}$$

We can now identify from (4.45), that the causative thermodynamic stimuli of the compositional and entropy fluxes are the chemical potential and temperature gradients divided by the temperature. Although the actual stimulus of the entropy flow is the temperature gradient and the material diffusion is stimulated by the chemical potential differences, in a binary alloy the material diffusion *can not* take place without simultaneous entropy transfer and thermal conduction is *necessarily* associated with simultaneous concentration flow. Thus each of the fluxes of $\xi$ and $s$ must depend on both of the gradients, i.e. the gradient of the chemical potential and the temperature gradient. If the gradients are small we may suppose the fluxes to be linearly related to $\nabla\mu_c$ and $\nabla T$,

$$\mathbf{j}_{\xi,\mathrm{mol}}=L_{\xi\xi}\frac{\nabla\mu_c}{T}+L_{\xi s}\frac{\nabla T}{T}, \tag{4.47}$$

$$\mathbf{j}_{s,\mathrm{mol}}=L_{s\xi}\frac{\nabla\mu_c}{T}+L_{ss}\frac{\nabla T}{T}, \tag{4.48}$$

where $L_{\xi\xi}$, $L_{\xi s}$, $L_{s\xi}$, $L_{ss}$ are called phenomenological coefficients,[5] independent of any thermodynamic forces and fluxes (the first lower index in the coefficients $L_{ij}$ is associated with the type of flux expressed by linear combination of $\nabla\mu_c/T$ and $\nabla T/T$ and the second lower index with the quantity, whose flux is multiplied by the relevant stimulus in the expression for $\sigma_s$). In the case when two or more fluxes share the same causative stimuli, the Onsager's reciprocity principle applies (cf. e.g. Landau and Lifshitz 1980, p. 365, chapter "The symmetry of the kinetic coefficients", or de Groot and Mazur 1984, p. 100), which stems from kinetic theory and utilizes the time reversal symmetry of the equations of motion of individual particles and

[5] Also termed "kinetic coefficients".

the assumption of only small departures from thermodynamic equilibrium[6]; it states, that the kinetic coefficients are symmetric

$$L_{\xi s} = L_{s\xi}. \tag{4.49}$$

The gradient of the chemical potential as a function of the pressure, the temperature and the mass fraction, $\mu_c(p, T, \xi)$, can be easily cast in the form

$$\nabla\mu_c = \frac{\chi_T}{\rho}\nabla p - \frac{h_{p,T}}{T}\nabla T + \Upsilon\nabla\xi. \tag{4.50}$$

On inserting the latter formula into expressions for the molecular compositional and entropy fluxes (4.47)–(4.48) one obtains

$$\mathbf{j}_{\xi,\mathrm{mol}} = -K\left(\nabla\xi + \frac{k_T}{T}\nabla T + \frac{k_p}{p}\nabla p\right), \tag{4.51}$$

$$T\mathbf{j}_{s,\mathrm{mol}} = -k\nabla T + \Lambda\mathbf{j}_{\xi,\mathrm{mol}}, \tag{4.52}$$

where the following new set of phenomenological coefficients was introduced

$$K = -L_{\xi\xi}\frac{\Upsilon}{T}, \qquad k = \frac{\Lambda}{T}L_{\xi s} - L_{ss}, \tag{4.53}$$

$$k_T = \frac{\Lambda - h_{p,T}}{\Upsilon}, \quad k_p = \frac{p\chi_T}{\rho\Upsilon}, \quad \Lambda = T\frac{L_{\xi s}}{L_{\xi\xi}}. \tag{4.54}$$

In the above $k$ is the coefficient of thermal conductivity, $K$ denotes the coefficient of material conductivity and

$$D = K/\rho \tag{4.55}$$

will be used to denote the material diffusion coefficient. Furthermore, $k_T$ is the Soret coefficient, which describes the effect of temperature gradient on the material flux, whereas $k_p$ describes an analogous effect of the pressure gradient. The coefficient $\Lambda$ (which has the units of energy mass density, J/kg) can be called the Dufour coefficient,

[6]In the current case the Eqs. (4.47)–(4.48) can be rewritten in the more general form utilized e.g. by Landau and Lifshitz (1987), Glansdorff and Prigogine (1971) or de Groot and Mazur (1984), $\mathrm{d}_t x_i = L_{ij}X_j$, where $\mathrm{d}_t x_1 = \mathrm{d}_t \mathring{m}_l = \mathbf{j}_{\xi,\mathrm{mol}}$, $\mathrm{d}_t x_2 = \mathrm{d}_t\mathring{S} = \mathbf{j}_{s,\mathrm{mol}}$, and $\mathring{m}_l$ expressed in [kg/m$^2$] is the mass of the light constituent transported through a unit surface of a fluid element in a time unit near thermodynamic equilibrium, whereas $\mathring{S}$ is the total entropy transported through a unit surface of a fluid element in a time unit near thermodynamic equilibrium. Inspection of the expression for the entropy sources (4.45) allows to write $\mathrm{d}_t\int_{V_0}\rho s\mathrm{d}V = -\int_{V_0}\mathbf{j}_{\xi,\mathrm{mol}}\cdot(\nabla\mu_c/T)\mathrm{d}V - \int_{V_0}\mathbf{j}_{s,\mathrm{mol}}\cdot(\nabla T/T)\mathrm{d}V + \cdots$, where the flux term, likewise the viscous and radiogenic terms have been omitted for brevity and $V_0$ denotes here the entire fluid volume. Thus we recognize $\rho\partial s/\partial\mathring{S} = \nabla T/T$ and $\rho\partial s/\partial\mathring{m}_l = \nabla\mu/T$, and the Onsager's reciprocity principle applies (cf. e.g. Landau and Lifshitz 1987, p. 230, chapter "Coefficients of mass transfer and thermal diffusion").

since it describes the effect of concentration gradient on the entropy flux. It is also of interest to observe, that the chemical potential gradient can be expressed by the material flux in the following way

$$\nabla \mu_c = -\frac{\Upsilon}{K} \mathbf{j}_{\xi,\mathrm{mol}} - \frac{\Lambda}{T} \nabla T. \tag{4.56}$$

We are now ready to write down the entropy balance (4.31) in a more explicit form

$$\rho T \left( \frac{\partial s}{\partial t} + \mathbf{u} \cdot \nabla s \right) = \nabla \cdot (k \nabla T) - \nabla \cdot \left( \Lambda \mathbf{j}_{\xi,\mathrm{mol}} \right) + \boldsymbol{\tau}_\nu : \mathbf{G}^s - \mathbf{j}_{\xi,\mathrm{mol}} \cdot \nabla \mu_c + Q, \tag{4.57}$$

where the flux of the light constituent is given by (4.51); the mass fraction balance, on the other hand, is given in (4.32).

Let us now utilize formula (4.46) for the total heat flux to write down the heat balance. By the use of the general local evolution law (1.4) the rate of variation of the heat exchanged via molecular processes between a fluid parcel and its surroundings is expressed by the heat flux divergence $-\nabla \cdot \mathbf{j}_q$ and the heat sources, which contain the viscous heating $\boldsymbol{\tau}_\nu : \mathbf{G}^s$ and other possible sources $Q$, e.g. radiogenic. Therefore the formulae (4.46), (4.57) and (4.32) yield

$$\begin{aligned}
\frac{đq_{\mathrm{tot}}}{\mathrm{d}t} &= - \nabla \cdot \mathbf{j}_q + \boldsymbol{\tau}_\nu : \mathbf{G}^s + Q \\
&= - \nabla \cdot \left( \mu_c \mathbf{j}_{\xi,\mathrm{mol}} \right) - \nabla \cdot \left( T \mathbf{j}_{s,\mathrm{mol}} \right) + \boldsymbol{\tau}_\nu : \mathbf{G}^s + Q \\
&= - \mu_c \nabla \cdot \mathbf{j}_{\xi,\mathrm{mol}} + \rho T \frac{Ds}{\partial t} \\
&= \rho T \frac{Ds}{\partial t} + \rho \mu_c \frac{D\xi}{\partial t}.
\end{aligned} \tag{4.58}$$

An important conclusion can be drawn from the latter equation, that the total infinitesimal heat delivered to the unit volume can not be expressed solely by $T \rho ds$, contrary to single-component fluids. In a binary alloy the effects of compositional variation (chemical potential equilibration) contribute significantly to the total molecular heat flux.

A significant simplification of the entropy balance (4.57) can be achieved through neglection of the Soret effect (the effect of temperature gradient on the material flux) and the effect of the pressure gradient on the mass fraction flux, so that effectively the binary alloy satisfies the Fick's law

$$\mathbf{j}_{\xi,\mathrm{mol}} = -K \nabla \xi. \tag{4.59}$$

This can be obtained by assuming e.g. $\Lambda \approx h_{p,T}$, so that $k_T \approx 0$ and only small values of the relative pressure gradient, $L \nabla p / p \ll 1$ (such as in the Boussinesq approximation). The entropy balance and the mass fraction equation then read

$$\rho T\left(\frac{\partial s}{\partial t}+\mathbf{u}\cdot\nabla s\right)=\nabla\cdot(k\nabla T)+\nabla\cdot(\Lambda K\nabla\xi)+\boldsymbol{\tau}_\nu:\mathbf{G}^s+K\nabla\xi\cdot\nabla\mu_c+Q, \tag{4.60}$$

$$\rho\left(\frac{\partial \xi}{\partial t}+\mathbf{u}\cdot\nabla\xi\right)=\nabla\cdot(K\nabla\xi), \tag{4.61}$$

where $\nabla\mu_c$ is given in (4.50). Further simplification is still possible when the Dufour coefficient $\Lambda$ is small, compared to some typical energy scale in the system, say the gravitational energy scale $\bar{g}L$ or thermal energy scale $c_{v,\xi}T$. Then the Dufour effect, that is the influence of the mass concentration flux on the entropy variation, can be neglected yielding

$$\rho T\left(\frac{\partial s}{\partial t}+\mathbf{u}\cdot\nabla s\right)=\nabla\cdot(k\nabla T)+\boldsymbol{\tau}_\nu:\mathbf{G}^s+K\nabla\xi\cdot\nabla\mu_c+Q. \tag{4.62}$$

It might be tempting to also provide some justification for neglection of the term $K\nabla\xi\cdot\nabla\mu_c$, so that the entropy balance directly corresponds to that of a single-component fluid. Unfortunately, it is hard to justify neglection of this term unless the parameter $\Upsilon$ is assumed small. However, if $\Upsilon$ were assumed small then the coefficients $k_T$ and $k_p$ describing the Soret effect and the pressure effect on the material flux become significant, unless $\chi_T$ and thus the compositional buoyancy is assumed negligible. The latter assumption means that only the thermal buoyancy plays a dynamical role (cf. Sect. 4.3) and the problem is reduced to purely thermal convection studied in previous chapters.

Therefore the only sensible way to neglect the term $K\nabla\xi\cdot\nabla\mu_c$ is on the grounds, that the ratio of material to thermal diffusion coefficients

$$\frac{D}{\kappa}=\frac{Pr}{Sc}\ll 1, \tag{4.63}$$

is small, where

$$Pr=\frac{\nu}{\kappa},\qquad Sc=\frac{\nu}{D}, \tag{4.64}$$

are the Prandtl and Schmidt numbers, respectively. Indeed, the ratios of the term $K\nabla\xi\cdot\nabla\mu_c$ to the thermal diffusion and entropy advection can be estimated as follows

$$\frac{|K\nabla\xi\cdot\nabla\mu_c|}{|\nabla\cdot(k\nabla T)|}\sim\frac{K\Upsilon\,|\nabla\xi|^2}{k\left|\nabla^2 T\right|}\sim\frac{Pr}{Sc}\frac{\Upsilon}{c_{p,\xi}}\frac{|\xi'|^2}{|T'|}, \tag{4.65}$$

$$\frac{|K\nabla\xi\cdot\nabla\mu_c|}{|\rho T\mathbf{u}\cdot\nabla s|}\sim\frac{K\Upsilon\,|\nabla\xi|^2}{|\rho|\,|T|\,\mathscr{U}\,|\nabla s|}\sim\frac{1}{ReSc}\frac{\Upsilon}{T_B}\frac{|\xi'|^2}{|s'|}, \tag{4.66}$$

hence the assumptions $D/\kappa\ll 1$ and $ReSc\gg 1$, or simply $Sc\gg 1$ allow to obtain the entropy equation in the single-component form (note, that in order for the thermal and compositional buoyancy forces to be comparable in magnitude we must require for the magnitudes of fluctuations $|\xi'|\sim|s'/c_{p,\xi}|$, which also implies $|\xi'|\sim|\alpha T'|$;

cf. the following Sects. 4.2 and 4.3). However, assumption such as $Sc \gg 1$ affects also the mass fraction equation (4.61). A non-dimensional form of that equation requires introduction of some time scale, say the viscous time scale $L^2/\nu$ or the inertial time scale $L/\mathscr{U}$. This leads to appearance of small parameter in front of the diffusive term which contains the highest-order derivative in the mass fraction equation, introducing compositional boundary layers into the dynamics (alternative choice of $L^2/D$ for the time scale leads to small, or large parameters in the remaining dynamical equations).

Perhaps a simplest way to neglect the Soret effect and the effect of the pressure gradient on the compositional flux is to assume large $\Upsilon \gg c_{v,\xi}\bar{T}$ (but *without* assuming the weak solution limit $\xi \ll 1$ explained in Sect. 4.5). In such a case the coefficients $k_T \approx 0$, $k_p \approx 0$ are negligibly small (cf. (4.54)), whereas the expression for the chemical potential gradient (4.50) simplifies to $\nabla\mu_c \approx \Upsilon\nabla\xi$. Additional assumption of a weak Dufour effect, $\Lambda \ll c_{v,\xi}\bar{T}$ allows to achieve the mass fraction and energy equations in greatly simplified forms of (4.61) and

$$\rho T\left(\frac{\partial s}{\partial t}+\mathbf{u}\cdot\nabla s\right)=\nabla\cdot(k\nabla T)+\boldsymbol{\tau}_\nu:\mathbf{G}^s+K\Upsilon\left(\nabla\xi\right)^2+Q. \tag{4.67}$$

Finally, it is of interest to note, that from the mass fraction equation (4.61) we may anticipate already, that the anelastic approximation for binary alloys requires

$$D\sim\delta^{1/2}\sqrt{\bar{g}L}L, \tag{4.68}$$

just as in the case of viscous and thermal diffusion coefficients, cf. (3.16) and (3.17).

### 4.1.1 Boundary Conditions on the Mass Fraction

The standard conditions that the mass fraction can be sensibly assumed to satisfy at boundaries are similar to those often assumed for temperature. Most frequently the flux of the light constituent through the boundaries is required to be constant, i.e.

$$K\left.\left(\frac{\partial\xi}{\partial z}+\frac{k_T}{T}\frac{\partial T}{\partial z}+\frac{k_p}{\tilde{p}}\frac{\partial p}{\partial z}\right)\right|_{z=0,L}=\text{const.} \tag{4.69}$$

Such a boundary condition corresponds to an exemplary situation, when the heavy constituent solidifies at the bottom, and the light constituent solidifies at the top of the domain; that way the light constituent is effectively supplied at the bottom and expelled at the top. Of course, when the Soret effect is negligible, $k_T = 0$, and the flux is independent of the pressure gradient, $k_p = 0$, this boundary condition simplifies to

$$K\left.\frac{\partial\xi}{\partial z}\right|_{z=0,L}=\text{const.} \tag{4.70}$$

Note, that in such a case the compositional convection at constant temperature, thus in the absence of thermal driving, is directly relevant to the Boussinesq Rayleigh-Bénard problem with fixed heat flux at the boundaries.

Another possibility is to assume, that the boundaries dissolve in the fluid (binary alloy) in such a way, that the saturation concentration $\xi_0$ (say) is rapidly established near their surfaces. In such a case the mass fraction can be assumed constant at the boundaries,

$$\xi|_{z=0,L} = \text{const.} \tag{4.71}$$

### 4.1.2 Production of the Total Entropy and the Second Law of Thermodynamics

The second law of thermodynamics states, that in a closed and adiabatically insulated system the production of the total entropy is always positive or null. The assumption of a closed system means, that there are no material interactions with the surroundings, i.e. the material (compositional) flux at the boundaries is zero

$$\mathbf{j}_{\xi,\text{mol}} \cdot \hat{\mathbf{e}}_z\big|_{z=0,L} = 0. \tag{4.72}$$

Adiabatic insulation on the other hand implies that the total heat flux (4.46), composed of the material and entropy fluxes, needs to vanish at the boundaries. By the use of the expression for the total entropy flux provided in (4.52) and the fact, that the material flux has already been assumed to vanish at the boundaries, the assumption of adiabatic insulation reduces to

$$-k\frac{\partial T}{\partial z}\bigg|_{z=0,L} = 0. \tag{4.73}$$

Impermeability of the boundaries additionally implies

$$\mathbf{u} \cdot \hat{\mathbf{e}}_z\big|_{z=0,L} = 0, \tag{4.74}$$

and to fix ideas let us assume periodicity in the horizontal directions. We can now divide the energy equation (4.57) by $T$ to get the entropy balance and rearrange the flux terms in the following way

$$\frac{1}{T}\nabla \cdot (k\nabla T) = \nabla \cdot \left(k\frac{\nabla T}{T}\right) + k\frac{(\nabla T)^2}{T^2}, \tag{4.75a}$$

$$-\frac{1}{T}\nabla \cdot \left(\Lambda \mathbf{j}_{\xi,\text{mol}}\right) = -\nabla \cdot \left(\Lambda\frac{\mathbf{j}_{\xi,\text{mol}}}{T}\right) - \Lambda\frac{\mathbf{j}_{\xi,\text{mol}} \cdot \nabla T}{T^2}. \tag{4.75b}$$

On integration of the entropy balance over the entire horizontally periodic volume and application of the boundary conditions (4.72), (4.73) and (4.74) one obtains

$$\frac{\partial}{\partial t}\int_V \rho s \mathrm{d}V = \int_V k\frac{(\nabla T)^2}{T^2}\mathrm{d}V - \int_V \Lambda\frac{\mathbf{j}_{\xi,\mathrm{mol}}\cdot\nabla T}{T^2}\mathrm{d}V + \int_V \frac{1}{T}\boldsymbol{\tau}_\nu : \mathbf{G}^s\mathrm{d}V$$
$$- \int_V \frac{\mathbf{j}_{\xi,\mathrm{mol}}\cdot\nabla\mu_c}{T}\mathrm{d}V + \int_V Q\mathrm{d}V, \tag{4.76}$$

where the mass conservation equation $\partial_t\rho + \nabla\cdot(\rho\mathbf{u}) = 0$ was used on the left hand side and

$$\int_V \nabla\cdot(\rho\mathbf{u}s)\mathrm{d}V = \int_{\partial V}\rho s\mathbf{u}\cdot\hat{\mathbf{n}}\mathrm{d}\Sigma = 0 \tag{4.77}$$

by the impermeability (and horizontal periodicity) conditions. Finally, introducing the expression (4.56) for the chemical potential gradient into the first term in the second line of (4.76) leads to

$$\frac{\partial}{\partial t}\int_V \rho s\mathrm{d}V = \int_V k\frac{(\nabla T)^2}{T^2}\mathrm{d}V + \int_V \frac{1}{T}\boldsymbol{\tau}_\nu : \mathbf{G}^s\mathrm{d}V$$
$$+ \int_V \frac{\Upsilon}{K}\frac{\mathbf{j}^2_{\xi,\mathrm{mol}}}{T}\mathrm{d}V + \int_V Q\mathrm{d}V. \tag{4.78}$$

It was shown in Sect. 1.4.1, that the viscous term

$$\int_V \frac{1}{T}\boldsymbol{\tau}_\nu : \mathbf{G}^s\mathrm{d}V \geq 0 \tag{4.79}$$

is positive definite and since by the second law of thermodynamics the entire production of the total entropy must be positive definite

$$\frac{\partial}{\partial t}\int_V \rho s\mathrm{d}V \geq 0, \tag{4.80}$$

this implies positivity of the material conductivity coefficient

$$K > 0. \tag{4.81}$$

## 4.2 Hydrostatic Reference State

At this stage we introduce the concept of the *anelastic approximation* for binary alloys; i.e. we list the additional assumptions with respect to those made for single-component fluids, that have to be made when binary alloys are considered. We proceed to derive the equations for the hydrostatic reference state. As in Chap. 3 we assume time independent boundary conditions and decompose the thermodynamic variables into the reference state contributions and the fluctuations induced by the convective flow (3.2a–3.2e), thus the mass fraction variable also takes the form

$$\xi(\mathbf{x}, t) = \tilde{\xi}(z) + \xi'(\mathbf{x}, t). \tag{4.82}$$

Within the anelastic approximation the convection is driven thermally by only small departures from the adiabatic profile, $s = \text{const}$, and compositionally through small departures from the well-mixed profile $\xi = \text{const}$ (the *first fundamental assumption* of the anelastic approximation). Therefore the assumption (3.11) still holds, but additionally we need to require

$$-L\left\langle \frac{\mathrm{d}\tilde{\xi}}{\mathrm{d}z} \right\rangle = \Delta\xi \sim \delta \ll 1, \tag{4.83}$$

where $\Delta\xi = \xi_B - \xi_T$ denotes the mass concentration jump across the fluid layer. Note, that similarly as in the case of the entropy variable (cf. discussion below (3.12)), the static state function $\tilde{\xi}$ can be split into two parts, $\tilde{\xi}_0 + \tilde{\tilde{\xi}}(z)$, where $\tilde{\xi}_0 = \text{const}$ is the uniform value of the mass concentration in the well-mixed, adiabatic state and $\tilde{\tilde{\xi}}(z) = \mathcal{O}(\delta)$ is the $z$-dependent part describing the departure of the reference state from the well-mixed one (i.e. in the hydrostatic equilibrium the variations in the mass fraction are only a weak correction to the mean, similarly as in the case of the entropy).

The *second fundamental assumption* of the anelastic approximation consists of (3.12) and an additional requirement, that the convective fluctuations of the mass fraction are small

$$\left|\xi'\right| \sim \mathcal{O}(\delta) \ll 1. \tag{4.84}$$

It is important to realize, that the mass fraction fluctuation need not be smaller than the equilibrium value $\tilde{\xi}$ and both $\left|\xi'\right|$ and $\tilde{\xi}$ can, in fact, be small (which means that the weak solution limit, $\xi \ll 1$, for a binary alloy is formally allowed, cf. Sect. 4.5.1).

It is now clear, that by inspection of the mass fraction equation (4.32) and the formula for the concentration flux (4.51) (or the simplified version (4.61)) we must require for consistency

$$K = \tilde{\rho} D = \mathcal{O}\left(\delta^{1/2} \rho_B \sqrt{\bar{g} L} L\right). \tag{4.85}$$

In consequence we obtain

$$j_\xi = \mathcal{O}\left(\delta K / \bar{\chi} L\right) = \mathcal{O}\left(\delta^{3/2} \rho_B \sqrt{\bar{g} L} / \bar{\chi}\right). \tag{4.86}$$

The general expression for the gradient of the chemical potential (4.50) implies

$$\frac{d\tilde{\mu}_c}{dz} = \tilde{\Upsilon}\frac{d\tilde{\xi}}{dz} - \frac{\tilde{h}_{p,T}}{\tilde{T}}\frac{d\tilde{T}}{dz} + \frac{\tilde{\chi}_T}{\tilde{p}}\frac{d\tilde{p}}{dz}$$
$$= -\tilde{\chi}\tilde{g} + \frac{\tilde{h}_{p,T}}{\tilde{T}}\Delta_S + \tilde{\Upsilon}\frac{d\tilde{\xi}}{dz}, \tag{4.87}$$

where

$$\Delta_S = -\left(\frac{d\tilde{T}}{dz} + \frac{\tilde{\alpha}\tilde{T}\tilde{g}}{\tilde{c}_{p,\xi}}\right), \tag{4.88}$$

therefore

$$\frac{L}{\tilde{h}_{p,T}}\left(\frac{d\tilde{\mu}_c}{dz} + \tilde{\chi}\tilde{g}\right) = \frac{\tilde{\Upsilon}L}{\tilde{h}_{p,T}}\frac{d\tilde{\xi}}{dz} + \frac{L}{\tilde{T}}\Delta_S. \tag{4.89}$$

From the latter equation and the assumptions (3.11) and (4.83) we see, that as long as $\tilde{\Upsilon}$ remains comparable with other thermodynamic properties of the system, thus e.g. $\tilde{\Upsilon} \sim \tilde{c}_{p,\xi}\tilde{\chi}/\tilde{\alpha} \sim \tilde{\chi}\tilde{h}_{p,T}$, then the quantity $L\left(d_z\tilde{\mu}_c + \tilde{\chi}\tilde{g}\right)/\tilde{h}_{p,T}$ must also be small, i.e.[7]

$$\frac{d\tilde{\mu}_c}{dz} = -\tilde{\chi}\tilde{g} + \mathcal{O}(\delta\tilde{g}). \tag{4.90}$$

The equations of the hydrostatic equilibrium, which include the hydrostatic momentum, mass fraction and entropy balances, the gravitational potential equation and the equations of state take the following, general form

$$\frac{d\tilde{p}}{dz} = -\tilde{\rho}\tilde{g}, \tag{4.91a}$$

$$\frac{d}{dz}\left[K\left(\frac{d\tilde{\xi}}{dz} + \frac{\tilde{k}_T}{\tilde{T}}\frac{d\tilde{T}}{dz} + \frac{\tilde{k}_p}{\tilde{p}}\frac{d\tilde{p}}{dz}\right)\right] = 0, \tag{4.91b}$$

$$\frac{d}{dz}\left[\left(k + \frac{\Lambda}{\tilde{T}}\tilde{k}_T K\right)\frac{d\tilde{T}}{dz} + \Lambda K\frac{d\tilde{\xi}}{dz} + \frac{\Lambda}{\tilde{p}}\tilde{k}_p K\frac{d\tilde{p}}{dz}\right]$$
$$+K\left(\frac{d\tilde{\xi}}{dz} + \frac{\tilde{k}_T}{\tilde{T}}\frac{d\tilde{T}}{dz} + \frac{\tilde{k}_p}{\tilde{p}}\frac{d\tilde{p}}{dz}\right)\left(\tilde{\Upsilon}\frac{d\tilde{\xi}}{dz} - \frac{\tilde{h}_{p,T}}{\tilde{T}}\frac{d\tilde{T}}{dz} + \frac{\tilde{\chi}_T}{\tilde{p}}\frac{d\tilde{p}}{dz}\right) = -\tilde{Q}, \tag{4.91c}$$

[7] However, as we have already remarked below the Eq. (4.28), it is to be emphasized, that in the case when the binary alloy is a weak solution of the light constituent, $\xi \ll 1$, the coefficient $\tilde{\Upsilon} \sim \tilde{c}_{p,\xi}\tilde{T}/\tilde{\xi}$ becomes large (cf. Landau and Lifshitz 1980, Eqs. 87.4–5, chapter "Weak solutions" and a comment below Eq. (96.7), chapter "Thermodynamic inequalities for solutions"). Then $\tilde{\Upsilon}/\tilde{h}_{p,T} \gg 1$ and the quantity $L\left(d_z\tilde{\mu}_c + \tilde{\chi}\tilde{g}\right)/\tilde{h}_{p,T}$ is *not* small, despite the fact, that the assumptions (3.11) and (4.83) still hold. The weak solutions are treated in Sect. 4.5.

$$\frac{d^2\tilde{\psi}}{dz^2} = 4\pi G \left[\tilde{\rho}(z)\left(\theta_H(z) - \theta_H(z-L)\right) + \rho_{in}(z)\theta_H(-z)\right], \tag{4.91d}$$

$$\tilde{\rho} = \rho(\tilde{p}, \tilde{T}, \tilde{\xi}), \quad \tilde{s} = s(\tilde{p}, \tilde{T}, \tilde{\xi}), \tag{4.91e}$$

where $\theta_H(z)$ is the Heaviside step function, $\tilde{g} = -\nabla\tilde{\psi}$ is the gravitational acceleration in the reference state, which consists of the contributions from the entire mass below the fluid layer of density, $\rho_{in}(z)$, and from the fluid layer itself, where the density is $\tilde{\rho}$. The assumption of small departures from the well-mixed state, (4.83) affects the above system of reference state equations. In the most general case, when the coefficients $\tilde{\Upsilon}$, $\tilde{k}_T$ and $\tilde{k}_p$ are finite, all the terms containing the mass fraction gradient $d_z\tilde{\xi}$ constitute $\mathcal{O}(\delta)$ corrections to the remaining terms. It follows from the Eq. (4.91b) that

$$K\left(\frac{\tilde{k}_T}{\tilde{T}}\frac{d\tilde{T}}{dz} + \frac{\tilde{k}_p}{\tilde{p}}\frac{d\tilde{p}}{dz}\right) = \text{const} + \mathcal{O}\left(\delta\frac{K}{L}\right), \tag{4.92}$$

and thus the entropy balance (4.91c) supplied by the force balance (4.91a) imply

$$\frac{d}{dz}\left(k\frac{d\tilde{T}}{dz} + \Lambda\,\text{const}\right) - \text{const}\left(\frac{\tilde{h}_{p,T}}{\tilde{T}}\frac{d\tilde{T}}{dz} + \tilde{\chi}_T\tilde{g}\right) = -\tilde{Q} + \mathcal{O}(\delta); \tag{4.93}$$

the latter determines the temperature gradient at leading order. On the other hand the pressure gradient can be expressed by the density, temperature and mass fraction gradients,

$$\begin{aligned}\frac{d\tilde{p}}{dz} &= \frac{1}{\tilde{\rho}\tilde{\beta}}\frac{d\tilde{\rho}}{dz} + \frac{\tilde{\alpha}}{\tilde{\beta}}\frac{d\tilde{T}}{dz} + \frac{\chi_T}{\beta}\frac{d\tilde{\xi}}{dz}\\ &= \frac{1}{\tilde{\rho}\tilde{\beta}}\frac{d\tilde{\rho}}{dz} + \frac{\tilde{\alpha}}{\tilde{\beta}}\frac{d\tilde{T}}{dz} + \mathcal{O}\left(\delta\bar{\rho}\bar{g}\right),\end{aligned} \tag{4.94}$$

and substitution of the latter to the hydrostatic force balance (4.91a) yields

$$\frac{1}{\tilde{\rho}}\frac{d\tilde{\rho}}{dz} = -\tilde{\beta}\tilde{\rho}\tilde{g} - \tilde{\alpha}\frac{d\tilde{T}}{dz} + \mathcal{O}\left(\frac{\delta}{L}\right), \tag{4.95}$$

which in turn determines the density gradient at leading order. Finally the state equation allows to determine the pressure. However, since the mass fraction gradient drops out of the leading order analysis, the so-determined reference state profiles $\tilde{\rho}$, $\tilde{T}$ and $\tilde{p}$ are in general inconsistent with the compositional flux balance (4.92), which is an independent equation that the temperature and pressure have to satisfy at leading order. In other words a height-dependent only hydrostatic state satisfying the assumption of weak mass fraction gradients does not exist. The situation may be rescued by assuming a slowly time-dependent basic state[8] and/or a basic state non-uniform in all three directions (that is dependent on $x$, $y$ and $z$), although, this introduces additional complications; nevertheless, slowly varying in time basic states are relevant to natural systems and commonly used in their quantitative description, since astrophysical objects such as planets and stars cool down and change chemical composition on long time scales. Furthermore, the well-mixed adiabatic state could also be chosen for the reference state, and then the convective flow would have to be driven by only weak departures from that state on boundaries. Since the reference state has to satisfy the equations (momentum, energy, mass fraction and mass balances and the gravitational and state equations), in general also in this case the adiabatic, well-mixed reference state would be required to be non-stationary.

The situation becomes greatly simplified, when the Soret effect and the term involving the pressure gradient in the expression for the compositional flux (4.51) are negligibly small compared to the mass fraction gradient. In such a case the material flux is given by the Fick's law (4.59) and the equations for the reference state take the following solvable form at the leading order

$$\frac{\mathrm{d}\tilde{p}}{\mathrm{d}z} = \tilde{\rho}\nabla\tilde{\psi}, \quad \frac{\mathrm{d}}{\mathrm{d}z}\left(K\frac{\mathrm{d}\tilde{\xi}}{\mathrm{d}z}\right) = 0, \quad \frac{\mathrm{d}}{\mathrm{d}z}\left(k\frac{\mathrm{d}\tilde{T}}{\mathrm{d}z}\right) = -\tilde{Q}, \tag{4.96a}$$

$$\tilde{\rho} = \rho(\tilde{p}, \tilde{T}, \tilde{\xi}), \quad \tilde{s} = s(\tilde{p}, \tilde{T}, \tilde{\xi}), \tag{4.96b}$$

$$\frac{\mathrm{d}^2\tilde{\psi}}{\mathrm{d}z^2} = 4\pi G\left[\tilde{\rho}(z)\left(\theta_H(z) - \theta_H(z-L)\right) + \rho_{in}(z)\theta_H(-z)\right], \tag{4.96c}$$

where, again, $\theta_H(z)$ is the Heaviside step function.

Finally, let us derive explicit relations between the thermodynamic fluctuations assuming, that the binary alloy is not a weak solution, which implies

$$\frac{|\xi'|}{\tilde{\xi}} \ll 1, \tag{4.97}$$

i.e. that the mean value of the concentration of the light constituent is much greater than its variations. The equations of state $\rho = \rho(p, T, \xi)$ and $s = s(p, T, \xi)$ now include the dependence on the mass fraction; expansion about the hydrostatic equilibrium at every height, similarly as in (3.13a, 3.13b) results in

$$\frac{\rho'}{\tilde{\rho}} = -\tilde{\alpha}T' - \tilde{\chi}_T\xi' + \tilde{\beta}p' + \mathcal{O}\left(\delta^2\right), \tag{4.98a}$$

[8] The dependence on time of the reference state can only be slow, in order not to violate the fundamental for the anelastic approximation leading order property $\nabla\cdot(\tilde{\rho}\mathbf{u}) = 0$, i.e. not to introduce the fast sound waves into the problem.

$$s' = -\tilde{\alpha}\frac{p'}{\tilde{\rho}} + \tilde{c}_{p,\xi}\frac{T'}{\tilde{T}} + \tilde{h}_{p,T}\frac{\xi'}{\tilde{T}} + \mathcal{O}\left(\tilde{c}_{p,\xi}\delta^2\right), \tag{4.98b}$$

which will be later utilized to cast the dynamical equations in a most suitable form. Additionally we may expand the chemical potential $\mu_c = \mu_c(p, T, \xi)$ about the reference state at every height, which yields

$$\mu_c' = \frac{\tilde{\chi}_T}{\tilde{\rho}}p' - \frac{\tilde{h}_{p,T}}{\tilde{T}}T' + \tilde{\Upsilon}\xi'. \tag{4.99}$$

However, we stress, that the assumption (4.97) is crucial in order for the expansions (4.98a, 4.98b) and (4.99) to be valid; in other words these expansions are valid only for solutions which are *not* weak. In particular, in the limit of a weak solution, when both $\tilde{\xi} \ll 1$ and $\xi' \ll 1$ are small the fluctuation $\xi'$ can be of comparable magnitude with the magnitude of the mass fraction in the reference state, i.e. $\xi'/\tilde{\xi} = \mathcal{O}(1)$. In such a case $\tilde{\Upsilon}$ is of the same order of magnitude as $\xi'^{-1}$ and the expressions for thermodynamic fluctuations need to be modified (cf. Sect. 4.5 on the weak solution limit).

### 4.2.1 Mixture of Ideal Gases

Let us now describe in detail one possible, and perhaps the simplest example of a binary alloy, which is formed when two ideal gases are mixed. In particular, this is the simplest example of what is termed an "ideal solution" in chemistry, that is a type of solution for which the entropy and internal energy are additive, i.e. the sum of entropies (or internal energies) of each constituent equals the total entropy (internal energy) of the solution; this means, that in the process of mixing of the constituents no heat is released. The heavy and light constituents, which occupy the same volume $V$ and have the same temperature $T$ are both described by the equations of state of a perfect gas,[9]

$$p^{(h)}V = N^{(h)}k_B T, \qquad p^{(l)}V = N^{(l)}k_B T, \tag{4.100a}$$

$$S^{(h)} = \frac{N^{(h)}}{N_A}\left(C_p^{(h)}\ln T - k_B N_A \ln p^{(h)} + S_0^{(h)}\right), \tag{4.100b}$$

$$S^{(l)} = \frac{N^{(l)}}{N_A}\left(C_p^{(l)}\ln T - k_B N_A \ln p^{(l)} + S_0^{(l)}\right), \tag{4.100c}$$

$$\mathcal{E}^{(h)} = \frac{N^{(h)}}{N_A}\left(C_v^{(h)}T + \mathcal{E}_0^{(h)}\right), \qquad \mathcal{E}^{(l)} = \frac{N^{(l)}}{N_A}\left(C_v^{(l)}T + \mathcal{E}_0^{(l)}\right), \tag{4.100d}$$

[9] See e.g. Guminski (1974), p. 167.

where $p^{(h)}$ and $p^{(l)}$ are the partial pressures of the heavy and light constituents, $k_B$ is the Boltzmann constant, $N_A$ is the Avogadro constant, uppercase $C_p$ and $C_v$ denote the molar specific heats ($C_p - C_v = k_B N_A$) and $S_0^{(h)}$, $S_0^{(l)}$, $\mathcal{E}_0^{(h)}$ and $\mathcal{E}_0^{(l)}$ are constants. In the above we have adopted for the time being the "canonical" thermodynamic variables, such as the volume $V$, the number of particles $N^{(h)} + N^{(l)} = N$, the actual entropy $S$ and internal energy $\mathcal{E}$,[10] pressure and temperature. To obtain the equations of state in terms of thermodynamic parameters of the mixture $\rho$, $p$, $T$, $s$ and $\varepsilon$ we first sum the two equations in (4.100a) and use the Dalton's law to write

$$p^{(h)} + p^{(l)} = p = \left(\frac{N^{(h)}m_h}{V}R^{(h)} + \frac{N^{(l)}m_l}{V}R^{(l)}\right)T, \tag{4.101}$$

where we have introduced the molecular masses of the light ($m_l$) and heavy ($m_h$) constituents and the specific gas constants for both constituents $R^{(h)} = k_B/m_h$ and $R^{(l)} = k_B/m_l$. Since the partial densities are

$$\rho^{(h)} = \frac{N^{(h)}m_h}{V}, \qquad \rho^{(l)} = \frac{N^{(l)}m_l}{V}, \tag{4.102}$$

and the total density

$$\rho = \frac{N^{(h)}m_h + N^{(l)}m_l}{V}, \tag{4.103}$$

whereas the mass fraction of the light constituent is defined as follows

$$\xi = \frac{\rho^{(l)}}{\rho} = \frac{N^{(l)}m_l}{N^{(h)}m_h + N^{(l)}m_l}, \tag{4.104}$$

we may rewrite the state equation (4.101) in the form

$$p = \rho R^{(h)}\left[1 + \xi\left(r_m - 1\right)\right]T, \tag{4.105}$$

where

$$r_m = \frac{m_h}{m_l} \tag{4.106}$$

is the ratio of molecular masses of the heavy to light constituent. Next we turn to the state equations for the entropy. According to the Gibbs law the total entropy of a mixture of ideal gases equals the sum of entropies of the individual constituents at the same temperature and volume; dividing by the total mass and using again the Dalton's law

$$p^{(h)} = \frac{N^{(h)}}{N}p, \qquad p^{(l)} = \frac{N^{(l)}}{N}p, \tag{4.107}$$

[10] As opposed to the mass densities of the entropy ($s$) and the internal energy ($\varepsilon$).

one obtains the formula for the mass density of the entropy of the mixture

$$
\begin{aligned}
s =& \frac{S^{(h)} + S^{(l)}}{N^{(h)} m_h + N^{(l)} m_l} \\
=& \left[\xi c_p^{(l)} + (1-\xi)\, c_p^{(h)}\right] \ln T - R^{(h)} \left[1 + \xi\,(r_m - 1)\right] \ln p \\
& - R^{(h)} (1-\xi) \ln \frac{N^{(h)}}{N} - r_m R^{(h)} \xi \ln \frac{N^{(l)}}{N} + s_0^{(l)} \xi + s_0^{(h)} (1-\xi), \qquad (4.108)
\end{aligned}
$$

where

$$
c_p^{(h)} = \frac{C_p^{(h)}}{m_h N_A}, \qquad c_p^{(l)} = \frac{C_p^{(l)}}{m_l N_A}, \qquad (4.109)
$$

$$
s_0^{(l)} = \frac{S_0^{(l)}}{m_l N_A}, \qquad s_0^{(h)} = \frac{S_0^{(h)}}{m_h N_A}. \qquad (4.110)
$$

The ratios $N^{(h)}/N$ and $N^{(l)}/N$ can be expressed in terms of the mass fraction $\xi$ by the use of (4.104),

$$
\frac{N^{(h)}}{N} = \frac{1-\xi}{1 + \xi\,(r_m - 1)}, \qquad \frac{N^{(l)}}{N} = \frac{r_m \xi}{1 + \xi\,(r_m - 1)}. \qquad (4.111)
$$

Finally, in a mixture of ideal gases the internal energy is also additive, thus the mass density of the internal energy of the mixture can be calculated in a straightforward way

$$
\varepsilon = \frac{\mathcal{E}^{(h)} + \mathcal{E}^{(l)}}{N^{(h)} m_h + N^{(l)} m_l} = \left[\xi c_v^{(l)} + (1-\xi)\, c_v^{(h)}\right] T + \varepsilon_0^{(l)} \xi + \varepsilon_0^{(h)} (1-\xi), \qquad (4.112)
$$

where

$$
c_v^{(h)} = \frac{C_v^{(h)}}{m_h N_A}, \qquad c_v^{(l)} = \frac{C_v^{(l)}}{m_l N_A}, \qquad (4.113)
$$

$$
\varepsilon_0^{(l)} = \frac{\mathcal{E}_0^{(l)}}{m_l N_A}, \qquad \varepsilon_0^{(h)} = \frac{\mathcal{E}_0^{(h)}}{m_h N_A}. \qquad (4.114)
$$

Directly from (4.108) and (4.112) and the definitions of the specific heats at constant volume and pressure for the mixture, it follows that

$$
c_{p,\xi} = T \left(\frac{\partial s}{\partial T}\right)_{p,\xi} = \xi c_p^{(l)} + (1-\xi)\, c_p^{(h)}, \qquad (4.115)
$$

$$
c_{v,\xi} = T \left(\frac{\partial s}{\partial T}\right)_{\rho,\xi} = \left(\frac{\partial \varepsilon}{\partial T}\right)_{\rho,\xi} = \xi c_v^{(l)} + (1-\xi)\, c_v^{(h)} \qquad (4.116)
$$

therefore the specific heats of a binary alloy are $\xi$-dependent. Note also, that for each constituent we have

$$c_p^{(h)} - c_v^{(h)} = R^{(h)} = \frac{k_B}{m_h}, \quad c_p^{(l)} - c_v^{(l)} = R^{(l)} = r_m R^{(h)} = \frac{k_B}{m_l}. \tag{4.117}$$

It is also useful to provide a formula for the chemical potential of the mixture of ideal gases. By the use of the general expression for the total differential of the internal energy

$$\mathrm{d}\varepsilon = T\mathrm{d}s + \frac{p}{\rho^2}\mathrm{d}\rho + \mu_c \mathrm{d}\xi, \tag{4.118}$$

the chemical potential can be calculated directly from the definition

$$\mu_c = \left(\frac{\partial \varepsilon}{\partial \xi}\right)_{\rho,s} = \left(\frac{\partial \varepsilon}{\partial \xi}\right)_{\rho,T} + \left(\frac{\partial \varepsilon}{\partial T}\right)_{\rho,\xi}\left(\frac{\partial T}{\partial \xi}\right)_{\rho,s}$$
$$\left(c_v^{(l)} - c_v^{(h)}\right)T + \varepsilon_0^{(l)} - \varepsilon_0^{(h)} - T\left(\frac{\partial s}{\partial \xi}\right)_{\rho,T}, \tag{4.119}$$

where we have used the implicit function theorem to get

$$\left(\frac{\partial T}{\partial \xi}\right)_{\rho,s} = -\frac{T}{c_{v,\xi}}\left(\frac{\partial s}{\partial \xi}\right)_{\rho,T}. \tag{4.120}$$

By analogy with $h_{p,T}$ defined in (4.18), the thermodynamic property of the alloy, $T\left(\partial_\xi s\right)_{\rho,T}$ can be denoted by $h_{v,T}$ and

$$\frac{h_{v,T}}{T} = \frac{h_{p,T}}{T} + \left(\frac{\partial s}{\partial p}\right)_{T,\xi}\left(\frac{\partial p}{\partial \xi}\right)_{\rho,T}. \tag{4.121}$$

With the aid of the Maxwell relation $\left(\partial_p s\right)_{T,\xi} = -\alpha/\rho$ and the equation of state (4.105), the latter equation can be transformed to

$$\frac{h_{v,T}}{T} = \frac{h_{p,T}}{T} - R^{(h)}\left(r_m - 1\right), \tag{4.122}$$

therefore the chemical potential can be alternatively expressed as follows

$$\mu_c = \left(c_p^{(l)} - c_p^{(h)}\right)T - h_{p,T} + \varepsilon_0^{(l)} - \varepsilon_0^{(h)}. \tag{4.123}$$

Note, that $h_{p,T} = T(\partial_\xi s)_{p,T}$ is a function of $p$, $T$ and $\xi$. Summarizing, we collect the expressions for $p(\rho, T, \xi)$, $s(p, T, \xi)$, $\varepsilon(T, \xi)$ and $\mu_c(p, T, \xi)$ and provide a complete set of equations of state for a mixture of ideal gases,

$$p = \rho R^{(h)}\left[1 + \xi\left(r_m - 1\right)\right]T, \tag{4.124a}$$

$$
\begin{aligned}
s = & c_{p,\xi} \ln T - R^{(h)} \left[1 + \xi \left(r_m - 1\right)\right] \ln p \\
& - R^{(h)} \left(1 - \xi\right) \ln \frac{1 - \xi}{1 + \xi \left(r_m - 1\right)} - r_m R^{(h)} \xi \ln \frac{r_m \xi}{1 + \xi \left(r_m - 1\right)} \\
& + s_0^{(l)} \xi + s_0^{(h)} \left(1 - \xi\right),
\end{aligned} \tag{4.124b}
$$

$$
\varepsilon = c_{v,\xi} T + + \varepsilon_0^{(l)} \xi + \varepsilon_0^{(h)} \left(1 - \xi\right), \tag{4.124c}
$$

$$
\mu_c = \left(c_p^{(l)} - c_p^{(h)}\right) T - h_{p,T} + \varepsilon_0^{(l)} - \varepsilon_0^{(h)}. \tag{4.124d}
$$

Formulae for any other thermodynamic potential, such as the free energy, the enthalpy, etc. can now be easily derived, if necessary. Having obtained the equations of state we can write down explicitly the expressions for all thermodynamic properties of interest, cf. their definitions in (4.30a–4.30e) and (4.54),

$$
\begin{aligned}
h_{p,T} = & \left(c_p^{(l)} - c_p^{(h)}\right) T \ln T - \left(r_m - 1\right) R^{(h)} T \ln p \\
& + R^{(h)} T \ln \frac{1 - \xi}{1 + \xi \left(r_m - 1\right)} - r_m R^{(h)} T \ln \frac{r_m \xi}{1 + \xi \left(r_m - 1\right)} \\
& + T \left(s_0^{(l)} - s_0^{(h)}\right),
\end{aligned} \tag{4.125a}
$$

$$
c_{p,\xi} = \xi c_p^{(l)} + \left(1 - \xi\right) c_p^{(h)}, \quad c_{v,\xi} = \xi c_v^{(l)} + \left(1 - \xi\right) c_v^{(h)}, \tag{4.125b}
$$

$$
c_{p,\xi} - c_{v,\xi} = R^{(h)} \left[1 + \xi \left(r_m - 1\right)\right] \tag{4.125c}
$$

$$
\chi_T = \frac{r_m - 1}{1 + \xi \left(r_m - 1\right)}, \quad \chi = \frac{r_m - 1}{1 + \xi \left(r_m - 1\right)} - \frac{h_{p,T}}{c_{p,\xi} T}, \tag{4.125d}
$$

$$
\Upsilon = \frac{r_m R^{(h)} T}{\xi \left(1 - \xi\right) \left[1 + \xi \left(r_m - 1\right)\right]}, \quad \alpha = \frac{1}{T}, \quad \beta = \frac{1}{p}, \tag{4.125e}
$$

$$
k_p = \frac{r_m - 1}{r_m} \xi \left(1 - \xi\right) \left[1 + \xi \left(r_m - 1\right)\right], \tag{4.125f}
$$

We recall, that $c_p^{(h)} - c_v^{(h)} = R^{(h)} = k_B / m_h$ and $c_p^{(l)} - c_v^{(l)} = r_m R^{(h)} = R^{(l)} = k_B / m_l$.

## 4.3 Buoyancy Force

In a similar manner as it was done in Chap. 3 (cf. Eqs. (3.85) and (3.86)) we will now express the buoyancy force in the Navier-Stokes equation solely in terms of the entropy and the mass fraction fluctuations. Elimination of the temperature fluctuation $T'$ from the Eqs. (4.98a, 4.98b) leads to

$$\frac{\rho'}{\tilde{\rho}} = -\frac{\tilde{\alpha}\tilde{T}}{\tilde{c}_{p,\xi}}s' - \left(\tilde{\chi}_T - \frac{\tilde{\alpha}\tilde{h}_{p,T}}{\tilde{c}_{p,\xi}}\right)\xi' + \tilde{\beta}\frac{\tilde{c}_{v,\xi}}{\tilde{c}_{p,\xi}}p', \tag{4.126}$$

therefore by the use of the isentropic compositional expansion coefficient (4.30c) one obtains

$$\begin{aligned}
-\frac{1}{\tilde{\rho}}\nabla p' - \nabla\psi' + \frac{\rho'}{\tilde{\rho}}\tilde{\mathbf{g}} &= -\frac{1}{\tilde{\rho}}\nabla p' - \nabla\psi' + \left(\frac{\tilde{\alpha}\tilde{T}}{\tilde{c}_{p,\xi}}s' + \tilde{\chi}\xi' - \tilde{\beta}\frac{\tilde{c}_{v,\xi}}{\tilde{c}_{p,\xi}}p'\right)\tilde{g}\hat{\mathbf{e}}_z \\
&= -\nabla\left(\frac{p'}{\tilde{\rho}} + \psi'\right) + \frac{\tilde{\alpha}\tilde{T}}{\tilde{c}_{p,\xi}}s'\tilde{g}\hat{\mathbf{e}}_z + \tilde{\chi}\xi'\tilde{g}\hat{\mathbf{e}}_z \\
&\quad - \left(\tilde{g}\tilde{\beta}\frac{\tilde{c}_{v,\xi}}{\tilde{c}_{p,\xi}} + \frac{1}{\tilde{\rho}^2}\frac{\mathrm{d}\tilde{\rho}}{\mathrm{d}z}\right)p'\hat{\mathbf{e}}_z.
\end{aligned} \tag{4.127}$$

Next, by the use of

$$\frac{\mathrm{d}\tilde{\rho}}{\mathrm{d}z} = \tilde{\rho}\tilde{\beta}\frac{\mathrm{d}\tilde{p}}{\mathrm{d}z} - \tilde{\rho}\tilde{\alpha}\frac{\mathrm{d}\tilde{T}}{\mathrm{d}z} - \tilde{\rho}\tilde{\chi}_T\frac{\mathrm{d}\tilde{\xi}}{\mathrm{d}z}, \tag{4.128}$$

the hydrostatic pressure balance $\mathrm{d}_z\tilde{p} = -\tilde{\rho}\tilde{g}$, the definition of the parameter $\delta$ in (3.11) together with (4.83) and $c_{p,\xi} - c_{v,\xi} = \alpha^2 T/\beta\rho$ (cf. (2.38)), one can neglect the entire term proportional to the pressure fluctuation $p'$, which is equal to

$$\left(-\tilde{\alpha}\Delta_S + \tilde{\chi}_T\frac{\mathrm{d}\tilde{\xi}}{\mathrm{d}z}\right)\frac{p'}{\tilde{\rho}}\hat{\mathbf{e}}_z = \left(\frac{\tilde{\alpha}\tilde{T}}{\tilde{c}_{p,\xi}}\frac{\mathrm{d}\tilde{s}}{\mathrm{d}z} + \tilde{\chi}\frac{\mathrm{d}\tilde{\xi}}{\mathrm{d}z}\right)\frac{p'}{\tilde{\rho}}\hat{\mathbf{e}}_z = \mathcal{O}\left(\bar{g}\delta^2\right); \tag{4.129}$$

in the above we have also used the expression (4.22a) for the basic entropy vertical gradient and we recall here $\Delta_S = -(\mathrm{d}_z\tilde{T} + \tilde{\alpha}\tilde{T}\tilde{g}/\tilde{c}_{p,\xi})$. This yields

$$-\frac{1}{\tilde{\rho}}\nabla p' - \nabla\psi' + \frac{\rho'}{\tilde{\rho}}\tilde{\mathbf{g}} = -\nabla\left(\frac{p'}{\tilde{\rho}} + \psi'\right) + \frac{\tilde{\alpha}\tilde{T}}{\tilde{c}_{p,\xi}}s'\tilde{g}\hat{\mathbf{e}}_z + \tilde{\chi}\xi'\tilde{g}\hat{\mathbf{e}}_z, \tag{4.130}$$

at the leading order.

## 4.4 Final Set of Equations

In the most general case, when the reference state is arbitrary, the thermodynamic variables take the form

$$\rho(\mathbf{x},t) = \tilde{\rho}(\mathbf{x},\varepsilon t) + \rho'(\mathbf{x},t), \quad p(\mathbf{x},t) = \tilde{p}(\mathbf{x},\varepsilon t) + p'(\mathbf{x},t), \tag{4.131a}$$

$$T(\mathbf{x},t) = \tilde{T}(\mathbf{x},\varepsilon t) + T'(\mathbf{x},t), \quad s(\mathbf{x},t) = \tilde{s}(\mathbf{x},\varepsilon t) + s'(\mathbf{x},t), \tag{4.131b}$$

$$\xi(\mathbf{x},t) = \tilde{\xi}(\mathbf{x},\varepsilon t) + \xi'(\mathbf{x},t), \quad \psi(\mathbf{x},t) = \tilde{\psi}(\mathbf{x},\varepsilon t) + \psi'(\mathbf{x},t), \tag{4.131c}$$

where $\varepsilon = \delta^n \ll 1, n \geq 1$ was introduced to stress, that within the anelastic approximation, which in particular requires $\nabla \cdot (\tilde{\rho}\mathbf{u}) = 0$ to be satisfied, the time dependence of the reference state can only be slow. The reference state could be chosen in various ways, e.g. as adiabatic, $\mathrm{d}_z\tilde{T} = -\tilde{\alpha}\tilde{T}\tilde{g}/\tilde{c}_{p,\xi}$ and well-mixed, $\mathrm{d}_z\tilde{\xi} = 0$, thus not satisfying the temperature and mass fraction boundary conditions which drive the flow or time-dependent and/or fully spatially inhomogeneous. Although a fully spatially dependent reference state may seem rather complicated at first sight, it can still be useful, since through satisfaction of the boundary conditions it can allow to impose homogeneous boundary conditions on temperature fluctuations (or entropy fluctuations if the entropy is fixed at boundaries) and mass fraction fluctuations. The equations for the fluctuations of the mass fraction and the entropy, in such a general case, can not be expressed in a simple, compact way. Hence we provide here only the most general form of those equations, however, we adopt the standard assumption that the hydrostatic force balance $\nabla\tilde{p} = -\tilde{\rho}\tilde{\mathbf{g}}$ holds in the reference state (thus flows in the basic state, even if present, are weak), which allows to simplify the Navier-Stokes equation,

$$\tilde{\rho}\left[\frac{\partial \mathbf{u}}{\partial t} + (\mathbf{u}\cdot\nabla)\mathbf{u}\right] = -\nabla p' - \tilde{\rho}\nabla\psi' + \rho'\tilde{\mathbf{g}} + \tilde{\rho}\nu\nabla^2\mathbf{u} + \tilde{\rho}\left(\frac{\nu}{3} + \nu_b\right)\nabla(\nabla\cdot\mathbf{u})$$
$$+ 2\nabla(\tilde{\rho}\nu)\cdot\mathbf{G}^s + \nabla\left(\tilde{\rho}\nu_b - \frac{2}{3}\tilde{\rho}\nu\right)\nabla\cdot\mathbf{u}, \tag{4.132a}$$

$$\nabla\cdot(\tilde{\rho}\mathbf{u}) = 0, \tag{4.132b}$$

$$\nabla^2\psi' = 4\pi G\rho', \tag{4.132c}$$

$$\tilde{\rho}\left(\frac{\partial\xi}{\partial t} + \mathbf{u}\cdot\nabla\xi\right) = -\nabla\cdot\mathbf{j}_{\xi,\mathrm{mol}}, \tag{4.132d}$$

$$\tilde{\rho}\tilde{T}\left(\frac{\partial s}{\partial t} + \mathbf{u}\cdot\nabla s\right) = \nabla\cdot(k\nabla T) - \nabla\cdot\left(\Lambda\mathbf{j}_{\xi,\mathrm{mol}}\right) + \boldsymbol{\tau}_\nu : \mathbf{G}^s - \mathbf{j}_{\xi,\mathrm{mol}}\cdot\nabla\mu_c + Q, \tag{4.132e}$$

$$\frac{\rho'}{\tilde{\rho}} = -\tilde{\alpha}T' - \tilde{\chi}_T\xi' + \tilde{\beta}p', \quad s' = -\tilde{\alpha}\frac{p'}{\tilde{\rho}} + \tilde{c}_{p,\xi}\frac{T'}{\tilde{T}} + \tilde{h}_{p,T}\frac{\xi'}{\tilde{T}}, \tag{4.132f}$$

$$\mu_c' = \frac{\tilde{\chi}_T}{\tilde{\rho}}p' - \frac{\tilde{h}_{p,T}}{\tilde{T}}T' + \tilde{\Upsilon}\xi'. \tag{4.132g}$$

where

$$\mathbf{j}_{\xi,\mathrm{mol}} = -K\left(\nabla\xi + \frac{k_T}{T}\nabla T + \frac{k_p}{p}\nabla p\right), \tag{4.133}$$

$$\begin{aligned}\nabla\mu_c =& \frac{\chi_T}{\rho}\nabla p - \frac{h_{p,T}}{T}\nabla T + \Upsilon\nabla\xi \\ =& -\frac{\Upsilon}{K}\mathbf{j}_{\xi,\mathrm{mol}} - \frac{\Lambda}{T}\nabla T.\end{aligned} \tag{4.134}$$

The list of definitions of the thermodynamic properties is provided in (4.30a–4.30e) and in (4.54) and the total molecular heat flux possesses contributions from both the thermal and compositional fluxes i.e. is given by

$$\mathbf{j}_q = -k\nabla T + (\Lambda + \mu_c)\,\mathbf{j}_{\xi,\mathrm{mol}}. \tag{4.135}$$

We note, that the equations for thermodynamic fluctuations take the form provided in (4.132f) and (4.132g) only when (cf. Eqs. (4.98a, 4.98b) and (4.99) and the comment below)

$$\frac{|\xi'|}{\tilde{\xi}} \ll 1, \tag{4.136}$$

and the limit of a weak solution, when $|\xi'|/\tilde{\xi} = \mathcal{O}(1)$ is allowed, is considered later in Sect. 4.5. A significant simplification is achieved when the Soret effect and the effect of the pressure gradient on the flux of the light constituent are weak,

$$\frac{k_T}{T}\nabla T + \frac{k_p}{p}\nabla p \ll \nabla\xi, \tag{4.137}$$

so that the Fick's law for the material flux is satisfied

$$\mathbf{j}_{\xi,\mathrm{mol}} = -K\nabla\xi. \tag{4.138}$$

This allows to introduce a time independent hydrostatic reference state (cf. Sect. 4.2 and the discussion below (4.96a–4.96c)), so that

$$\rho(\mathbf{x},t) = \tilde{\rho}(z) + \rho'(\mathbf{x},t), \quad p(\mathbf{x},t) = \tilde{p}(z) + p'(\mathbf{x},t), \tag{4.139a}$$

$$T(\mathbf{x},t) = \tilde{T}(z) + T'(\mathbf{x},t), \quad s(\mathbf{x},t) = \tilde{s}(z) + s'(\mathbf{x},t), \tag{4.139b}$$

$$\xi(\mathbf{x},t) = \tilde{\xi}(z) + \xi'(\mathbf{x},t), \quad \psi(\mathbf{x},t) = \tilde{\psi}(z) + \psi'(\mathbf{x},t). \tag{4.139c}$$

Note, that just like for the thermal conductivity coefficient $k$ we assume, that the material conductivity coefficient is also a function of height only, $K = K(z)$ and the coefficients are of the order $k = \mathcal{O}(\delta^{1/2}\rho_B\bar{c}_{p,\xi}\sqrt{\bar{g}L}L)$ and $K = \mathcal{O}(\delta^{1/2}\rho_B\sqrt{\bar{g}L}L)$. Under the above assumptions the material flux can be approximated as follows

$$\tilde{\mathbf{j}}_{\xi,\mathrm{mol}} \cdot \hat{\mathbf{e}}_z = -K \frac{\mathrm{d}\tilde{\xi}}{\mathrm{d}z} = \text{const}, \tag{4.140a}$$

$$\tilde{\mathbf{j}}_{\xi,\mathrm{mol}} \cdot \hat{\mathbf{e}}_x = \tilde{\mathbf{j}}_{\xi,\mathrm{mol}} \cdot \hat{\mathbf{e}}_y = 0 \tag{4.140b}$$

$$\mathbf{j}'_{\xi,\mathrm{mol}} = -K\nabla\xi'. \tag{4.140c}$$

Furthermore, under the anelastic approximation it is demanded, that the material flux is weak, i.e.

$$-K\frac{\mathrm{d}\tilde{\xi}}{\mathrm{d}z} = \mathcal{O}\left(\delta^{3/2}\rho_B\sqrt{\bar{g}L}\right), \qquad -K\nabla\xi' = \mathcal{O}\left(\delta^{3/2}\rho_B\sqrt{\bar{g}L}\right), \tag{4.141}$$

so that the departure from the well-mixed state, which drives convection is small (cf. the definition of $\delta \ll 1$ in (3.11) and (4.83)). In consequence the heat flux in the reference state is dominated by the thermal flux $-k\mathrm{d}_z\tilde{T} = \mathcal{O}(\delta^{1/2}\rho_B\bar{c}_{p,\xi}\Delta T\sqrt{\bar{g}L})$, which is $\delta^{-1}$ times greater than the material flux contribution $\Lambda K\mathrm{d}_z\tilde{\xi} = \mathcal{O}(\delta^{3/2}\rho_B\bar{c}_{p,\xi}\Delta T\sqrt{\bar{g}L})$.

Therefore in the light of (4.90) we can make the following estimate

$$\begin{aligned}-\mathbf{j}_{\xi,\mathrm{mol}} \cdot \nabla\mu_c &= -\tilde{\chi}\bar{g}K\frac{\mathrm{d}\tilde{\xi}}{\mathrm{d}z} - \tilde{\chi}\bar{g}K\frac{\partial\xi'}{\partial z} + K\frac{\mathrm{d}\tilde{\xi}}{\mathrm{d}z}\nabla\mu'_c + \mathcal{O}\left(\delta^{5/2}\rho_B\bar{g}\sqrt{\bar{g}L}\right)\\ &= -\tilde{\chi}\bar{g}K\frac{\mathrm{d}\tilde{\xi}}{\mathrm{d}z} - \tilde{\chi}\bar{g}K\frac{\partial\xi'}{\partial z} + \mathcal{O}\left(\delta^{5/2}\rho_B\bar{g}\sqrt{\bar{g}L}\right),\end{aligned} \tag{4.142}$$

which is valid as long as the assumption of a *not* weak solution (4.136) holds and consequently the convective fluctuation of the chemical potential $\mu'_c$ and its gradient are small ($\Upsilon = \mathcal{O}(\bar{g}L)$), so that

$$\begin{aligned}K\frac{\mathrm{d}\tilde{\xi}}{\mathrm{d}z}\nabla\mu'_c &= K\frac{\mathrm{d}\tilde{\xi}}{\mathrm{d}z}\nabla\left(\frac{\tilde{\chi}_T}{\tilde{\rho}}p' - \frac{\tilde{h}_{p,T}}{\tilde{T}}T' + \tilde{\Upsilon}\xi'\right)\\ &= K\mathcal{O}\left(\frac{\delta}{L}\right)\mathcal{O}(\delta\bar{g}) = \mathcal{O}\left(\delta^{5/2}\rho_B\bar{g}\sqrt{\bar{g}L}\right).\end{aligned} \tag{4.143}$$

Of course the term $-\tilde{\chi}\bar{g}K\mathrm{d}_z\tilde{\xi}$ in (4.142) belongs to the basic state energy (entropy) balance, hence it does not contribute to the dynamical equation for the evolution of the entropy fluctuation.

Finally by the use of (4.130) the Navier-Stokes equation can be written in the form which emphasizes, that the buoyancy force is created by departures from the isentropic and well-mixed state. Therefore the leading-order anelastic equations under the additional assumptions that the mean concentration of the light constituent greatly exceeds its convective fluctuation (4.136) and that the Soret and pressure gradient effects on the material flux are negligible (4.137) can be cast in the following, simplified form

$$\frac{\partial \mathbf{u}}{\partial t}+(\mathbf{u}\cdot\nabla)\,\mathbf{u}=-\nabla\left(\frac{p'}{\tilde{\rho}}+\psi'\right)+\frac{\tilde{\alpha}\tilde{T}}{\tilde{c}_{p,\xi}}s'\tilde{g}\hat{\mathbf{e}}_z+\tilde{\chi}\xi'\tilde{g}\hat{\mathbf{e}}_z$$
$$+\nu\nabla^2\mathbf{u}+\left(\frac{\nu}{3}+\nu_b\right)\nabla\left(\nabla\cdot\mathbf{u}\right)$$
$$+\frac{2}{\tilde{\rho}}\nabla\left(\tilde{\rho}\nu\right)\cdot\mathbf{G}^s+\frac{1}{\tilde{\rho}}\nabla\left(\tilde{\rho}\nu_b-\frac{2}{3}\tilde{\rho}\nu\right)\nabla\cdot\mathbf{u}, \quad (4.144a)$$

$$\nabla\cdot(\tilde{\rho}\mathbf{u})=0, \quad (4.144b)$$

$$\nabla^2\psi'=4\pi G\rho', \quad (4.144c)$$

$$\tilde{\rho}\left(\frac{\partial\xi'}{\partial t}+\mathbf{u}\cdot\nabla\xi'\right)+\tilde{\rho}u_z\frac{\mathrm{d}\tilde{\xi}}{\mathrm{d}z}=\nabla\cdot\left(K\nabla\xi'\right), \quad (4.144d)$$

$$\tilde{\rho}\tilde{T}\left(\frac{\partial s'}{\partial t}+\mathbf{u}\cdot\nabla s'\right)-\tilde{\rho}\tilde{c}_{p,\xi}u_z\Delta_S+\tilde{\rho}\tilde{h}_{p,T}u_z\frac{\mathrm{d}\tilde{\xi}}{\mathrm{d}z}$$
$$=\nabla\cdot\left(k\nabla T'\right)+\nabla\cdot\left(\Lambda K\nabla\xi'\right)-\tilde{\chi}\tilde{g}K\frac{\partial\xi'}{\partial z}$$
$$+2\tilde{\rho}\nu\mathbf{G}^s:\mathbf{G}^s+\tilde{\rho}\left(\nu_b-\frac{2}{3}\nu\right)(\nabla\cdot\mathbf{u})^2+Q', \quad (4.144e)$$

$$\frac{\rho'}{\tilde{\rho}}=-\tilde{\alpha}T'-\tilde{\chi}_T\xi'+\tilde{\beta}p', \qquad s'=-\tilde{\alpha}\frac{p'}{\tilde{\rho}}+\tilde{c}_{p,\xi}\frac{T'}{\tilde{T}}+\tilde{h}_{p,T}\frac{\xi'}{\tilde{T}}, \quad (4.144f)$$

where we have used

$$\frac{\mathrm{d}\tilde{s}}{\mathrm{d}z}=\frac{\tilde{c}_{p,\xi}}{\tilde{T}}\left(\frac{\mathrm{d}\tilde{T}}{\mathrm{d}z}+\frac{\tilde{\alpha}\tilde{T}\tilde{g}}{\tilde{c}_{p,\xi}}\right)+\frac{\tilde{h}_{p,T}}{\tilde{T}}\frac{\mathrm{d}\tilde{\xi}}{\mathrm{d}z}, \quad (4.145)$$

and $\Delta_S=-(\mathrm{d}_z\tilde{T}+\tilde{\alpha}\tilde{T}\tilde{g}/\tilde{c}_{p,\xi})$. Once the thermodynamic properties of the binary alloy $\tilde{\alpha}$, $\tilde{\beta}$, $\tilde{\chi}_T$, $\tilde{c}_{p,\xi}$ and $\tilde{h}_{p,T}$, together with $r_m$ and the density profile for the bottom body $\rho_{in}(z)$ are specified, the system of Eqs. (4.144a–4.144f), supplied by adequate boundary conditions becomes fully determined and can be solved. One of the simplest examples of an equation of state for a mixture of two constituents corresponds to a mixture of ideal gases for which all the thermodynamic properties were provided in Sect. 4.2.1.

### 4.4.1 Global Balance

In order to obtain a global force balance we multiply the Navier-Stokes equation (4.144a) by $\tilde{\rho}\mathbf{u}$, average over the entire volume (with assumed periodicity in the '$x$' and '$y$' directions) and for a (statistically) stationary state this yields the following balance between the total work per unit volume of the buoyancy forces averaged over the horizontal directions and the total viscous dissipation in the fluid volume

$$\left\langle \frac{\tilde{\alpha}\tilde{T}\tilde{g}\tilde{\rho}}{\tilde{c}_{p,\xi}} u_z s' \right\rangle + \left\langle \tilde{\chi}\tilde{g}\tilde{\rho} u_z \xi' \right\rangle = 2\left\langle \mu \mathbf{G}^s : \mathbf{G}^s \right\rangle - \left\langle \left(\frac{2}{3}\mu - \mu_b\right)(\nabla \cdot \mathbf{u})^2 \right\rangle. \tag{4.146}$$

To derive the latter relation we have used the impermeable and either no-slip or stress-free boundary conditions, as in (2.65a, 2.65b).

Next we investigate the mean energy (entropy) and material fluxes and for simplicity we assume no additional (such as e.g. radiogenic or radiational) heat sources $Q = 0$. First we average over a horizontal plane and integrate from 0 to $z$ the stationary energy equation (4.132e) and the stationary mass fraction equation (4.132d),

$$\begin{aligned} 0 = &\int_0^z \tilde{\rho}\frac{\mathrm{d}\tilde{T}}{\mathrm{d}z}\left\langle u_z s' \right\rangle_h \mathrm{d}z - \tilde{\rho}\tilde{T}\left\langle u_z s' \right\rangle_h - k\frac{\mathrm{d}\langle T \rangle_h}{\mathrm{d}z}\bigg|_{z=0} + k\frac{\mathrm{d}\langle T \rangle_h}{\mathrm{d}z} \\ &+ (\Lambda + \mu_c)\left\langle \mathbf{j}_{\xi,\mathrm{mol}} \cdot \hat{\mathbf{e}}_z \right\rangle_h\big|_{z=0} - (\Lambda + \mu_c)\left\langle \mathbf{j}_{\xi,\mathrm{mol}} \cdot \hat{\mathbf{e}}_z \right\rangle_h \\ &+2\int_0^z \left\langle \mu \mathbf{G}^s : \mathbf{G}^s \right\rangle_h \mathrm{d}z - \int_0^z \left\langle \left(\frac{2}{3}\mu - \mu_b\right)(\nabla \cdot \mathbf{u})^2 \right\rangle_h \mathrm{d}z \\ &+ \int_0^z \left\langle \mu_c \nabla \cdot \mathbf{j}_{\xi,\mathrm{mol}} \right\rangle_h \mathrm{d}z, \end{aligned} \tag{4.147a}$$

$$\tilde{\rho}\left\langle u_z \xi' \right\rangle_h = \left\langle \mathbf{j}_{\xi,\mathrm{mol}} \cdot \hat{\mathbf{e}}_z \right\rangle_h\big|_{z=0} - \left\langle \mathbf{j}_{\xi,\mathrm{mol}} \cdot \hat{\mathbf{e}}_z \right\rangle_h, \tag{4.147b}$$

where we have used the fact, that the impermeable boundary conditions at $z = 0,\ L$ and the mass conservation imply $\langle u_z \rangle_h = 0$ in the fluid volume (cf. (3.101)).

By the use of stationary form of the Eq. (4.132d) and the fact that $\nabla\xi = \mathcal{O}(\delta/L)$

$$\begin{aligned} \int_0^z \left\langle \mu_c \nabla \cdot \mathbf{j}_{\xi,\mathrm{mol}} \right\rangle_h \mathrm{d}z &= -\int_0^z \left\langle \tilde{\mu}_c \tilde{\rho}\mathbf{u} \cdot \nabla \xi' \right\rangle_h \mathrm{d}z + \mathcal{O}\left(\delta^{5/2}\frac{k\Delta T}{L}\right) \\ &= -\int_0^z \left\langle \tilde{\mu}_c \nabla \cdot (\tilde{\rho}\mathbf{u}\xi') \right\rangle_h \mathrm{d}z + \mathcal{O}\left(\delta^{5/2}\frac{k\Delta T}{L}\right) \\ &= -\tilde{\mu}_c\tilde{\rho}\left\langle u_z \xi' \right\rangle_h + \int_0^z \frac{\mathrm{d}\tilde{\mu}_c}{\mathrm{d}z}\tilde{\rho}\left\langle u_z \xi' \right\rangle_h \mathrm{d}z + \mathcal{O}\left(\delta^{5/2}\frac{k\Delta T}{L}\right) \end{aligned} \tag{4.148}$$

where we have used again $\langle u_z \rangle_h = 0$, the anelastic mass balance $\nabla \cdot (\tilde{\rho}\mathbf{u}) = 0$ and finally the impermeability condition at the bottom boundary. On introduction of (4.148) into the global balance (4.147a), together with $\mathrm{d}_z \tilde{T} = -\tilde{g}\tilde{\alpha}\tilde{T}/\tilde{c}_p + \mathcal{O}(\delta\Delta T/L)$ and $\mathrm{d}_z \tilde{\mu}_c = -\tilde{g}\tilde{\chi} + \mathcal{O}(\delta\bar{g})$ we get the following expression for the total horizontally averaged heat flux entering the system at the bottom

$$
\begin{aligned}
F_{total}(z=0) =& -k\frac{\mathrm{d}}{\mathrm{d}z}\left(\tilde{T} + \langle T' \rangle_h\right)\Big|_{z=0} + (\Lambda + \mu_c)\left\langle \mathbf{j}_{\xi,\mathrm{mol}} \cdot \hat{\mathbf{e}}_z \right\rangle_h \Big|_{z=0} \\
=& -k\frac{\mathrm{d}}{\mathrm{d}z}\left(\tilde{T} + \langle T' \rangle_h\right) + (\Lambda + \mu_c)\left\langle \mathbf{j}_{\xi,\mathrm{mol}} \cdot \hat{\mathbf{e}}_z \right\rangle_h \\
&+ \tilde{\rho}\tilde{T}\left\langle u_z s' \right\rangle_h + \tilde{\mu}_c \tilde{\rho}\left\langle u_z \xi' \right\rangle_h \\
&+ \int_0^z \frac{\tilde{\rho}\tilde{g}\tilde{\alpha}\tilde{T}}{\tilde{c}_{p,\xi}}\left\langle u_z s' \right\rangle_h \mathrm{d}z + \int_0^z \tilde{g}\tilde{\chi}\tilde{\rho}\left\langle u_z \xi' \right\rangle_h \mathrm{d}z \\
&- 2\int_0^z \left\langle \mu \mathbf{G}^s : \mathbf{G}^s \right\rangle_h \mathrm{d}z + \int_0^z \left\langle \left(\frac{2}{3}\mu - \mu_b\right)(\nabla \cdot \mathbf{u})^2 \right\rangle_h \mathrm{d}z. \quad (4.149)
\end{aligned}
$$

The latter formula expresses the fact, that the horizontally averaged vertical heat flux at any height $z$ is significantly influenced (could be either increased or decreased) by the work of the buoyancy forces and the viscous heating, i.e.

$$
\begin{aligned}
F_{total}(z) =& -k\frac{\mathrm{d}}{\mathrm{d}z}\left(\tilde{T} + \langle T' \rangle_h\right) + (\Lambda + \mu_c)\left\langle \mathbf{j}_{\xi,\mathrm{mol}} \cdot \hat{\mathbf{e}}_z \right\rangle_h \\
&+ \tilde{\rho}\tilde{T}\left\langle u_z s' \right\rangle_h + \tilde{\mu}_c \tilde{\rho}\left\langle u_z \xi' \right\rangle_h \\
=& F_{total}(z=0) - \int_0^z \frac{\tilde{\rho}\tilde{g}\tilde{\alpha}\tilde{T}}{\tilde{c}_{p,\xi}}\left\langle u_z s' \right\rangle_h \mathrm{d}z - \int_0^z \tilde{g}\tilde{\chi}\tilde{\rho}\left\langle u_z \xi' \right\rangle_h \mathrm{d}z \\
&+ 2\int_0^z \left\langle \mu \mathbf{G}^s : \mathbf{G}^s \right\rangle_h \mathrm{d}z - \int_0^z \left\langle \left(\frac{2}{3}\mu - \mu_b\right)(\nabla \cdot \mathbf{u})^2 \right\rangle_h \mathrm{d}z. \quad (4.150)
\end{aligned}
$$

In an analogous way, from (4.147b) we can derive the expression for the total horizontally averaged material flux, denoted by $G_{total}(z)$, entering the system at the bottom,

$$
G_{total}(z=0) = \left\langle \mathbf{j}_{\xi,\mathrm{mol}} \cdot \hat{\mathbf{e}}_z \right\rangle_h \Big|_{z=0} = \tilde{\rho}\left\langle u_z \xi' \right\rangle_h + \left\langle \mathbf{j}_{\xi,\mathrm{mol}} \cdot \hat{\mathbf{e}}_z \right\rangle_h = G_{total}(z), \quad (4.151)
$$

which is the same at every horizontal plane, thus independent of height, $G_{total}(z) = G_{total}(z=0)$.[11]

Next, by setting the upper limit of the vertical integration in (4.147a) and (4.148) to $z = L$ (i.e. integrating over the entire fluid volume), applying the boundary condition of impermeability at the top $u_z(z=L) = 0$ and utilizing (4.146) one obtains

[11] Similarly as the heat flux in the Boussinesq approximation; see Sect. 2.1.3, Eq. (2.55).

$$L\left\langle\frac{\tilde{g}\tilde{\alpha}\tilde{T}\tilde{\rho}}{\tilde{c}_p}u_z s'\right\rangle + L\left\langle\tilde{\rho}\frac{\mathrm{d}\tilde{T}}{\mathrm{d}z}u_z s'\right\rangle + L\left\langle\tilde{\chi}\tilde{g}\tilde{\rho}u_z\xi'\right\rangle + L\left\langle\frac{\mathrm{d}\tilde{\mu}_c}{\mathrm{d}z}\tilde{\rho}u_z\xi'\right\rangle$$
$$= -k\frac{\mathrm{d}\langle T\rangle_h}{\mathrm{d}z}\bigg|_{z=L} + k\frac{\mathrm{d}\langle T\rangle_h}{\mathrm{d}z}\bigg|_{z=0} + (\Lambda+\mu_c)\left\langle\mathbf{j}_{\xi,\mathrm{mol}}\cdot\hat{\mathbf{e}}_z\right\rangle_h\Big|_{z=L}$$
$$-(\Lambda+\mu_c)\left\langle\mathbf{j}_{\xi,\mathrm{mol}}\cdot\hat{\mathbf{e}}_z\right\rangle_h\Big|_{z=0}. \tag{4.152}$$

Of course the left hand side of the latter equation vanishes at leading order since $\mathrm{d}_z\tilde{T} = -\tilde{g}\tilde{\alpha}\tilde{T}/\tilde{c}_p + \mathcal{O}(\delta\Delta T/L)$ and $\mathrm{d}_z\tilde{\mu}_c = -\tilde{g}\tilde{\chi} + \mathcal{O}(\delta\bar{g})$. Therefore we can conclude, that in a stationary state the total heat flux which enters the system at the bottom must be equal to the total heat flux which leaves the system at the top,

$$-k\frac{\mathrm{d}\langle T\rangle_h}{\mathrm{d}z}\bigg|_{z=0} + (\Lambda+\mu_c)\left\langle\mathbf{j}_{\xi,\mathrm{mol}}\cdot\hat{\mathbf{e}}_z\right\rangle_h\Big|_{z=0}$$
$$= -k\frac{\mathrm{d}\langle T\rangle_h}{\mathrm{d}z}\bigg|_{z=L} + (\Lambda+\mu_c)\left\langle\mathbf{j}_{\xi,\mathrm{mol}}\cdot\hat{\mathbf{e}}_z\right\rangle_h\Big|_{z=L}. \tag{4.153}$$

The same is true for the total compositional flux

$$\left\langle\mathbf{j}_{\xi,\mathrm{mol}}\cdot\hat{\mathbf{e}}_z\right\rangle_h\Big|_{z=0} = \left\langle\mathbf{j}_{\xi,\mathrm{mol}}\cdot\hat{\mathbf{e}}_z\right\rangle_h\Big|_{z=L}, \tag{4.154}$$

which has been obtained by setting $z = L$ in (4.147b) and application of the impermeability condition at the top boundary.

### 4.4.2 Definitions of the Rayleigh and Nusselt Numbers

For simplicity let us assume, that the influence of the temperature and pressure gradients on the material flux is negligible, i.e. $\mathbf{j}_{\xi,\mathrm{mol}} = -K\nabla\xi$. Then the total local heat flux is given by (cf. Eq. (4.135))

$$\mathbf{j}_q = -k\nabla T - (\Lambda+\mu_c)\,K\nabla\xi. \tag{4.155}$$

To properly define the Nusselt number we must first identify the heat flux associated with the physical driving of the convective flow. Since buoyancy effects are allowed only when the adiabatic temperature gradient is exceeded and/or when the gradient of the mass fraction $\xi$ becomes negative, the driving comes from the superadiabatic excess in the temperature gradient and the magnitude of the negative mass fraction gradient, both imposed by the boundary conditions. Therefore to describe heat transfer by convection it is suitable to use the total superadiabatic heat flux composed of contributions from the superadiabatic thermal molecular flux, compositional molecular flux and the heat flux resulting from advection of the entropy and concentration, defined in the following way

$$F_S(z) = -k\frac{\mathrm{d}}{\mathrm{d}z}\left(\tilde{T} + \langle T'\rangle_h - T_{ad}\right) - (\Lambda + \mu_c) K \frac{\mathrm{d}}{\mathrm{d}z}\left(\tilde{\xi} + \langle \xi'\rangle_h\right)$$
$$+ \tilde{\rho}\tilde{T}\left\langle u_z s'\right\rangle_h + \tilde{\mu}_c\tilde{\rho}\left\langle u_z \xi'\right\rangle_h . \tag{4.156}$$

In such a way the heat flux conducted down the adiabat, unrelated to the convective motions, is excluded. This allows to define the Nusselt number $Nu$ as a ratio of the total superadiabatic heat flux which enters the system at the bottom in a convective state, $F_S(z = 0)$ to the total superadiabatic heat flux in the hydrostatic reference state, for which the temperature $\tilde{T}$ and mass fraction $\tilde{\xi}$ satisfy the non-homogeneous boundary conditions responsible for driving (the fluctuations $T'$ and $\xi'$ satisfy homogeneous boundary conditions),

$$Nu = \frac{F_S\,(z = 0)}{k\Delta_S - (\Lambda + \mu_c)\, K \frac{\mathrm{d}\tilde{\xi}}{\mathrm{d}z}}$$
$$= \frac{-k\,\frac{\mathrm{d}}{\mathrm{d}z}\left(\tilde{T} + \langle T'\rangle_h - T_{ad}\right)\Big|_{z=0} - (\Lambda + \mu_c)\, K \frac{\mathrm{d}}{\mathrm{d}z}\left(\tilde{\xi} + \langle \xi'\rangle_h\right)\Big|_{z=0}}{k\Delta_S - (\Lambda + \mu_c)\, K \frac{\mathrm{d}\tilde{\xi}}{\mathrm{d}z}}. \tag{4.157}$$

Furthermore, the most convenient definition of the Rayleigh numbers associated with thermal and compositional buoyancies depends on a particular application. If the fluid properties are spatially non-uniform a useful Rayleigh number definition can involve spatial averaging. However, when one assumes uniform dynamical viscosity $\mu$, thermal conductivity $k$, specific heat $c_{p,\xi}$ and gravity $g = \mathrm{const}$ (which is allowed e.g. in the case of a weak solution considered in Sect. 4.5), the following definitions of the thermal and compositional Rayleigh numbers can be proposed

$$Ra_{th} = \frac{g\,\Delta_S\, L^4 \rho_B^2 c_{p,\xi}}{T_B \mu k}, \tag{4.158}$$

$$Ra_{comp} = \frac{g\left(-\frac{\mathrm{d}\tilde{\xi}}{\mathrm{d}z}\right) L^4 \rho_B^2 c_{p,\xi}}{\mu k}, \tag{4.159}$$

where $k\Delta_S = k(\Delta T/L - g/c_p)$ is the superadiabatic thermal conductive heat flux in the hydrostatic basic state.

### 4.4.3 Boussinesq Equations

In order to reduce the system of anelastic dynamical equations into the Boussinesq one it is necessary to assume that the scale heights associated with density, temperature and pressure are large compared to the fluid layer thickness, i.e.

$$\frac{\mathrm{d}\tilde{\rho}}{\mathrm{d}z} \ll \frac{\bar{\rho}}{L}, \quad \frac{\mathrm{d}\tilde{T}}{\mathrm{d}z} \ll \frac{\bar{T}}{L}, \quad \frac{\mathrm{d}\tilde{p}}{\mathrm{d}z} \ll \frac{\bar{p}}{L}. \tag{4.160}$$

Since this limit is mainly applicable to laboratory systems we assume for simplicity, that the gravitational acceleration is constant, $g = \text{const}$. The Boussinesq approximation is characterized by small departures of the thermodynamic variables from their mean values, thus we split the hydrostatic basic state into the mean and the small vertically varying correction, e.g. $\tilde{T} = \bar{T} + \tilde{\tilde{T}}$ and $\tilde{\tilde{T}}/\bar{T} \ll 1$, cf. Chap. 2 on the Boussinesq convection. Recall, that the small Boussinesq parameter was defined as $\epsilon = \Delta\tilde{\tilde{\rho}}/\bar{\rho} \ll 1$ and therefore the Boussinesq limit of the anelastic approximation implies $\delta \lesssim \mathcal{O}(\epsilon)$ and $\mathrm{d}_z\tilde{\xi} \leq \mathcal{O}(\epsilon)$. In the equations we can substitute $\tilde{\rho} \approx \bar{\rho}, \tilde{T} \approx \bar{T}$, $\tilde{p} \approx \bar{p}$, $\mathrm{d}\tilde{\rho}/\mathrm{d}z = \mathrm{d}\tilde{\tilde{\rho}}/\mathrm{d}z$, $\mathrm{d}\tilde{T}/\mathrm{d}z = \mathrm{d}\tilde{\tilde{T}}/\mathrm{d}z$, $\mathrm{d}\tilde{p}/\mathrm{d}z = \mathrm{d}\tilde{\tilde{p}}/\mathrm{d}z$.

It can already be seen, that in the continuity equation the term $u_z\mathrm{d}\tilde{\rho}/\mathrm{d}z = \mathcal{O}(\bar{\rho}\sqrt{g/L}\epsilon^{3/2})$ becomes negligible compared to $\bar{\rho}\nabla \cdot \mathbf{u} = \mathcal{O}(\bar{\rho}\sqrt{g/L}\epsilon^{1/2})$ and hence the mass conservation at leading order is expressed by $\nabla \cdot \mathbf{u} = 0$; we recall, that in the Boussinesq limit $|\mathbf{u}| = \mathcal{O}(\epsilon^{1/2}\sqrt{gL})$.

Next, by the use of the formula

$$\frac{\partial s}{\partial z} = \frac{c_{p,\xi}}{T}\frac{\partial T}{\partial z} - \frac{\alpha}{\rho}\frac{\partial p}{\partial z} + \frac{h_{p,T}}{T}\frac{\partial \xi}{\partial z}, \tag{4.161}$$

and with the aid of the hydrostatic force balance $\partial_z p \sim -g\rho$ we can provide the following estimates

$$c_{p,\xi} = \mathcal{O}\left(-\frac{\bar{\alpha}\bar{T}gL}{L\frac{\mathrm{d}\tilde{\tilde{T}}}{\mathrm{d}z}}\right) = \mathcal{O}\left(\epsilon^{-1}gL/\bar{T}\right), \qquad h_{p,T} = \mathcal{O}\left(\epsilon^{-1}gL\right). \tag{4.162}$$

The compositional and thermal conduction and diffusion coefficients satisfy

$$D = \frac{K}{\bar{\rho}} = \mathcal{O}\left(\epsilon^{1/2}\sqrt{gL}L\right), \quad \kappa = \frac{k}{\bar{\rho}\bar{c}_{p,\xi}} = \mathcal{O}\left(\epsilon^{1/2}\sqrt{gL}L\right), \tag{4.163}$$

hence for the compositional heat flux we get

$$\mathbf{j}_{\xi,\mathrm{mol}} \sim -K\nabla\xi = \mathcal{O}\left(\epsilon^{3/2}\bar{\rho}\sqrt{gL}\right). \tag{4.164}$$

Since the full expression for the compositional flux (4.133) involves gradients of temperature and pressure and the total heat flux (4.135) involves a contribution from the material flux, we require for consistency

$$\Lambda = \mathcal{O}\left(\epsilon^{-1}gL\right), \qquad \Upsilon = \mathcal{O}\left(\epsilon^{-1}gL\right), \tag{4.165}$$

$$k_T = \frac{\Lambda - h_{p,T}}{\Upsilon} \lesssim \mathcal{O}(1) \quad k_p = \frac{p\chi_T}{\rho\Upsilon} = \chi_T \frac{p}{\rho g L} \frac{gL}{\Upsilon} = \mathcal{O}(1), \tag{4.166}$$

where we have utilized (2.28d) to estimate $p/\rho g L$. Furthermore, by the use of (2.28c) the pressure fluctuation in the Boussinesq limit is so small, that the effect of $\nabla p'$ on the material flux can be neglected to yield

$$\begin{aligned} \mathbf{j}_{\xi,\text{mol}} &= -K\left(\nabla\xi + \frac{k_T}{T}\nabla T + \frac{k_p}{p}\nabla p\right) \\ &= -K\left(\frac{\mathrm{d}\tilde{\bar{\xi}}}{\mathrm{d}z} + \frac{\bar{k}_T}{\bar{T}}\frac{\mathrm{d}\tilde{\bar{T}}}{\mathrm{d}z} + \frac{\bar{k}_p}{\bar{p}}\frac{\mathrm{d}\tilde{\bar{p}}}{\mathrm{d}z}\right) - K\left(\nabla\xi' + \frac{\bar{k}_T}{\bar{T}}\nabla T'\right). \end{aligned} \tag{4.167}$$

On the basis of (4.134) the gradient of the chemical potential is of the order $\epsilon^0$, i.e. $\nabla\mu_c = \mathcal{O}(g)$ and this implies

$$-\frac{1}{\bar{\rho}\bar{c}_{p,\xi}}\mathbf{j}_{\xi,\text{mol}} \cdot \nabla\mu_c = \mathcal{O}\left(\epsilon^{5/2}\bar{T}\sqrt{g/L}\right), \tag{4.168}$$

which will allow to neglect this term in the temperature equation, since it is of the same order of magnitude as the viscous heating, cf. (2.42). The following expression for the entropy differential

$$\mathrm{d}s = \frac{c_{p,\xi}}{T}\mathrm{d}T - \frac{\alpha}{\rho}\mathrm{d}p + \frac{h_{p,T}}{T}\mathrm{d}\xi, \tag{4.169}$$

supplied by the mass fraction balance (4.132d) allows to transform the entropy balance (4.132e) into the temperature equation

$$\begin{aligned} &\bar{\rho}\bar{c}_{p,\xi}\left(\frac{\partial T}{\partial t} + \mathbf{u}\cdot\nabla T\right) - \bar{\alpha}\bar{T}\left(\frac{\partial p}{\partial t} + \mathbf{u}\cdot\nabla p\right) \\ &= \nabla\cdot(k\nabla T) - \nabla\cdot\left[\left(\Lambda + \bar{h}_{p,T}K\right)\mathbf{j}_{\xi,\text{mol}}\right] + \boldsymbol{\tau}_\nu : \mathbf{G}^s - \mathbf{j}_{\xi,\text{mol}}\cdot\nabla\mu_c + Q. \end{aligned} \tag{4.170}$$

The pressure material derivative can be greatly simplified using the fact, that the pressure fluctuation is small, as in (2.28c), so that

$$\frac{\partial p}{\partial t} + \mathbf{u}\cdot\nabla p = \frac{\partial p'}{\partial t} + u_z\frac{\mathrm{d}\tilde{\bar{p}}}{\mathrm{d}z} + \mathbf{u}\cdot\nabla p' = -\bar{\rho}g u_z + \mathcal{O}\left(\epsilon^{3/2}\bar{\rho}\sqrt{g^3L}\right). \tag{4.171}$$

Finally the smallness of the pressure fluctuation implies also $\left|\tilde{\beta}p'\right| \ll \left|\tilde{\alpha}T'\right|$ (cf. discussion below (2.19)), thus the buoyancy force in the Navier-Stokes equation (4.132a) takes the leading order form

$$\mathbf{g}\frac{\rho'}{\bar{\rho}} = -\mathbf{g}\bar{\alpha}T' - \mathbf{g}\bar{\chi}_T\xi' + \mathcal{O}\left(\mathbf{g}\epsilon^2\right). \tag{4.172}$$

We are now ready to write down the final form of the dynamical equations under the Boussinesq approximation which reads

$$\frac{\partial \mathbf{u}}{\partial t} + (\mathbf{u}\cdot\nabla)\,\mathbf{u} = -\frac{1}{\bar{\rho}}\nabla p' + g\bar{\alpha}T'\hat{\mathbf{e}}_z + g\bar{\chi}_T\xi'\hat{\mathbf{e}}_z + \nu\nabla^2\mathbf{u} + 2\nabla\nu\cdot\mathbf{G}^s, \tag{4.173a}$$

$$\nabla\cdot\mathbf{u} = 0, \tag{4.173b}$$

$$\frac{\partial \xi'}{\partial t} + \mathbf{u}\cdot\nabla\xi' + u_z\frac{\mathrm{d}\tilde{\tilde{\xi}}}{\mathrm{d}z} = \nabla\cdot\left[D\left(\nabla\xi' + \frac{k_T}{\bar{T}}\nabla T'\right)\right], \tag{4.173c}$$

$$\begin{aligned}\frac{\partial T'}{\partial t} + \mathbf{u}\cdot\nabla T' + u_z\left(\frac{\mathrm{d}\tilde{\tilde{T}}}{\mathrm{d}z} + \frac{g\bar{\alpha}\bar{T}}{\bar{c}_{p,\xi}}\right) =& \nabla\cdot\left[\left(\kappa + D\frac{\bar{\Upsilon}k_T^2}{\bar{c}_{p,\xi}\bar{T}}\right)\nabla T'\right] \\ &+ \frac{\bar{\Upsilon}}{\bar{c}_{p,\xi}}\nabla\cdot\left(k_T D\nabla\xi'\right) + \frac{Q'}{\bar{\rho}\bar{c}_{p,\xi}}.\end{aligned} \tag{4.173d}$$

Note, that we did not utilize the smallness of the ratios $\xi'/\bar{\xi}$ nor $s'/\bar{s}$ at any point in the derivation of the Boussinesq equation and in fact in the Boussinesq limit those ratios need not be small.

Moreover, we observe that when the Soret coefficient $k_T$ is negligibly small both the Soret and Dufour effects are excluded at once and the mass fraction and temperature equations simplify to

$$\frac{\partial \xi'}{\partial t} + \mathbf{u}\cdot\nabla\xi' + u_z\frac{\mathrm{d}\tilde{\tilde{\xi}}}{\mathrm{d}z} = \nabla\cdot\left(D\nabla\xi'\right), \tag{4.174}$$

$$\frac{\partial T'}{\partial t} + \mathbf{u}\cdot\nabla T' + u_z\left(\frac{\mathrm{d}\tilde{\tilde{T}}}{\mathrm{d}z} + \frac{g\bar{\alpha}\bar{T}}{\bar{c}_{p,\xi}}\right) = \nabla\cdot\left(\kappa\nabla T'\right) + \frac{Q'}{\bar{\rho}\bar{c}_{p,\xi}}. \tag{4.175}$$

When there are no heating sources, $Q = 0$, the equations for the mass fraction and the temperature are of exactly the same type and since the compositional and thermal contributions to the total buoyancy force $g\bar{\alpha}T'\hat{\mathbf{e}}_z + g\bar{\chi}_T\xi'\hat{\mathbf{e}}_z$ are alike, if additionally the boundary conditions for $T'$ and $\xi'$ are of the same type, the physical effect of both is qualitatively the same. If, however, some radiogenic or radioactive heat sources are present the effect of thermal driving significantly differs from that of the compositional driving. In general the boundary conditions for $T'$ and $\xi'$ are also not of the same type, making their effects distinguishable.

## 4.5 Weak Solution Limit, $\xi \ll 1$

When the solution of the light constituent is weak, that is the mass fraction

$$\xi \ll 1 \tag{4.176}$$

is small, the coefficient $\Upsilon$, which describes the variation of the chemical potential with concentration at constant pressure and temperature satisfies

$$\Upsilon \approx \frac{k_B T}{\xi m_l} = \mathcal{O}\left(\bar{c}_{p,\xi}\bar{T}/\bar{\xi}\right) \gg 1, \tag{4.177}$$

where $k_B$ is the Boltzmann constant (cf. Landau and Lifshitz 1980, Eqs. 87.4–5, chapter "Weak solutions" and Chap. 96 on "Thermodynamic inequalities for solutions").[12] Therefore $\alpha\Upsilon/\chi_T \gg c_{p,\xi} \sim h_{p,T}/T$ is large and by (4.27) the quantity

$$-\frac{L}{Tc_{p,\xi}}\left(\frac{\mathrm{d}\mu_c}{\mathrm{d}z} + \chi g\right) = \mathcal{O}(1) \tag{4.178}$$

is, in general *not* small and much greater than the superadiabatic gradient $\alpha L(\mathrm{d}_z T + \alpha T g/c_{p,\xi})$, despite the smallness of $\mathrm{d}_z\xi$.

A hydrostatic reference state for anelastic convection in the case of a weak solution, $\tilde{\xi} \ll 1$ and $\xi' \ll 1$, can be obtained in the following way. In such a case the coefficient $\Upsilon \gg \chi c_{p,\xi}/\alpha$ is large in comparison with other standard thermodynamic properties of the binary alloy, which by the use of (4.54) implies that the coefficients $k_T$ and $k_p$ are small, hence the Soret effect and $\nabla p$-dependence of the mass concentration flux are weak. The following assumption

$$k_T \sim k_p \sim \frac{\bar{p}}{\bar{\rho}\Upsilon} \sim \delta, \tag{4.179}$$

where $\delta$ is defined in (3.11) (cf. also (4.83)), allows to simplify the reference state equations, so that at the leading order they read

$$\frac{\mathrm{d}\tilde{p}}{\mathrm{d}z} = -\tilde{\rho}\tilde{g}, \tag{4.180a}$$

$$\frac{\mathrm{d}}{\mathrm{d}z}\left[K\left(\frac{\mathrm{d}\tilde{\xi}}{\mathrm{d}z} + \frac{\tilde{k}_T}{\tilde{T}}\frac{\mathrm{d}\tilde{T}}{\mathrm{d}z} + \frac{\tilde{k}_p}{\tilde{p}}\frac{\mathrm{d}\tilde{p}}{\mathrm{d}z}\right)\right] = 0, \tag{4.180b}$$

[12] Note, that in the book of Landau and Lifshitz (1980) the temperature is expressed in energy units, thus their temperature is in fact $k_B T$, where $T$ is expressed in degrees Kelvin. Since in the limit of $\xi \ll 1$ and in the notation of the footnote (1) we get $\xi \approx N^{(l)}m_l/N^{(h)}m_h$, therefore Landau's $c = N^{(l)}/N^{(h)} \approx \xi m_h/m_l$ is equivalent to $\xi$ up to a constant factor and hence the parameter $\Upsilon \approx k_B T/\xi m_l$.

$$\frac{\mathrm{d}}{\mathrm{d}z}\left(k\frac{\mathrm{d}\tilde{T}}{\mathrm{d}z}\right) = -\tilde{Q}, \tag{4.180c}$$

$$\frac{\mathrm{d}^2\tilde{\psi}}{\mathrm{d}z^2} = 4\pi G\left[\tilde{\rho}(z)\left(\theta_H(z) - \theta_H(z-L)\right) + \rho_{in}(z)\theta_H(-z)\right], \tag{4.180d}$$

$$\tilde{\rho} = \rho(\tilde{p}, \tilde{T}, \tilde{\xi}), \quad \tilde{s} = s(\tilde{p}, \tilde{T}, \tilde{\xi}). \tag{4.180e}$$

In the above system the reference temperature, density and pressure profiles are determined from the Eqs. (4.180a), (4.180c) and (4.180e) and then the Eq. (4.180b) allows to calculate the mass fraction reference profile. $\theta_H(z)$ is the Heaviside step function.

### 4.5.1 Mixture of Ideal Gases in the Weak Solution Limit

It is of interest to consider the simplifying case when the binary alloy is a mixture of two ideal gases (cf. Sect. 4.2.1), but it remains a weak solution of the light constituent, more precisely

$$\xi \sim \delta \ll 1. \tag{4.181}$$

Under this assumption the thermodynamic properties of the alloy significantly simplify at the leading order to

$$\begin{aligned} h_{p,T} \approx & \left(c_p^{(l)} - c_p^{(h)}\right) T \ln T - (r_m - 1)\, R^{(h)} T \ln p - r_m R^{(h)} T \ln (r_m \xi) \\ & + T\left(s_0^{(l)} - s_0^{(h)}\right), \end{aligned} \tag{4.182a}$$

$$c_{p,\xi} \approx c_p^{(h)}, \qquad c_{v,\xi} \approx c_v^{(h)}, \qquad c_{p,\xi} - c_{v,\xi} \approx R^{(h)}, \tag{4.182b}$$

$$\chi_T \approx r_m - 1, \quad \Upsilon \approx \frac{r_m R^{(h)} T}{\xi}, \qquad k_p \approx \frac{r_m - 1}{r_m}\xi, \tag{4.182c}$$

$$\begin{aligned} \chi \approx & \left(1 - \frac{c_p^{(l)}}{c_p^{(h)}}\right) \ln T + (r_m - 1)\frac{\gamma^{(h)} - 1}{\gamma^{(h)}} \ln p + r_m \frac{\gamma^{(h)} - 1}{\gamma^{(h)}} \ln (r_m \xi) \\ & + r_m - 1 - \frac{s_0^{(l)} - s_0^{(h)}}{c_p^{(h)}}. \end{aligned} \tag{4.182d}$$

Therefore $\Upsilon \sim R^{(h)} T \delta^{-1}$ is large, whereas $k_p \sim \delta$ likewise $k_T = (\Lambda - h_{p,T})/\Upsilon \sim \delta$ are small.[13] In other words we can write

$$\xi = \mathcal{O}(\delta) \Rightarrow \Upsilon = \mathcal{O}\left(\delta^{-1} \bar{g} L\right), \quad k_T = \mathcal{O}(\delta), \quad k_p = \mathcal{O}(\delta), \tag{4.183}$$

where $\delta \ll 1$ is the measure of the departure of the system from the adiabatic well-mixed state, defined in (3.11) and (4.83). The equations describing the hydrostatic reference state (4.180a–4.180e), in the current situation take the form

$$\frac{\mathrm{d}\tilde{p}}{\mathrm{d}z} = -\tilde{\rho}\tilde{g}, \tag{4.184a}$$

$$\frac{\mathrm{d}}{\mathrm{d}z}\left[K\left(\frac{\mathrm{d}\tilde{\xi}}{\mathrm{d}z} + \frac{\tilde{k}_T}{\tilde{T}}\frac{\mathrm{d}\tilde{T}}{\mathrm{d}z} - \frac{(r_m - 1)\,\tilde{g}\tilde{\xi}}{r_m R^{(h)} \tilde{T}}\right)\right] = 0, \tag{4.184b}$$

$$\frac{\mathrm{d}}{\mathrm{d}z}\left(k\frac{\mathrm{d}\tilde{T}}{\mathrm{d}z}\right) = -\tilde{Q}, \tag{4.184c}$$

$$\frac{\mathrm{d}^2\tilde{\psi}}{\mathrm{d}z^2} = 4\pi G\left[\tilde{\rho}(z)\left(\theta_H(z) - \theta_H(z-L)\right) + \rho_{in}(z)\theta_H(-z)\right], \tag{4.184d}$$

$$\tilde{p} = \tilde{\rho} R^{(h)} \tilde{T}, \quad \tilde{s} = c_p^{(h)} \ln \tilde{T} - R^{(h)} \ln \tilde{p} + s_0^{(h)}. \tag{4.184e}$$

The temperature in the reference state $\tilde{T}$ can now be explicitly calculated from (4.184c), whereas the pressure $\tilde{p}$ and density $\tilde{\rho}$ from the hydrostatic force balance (4.184a) and the first of the equations of state in (4.184e). Let us now consider the special case when the transport coefficients $K$ and $k$, the Soret coefficient $k_T$ and the specific heats $c_{p,\xi} \approx c_p^{(h)}$ and $c_{v,\xi} \approx c_v^{(h)}$ can be assumed constant, and moreover the total mass of the fluid layer is negligibly small compared to the mass of the body below the layer, which implies $\mathrm{d}_z^2\tilde{\psi} = 4\pi G \rho_{in}(z)\theta_H(-z)$ and consequently $g \approx \tilde{g} = \mathrm{const}$ within the region of the fluid. Additionally we assume, that the radiogenic heat can be neglected $\tilde{Q} = 0$. The temperature, density, pressure and entropy in the reference state are then described at the leading order by the Eqs. (3.68a–3.68c) with $c_p$, $\gamma = c_p/c_v$ and $R$ replaced by $c_p^{(h)}$, $\gamma^{(h)} = c_p^{(h)}/c_v^{(h)}$ and $R^{(h)}$ respectively, i.e. the equations of state of a single-component ideal gas, that is the heavy constituent only. These expressions are now supplied by the solution of the mass fraction equation (4.184c) which in the case at hand takes the following, fairly simple form

$$\tilde{\xi} = \frac{C_0}{\left(1 - \theta\frac{z}{L}\right)^l} + \frac{\tilde{j}_\xi}{K}\frac{L}{\theta}\frac{1}{l+1}\left(1 - \theta\frac{z}{L}\right) - \frac{k_T}{l}, \tag{4.185}$$

[13]Note, that $h_{p,T}$ has a logarithmic dependence on the mass fraction $\xi$, however, since $\xi \ln \xi \overset{\xi\to 0}{\longrightarrow} -\xi$ ($\Lambda$ is a phenomenological material property), the assumption $k_T \sim \delta$ is justified.

where $\theta = \Delta T / T_B$, $C_0 = \text{const}$,

$$\tilde{j}_\xi = -K\left(\frac{\mathrm{d}\tilde{\xi}}{\mathrm{d}z} + \frac{\tilde{k}_T}{\tilde{T}}\frac{\mathrm{d}\tilde{T}}{\mathrm{d}z} - \frac{(r_m - 1)\, g\tilde{\xi}}{r_m R^{(h)}\tilde{T}}\right), \tag{4.186}$$

is the constant flux of the mass concentration and

$$l = \frac{r_m - 1}{r_m}\frac{gL}{R^{(h)}\Delta T} = \frac{r_m - 1}{r_m}(m+1). \tag{4.187}$$

The constant $C_0$ can be determined by application of the boundary conditions when the values of the mass fraction are specified at the boundaries, however, when the mass concentration flux is held fixed at the boundaries, this constant remains undetermined. In the limit of weak stratification, $\theta \ll 1$, the $\tilde{\xi}$ profile becomes linear.

It is sometimes useful to possess also the explicit formula for the chemical potential in the reference state, which in the limit $\xi \ll 1$ reads (cf. (4.123) and (4.182a))

$$\begin{aligned}\tilde{\mu}_c = {} & \left(c_p^{(l)} - c_p^{(h)}\right)\tilde{T}\left(1 - \ln\tilde{T}\right) + (r_m - 1)\, R^{(h)}\tilde{T}\ln\tilde{p} \\ & + r_m R^{(h)}\tilde{T}\ln\left(r_m\tilde{\xi}\right) - \tilde{T}\left(s_0^{(l)} - s_0^{(h)}\right) + \varepsilon_0^{(l)} - \varepsilon_0^{(h)}.\end{aligned} \tag{4.188}$$

Let us stress, that there is, in fact, a large degree of freedom in the choice of the reference state, which depends on the particular application. The reference state must satisfy the dynamical equations, and therefore may be required to be non-stationary if e.g. the thermal diffusion in the basic state does not vanish $\nabla \cdot \left(k\nabla\tilde{T}\right) \neq 0$. However, when specifying a reference state one must take good care to specify also the boundary conditions for the fluctuations and realize precisely what does the assumption of small departures from the adiabatic well-mixed state mean in the particular situation. For example, as already noted, the well-mixed adiabatic state itself can be chosen for the reference state, in which case the boundary conditions on the fluctuations *must be* non-zero in order to drive the flow; in such a case the boundary conditions define the departure from adiabaticity and the well-mixed state, which must be weak.

Next we proceed to derive the final form of the dynamical equations in the considered limit. A significant simplification is achieved when the Dufour effect is weak

$$\Lambda \lesssim \mathcal{O}\left(\delta\bar{g}L\right), \tag{4.189}$$

where $\bar{g}L$ is the gravitational energy scale, since the chemical potential gradient, which is an order unity quantity, by the use of (4.134) and (4.189) is proportional to the material flux

$$\nabla\mu_c = -\frac{\Upsilon}{K}\mathbf{j}_{\xi,\mathrm{mol}} + \mathcal{O}\left(\delta\bar{g}\right). \tag{4.190}$$

It is important to emphasize, that in the weak solution limit both, the mass fraction in the reference state, $\tilde{\xi}$ and its fluctuation $\xi'$ are small and of the same order of magnitude in terms of the small parameter $\delta$, that is $\tilde{\xi} = \mathcal{O}(\delta)$ and $\xi' = \mathcal{O}(\delta)$, but the compositional expansion coefficients $\chi$ and $\chi_T$ are both of the order of unity. Under the above assumptions the terms in the material flux can be approximated as follows

$$\begin{aligned}\mathbf{j}_{\xi,\mathrm{mol}} &= -K\left(\nabla\xi' + \frac{\mathrm{d}\tilde{\xi}}{\mathrm{d}z}\hat{\mathbf{e}}_z + \frac{\tilde{k}_T}{\tilde{T}}\frac{\mathrm{d}\tilde{T}}{\mathrm{d}z}\hat{\mathbf{e}}_z + \frac{\tilde{k}_p}{\tilde{p}}\frac{\mathrm{d}\tilde{p}}{\mathrm{d}z}\hat{\mathbf{e}}_z\right) + \mathcal{O}\left(\delta^{5/2}\rho_B\sqrt{\bar{g}L}\right)\\ &= -K\nabla\xi' + \tilde{j}_\xi\hat{\mathbf{e}}_z + \mathcal{O}\left(\delta^{5/2}\rho_B\sqrt{\bar{g}L}\right),\end{aligned} \tag{4.191}$$

where the material flux in the reference state

$$\tilde{j}_\xi = -K\left(\frac{\mathrm{d}\tilde{\xi}}{\mathrm{d}z} + \frac{\tilde{k}_T}{\tilde{T}}\frac{\mathrm{d}\tilde{T}}{\mathrm{d}z} + \frac{\tilde{k}_p}{\tilde{p}}\frac{\mathrm{d}\tilde{p}}{\mathrm{d}z}\right) = \mathrm{const} \tag{4.192}$$

is uniform (recall, that the material conductivity coefficient is a function of height only, $K = K(z)$). In the light of (4.190) and (4.191) the term $-\mathbf{j}_{\xi,\mathrm{mol}}\cdot\nabla\mu_c$ in the energy equation takes the form

$$\begin{aligned}-\mathbf{j}_{\xi,\mathrm{mol}}\cdot\nabla\mu_c = \frac{\Upsilon}{K}\mathbf{j}^2_{\xi,\mathrm{mol}} &= \frac{\Upsilon}{K}\left(K\nabla\xi' - \tilde{j}_\xi\hat{\mathbf{e}}_z\right)^2\\ &= \Upsilon K\left(\nabla\xi'\right)^2 - 2\tilde{j}_\xi\Upsilon\frac{\partial\xi'}{\partial z} - \frac{\tilde{j}^2_\xi}{K}\frac{r_m R^{(h)}\tilde{T}}{\tilde{\xi}+\xi'}\frac{\xi'}{\tilde{\xi}}\\ &\quad + \frac{\tilde{j}^2_\xi}{K}\frac{r_m R^{(h)}\tilde{T}}{\tilde{\xi}} + \mathcal{O}\left(\delta^{5/2}\rho_B\bar{g}\sqrt{\bar{g}L}\right),\end{aligned} \tag{4.193}$$

where in the last term we have substituted

$$\Upsilon = \frac{r_m R^{(h)}\left(\tilde{T}+T'\right)}{\tilde{\xi}+\xi'} = \frac{r_m R^{(h)}\tilde{T}}{\tilde{\xi}+\xi'} + \mathcal{O}\left(\bar{g}L\right) = \frac{r_m R^{(h)}\tilde{T}}{\tilde{\xi}} - \frac{r_m R^{(h)}\tilde{T}}{\tilde{\xi}+\xi'}\frac{\xi'}{\tilde{\xi}} + \mathcal{O}\left(\bar{g}L\right), \tag{4.194}$$

to clearly separate the last term $\tilde{j}^2_\xi r_m R^{(h)}\tilde{T}/K\tilde{\xi}$ in (4.193) which belongs to the basic state energy (entropy) balance. It can now be clearly seen, that as a result of (4.178), in the limit of a weak solution the term $-\mathbf{j}_{\xi,\mathrm{mol}}\cdot\nabla\mu_c$ takes a significantly different form than (4.142) obtained for the case of $|\xi'| \ll \tilde{\xi}$. Consequently the entropy equation (4.144e) at the leading order is modified to

$$\tilde{\rho}\tilde{T}\left(\frac{\partial s'}{\partial t}+\mathbf{u}\cdot\nabla s'\right)-\tilde{\rho}\tilde{c}_{p,\xi}u_z\Delta_S+\tilde{\rho}\tilde{h}_{p,T}u_z\frac{\mathrm{d}\tilde{\xi}}{\mathrm{d}z}$$
$$=\nabla\cdot\left(k\nabla T'\right)+K\Upsilon\left(\nabla\xi'\right)^2-2\tilde{j}_\xi\Upsilon\frac{\partial\xi'}{\partial z}-\frac{\tilde{j}_\xi^2}{K}\frac{r_m R^{(h)}\tilde{T}}{\tilde{\xi}+\xi'}\frac{\xi'}{\tilde{\xi}}$$
$$+2\tilde{\rho}\nu\mathbf{G}^s:\mathbf{G}^s+\tilde{\rho}\left(\nu_b-\frac{2}{3}\nu\right)(\nabla\cdot\mathbf{u})^2+Q', \qquad (4.195)$$

where $\Upsilon$ is given by the full expression in (4.194). The Eqs. (4.144b–4.144d) remain unaltered. The Navier-Stokes equation can be conveniently left in the most general form (4.132a). The full equations of state of a mixture of ideal gases and the thermodynamic properties are provided in (4.124a–4.124d) and (4.125a–4.125f) whereas the leading order form of the thermodynamic properties in the limit $\xi \ll 1$ can be found in (4.182a–4.182d). Since the parameters $\alpha$, $\beta$ and $\chi_T$ for weak solutions remain order unity quantities in terms of $\delta$, the density fluctuation is still expressed by

$$\frac{\rho'}{\tilde{\rho}}=\frac{p'}{\tilde{p}}-\frac{T'}{\tilde{T}}-(r_m-1)\,\xi', \qquad (4.196)$$

and is of the order $\rho' = \mathcal{O}(\delta\tilde{\rho})$. The entropy fluctuation on the other hand can no longer be expressed as in (4.144f) because $\xi'/\tilde{\xi}$ is no longer required to be small and the parameter $h_{p,T}$ is irregular when $\xi \to 0$. Hence the entropy fluctuation must be calculated explicitly from (4.124b) and (4.125a–4.125c)

$$s'=s-\tilde{s}$$
$$=c_p^{(h)}\frac{T'}{\tilde{T}}-\frac{p'}{\tilde{\rho}\tilde{T}}+\frac{\tilde{h}_{p,T}}{\tilde{T}}\xi'$$
$$+r_m R^{(h)}\tilde{\xi}\left[\frac{\xi'}{\tilde{\xi}}-\left(1+\frac{\xi'}{\tilde{\xi}}\right)\ln\left(1+\frac{\xi'}{\tilde{\xi}}\right)\right], \qquad (4.197)$$

where $\tilde{h}_{p,T}$ is taken from (4.182a). However, $s'$ still remains an order $\mathcal{O}\left(\delta\tilde{c}_{p,\xi}\right)$ quantity because $\tilde{\xi}=\mathcal{O}(\delta)$ and as remarked in the footnote (13), $\tilde{h}_{p,T}\xi'/\tilde{T}=\mathcal{O}(\delta\tilde{h}_{p,T}/\tilde{T})=\mathcal{O}(\delta\tilde{c}_{p,\xi})$. The two latter expressions for the density and entropy fluctuations can be used to express the buoyancy force in terms of the entropy and mass fraction fluctuations, similarly as in Sect. 4.3; in an analogous way one obtains at the leading order

$$-\frac{1}{\tilde{\rho}}\nabla p'-\nabla\psi'+\frac{\rho'}{\tilde{\rho}}\tilde{\mathbf{g}}=-\nabla\left(\frac{p'}{\tilde{\rho}}+\psi'\right)+\frac{s'}{c_p^{(h)}}\tilde{g}\hat{\mathbf{e}}_z+\tilde{\chi}\xi'\tilde{g}\hat{\mathbf{e}}_z$$
$$-r_m g\frac{\gamma^{(h)}-1}{\gamma^{(h)}}\tilde{\xi}\left[\frac{\xi'}{\tilde{\xi}}-\left(1+\frac{\xi'}{\tilde{\xi}}\right)\ln\left(1+\frac{\xi'}{\tilde{\xi}}\right)\right]\hat{\mathbf{e}}_z, \qquad (4.198)$$

where $\tilde{\chi}$ can be taken from (4.182d). Such an expression for the buoyancy force, although not as simple as for the case of $|\xi'|/\tilde{\xi} \ll 1$ may still be useful, since it does not involve the density and temperature fluctuations and the pressure fluctuation could be easily removed from the Navier-Stokes equation by taking its curl.

For the sake of completeness we also provide the expressions for the chemical potential in the current case. The leading order form of the chemical potential in the reference state was already derived in (4.188). The convective fluctuation has to be calculated directly from (4.124d) and (4.125a)

$$\begin{aligned} \mu_c' =& \mu_c - \tilde{\mu}_c \\ =& (r_m - 1)\, R^{(h)} \tilde{T} \frac{p'}{\tilde{p}} - \tilde{h}_{p,T} \frac{T'}{\tilde{T}} \\ &+ r_m R^{(h)} \tilde{T} \left( \frac{T'}{\tilde{T}} + 1 \right) \ln \left( 1 + \frac{\xi'}{\tilde{\xi}} \right). \end{aligned} \quad (4.199)$$

Observe, that the fluctuation of the chemical potential is an order unity quantity in terms of the small parameter $\delta$, i.e. $\mu_c' = \mathcal{O}(\bar{g}L)$ and thus the magnitude of the reference state profile $\tilde{\mu}_c(z)$ is not expected to exceed the magnitude of the fluctuation.

#### 4.5.1.1 Entropy Formulation with Compositional Effects for Systems with Volume Cooling

When the binary alloy is a mixture of ideal, light and heavy gases, but the solution of the light constituent is weak, i.e. we hold $\xi = \mathcal{O}(\delta) \ll 1$, and there are volume heat sinks which can be modelled by $\tilde{Q} = \kappa g \mathrm{d}_z \tilde{p} < 0$, it is possible to express the dynamical equations in terms of only three thermodynamic variables describing fluctuations, namely $p'$, $s'$ and $\xi'$. Moreover, the pressure then, appears solely in the Navier-Stokes equation and can be easily removed by taking a curl of this equation. The remaining variables $\rho'$ and $T'$ are entirely eliminated and if necessary can be calculated afterwards, when the system of dynamical equations is solved and $\mathbf{u}$, $p'$, $s'$ and $\xi'$ are determined. This is called the *entropy formulation* (cf. Sect. 3.3 for comparison with the case of a single-component fluid). This formulation can be achieved only, when additionally a constant thermal diffusivity is assumed and the fluid's contributions to the total gravity are neglected, so that the gravitational acceleration is effectively constant within the fluid, i.e.

$$\kappa = \frac{k}{\tilde{\rho} c_p^{(h)}} = \text{const}, \qquad g = \text{const}, \quad (4.200)$$

where $c_p^{(h)}$ is the specific heat at constant pressure of the heavy constituent alone; the specific heats $c_p^{(h)}$ and $c_v^{(h)}$ are also assumed constant. The properties of the binary alloy are described by (4.182a–4.182d) and we assume, that the Dufour effect is negligible

$$\Lambda \lesssim \mathcal{O}(\delta \bar{g} L), \tag{4.201}$$

thus the Soret coefficient simplifies to

$$\tilde{k}_T = -\frac{\tilde{h}_{p,T}}{\tilde{\Upsilon}}. \tag{4.202}$$

The hydrostatic reference state is determined by the Eqs. (4.184a–4.184c, 4.184e) with $g = \text{const}$ (the chemical potential is given in (4.188)).

The Eq. (4.197) allows to express the temperature fluctuations in terms of the entropy, pressure and mass fraction fluctuations

$$\begin{aligned} T' =& \frac{\tilde{T}}{c_p^{(h)}} s' + \frac{p'}{\tilde{\rho} c_p^{(h)}} - \frac{\tilde{h}_{p,T}}{c_p^{(h)}} \xi' \\ & - r_m \frac{\gamma^{(h)} - 1}{\gamma^{(h)}} \tilde{T} \tilde{\xi} \left[ \frac{\xi'}{\tilde{\xi}} - \left(1 + \frac{\xi'}{\tilde{\xi}}\right) \ln \left(1 + \frac{\xi'}{\tilde{\xi}}\right) \right], \end{aligned} \tag{4.203}$$

where $\gamma^{(h)} = c_p^{(h)} / c_v^{(h)}$. On inserting the latter expression into the thermal diffusion term in the entropy equation one obtains

$$\begin{aligned} \nabla \cdot (k \nabla T') =& \kappa \nabla \cdot \left[ \tilde{\rho} \nabla \left( \tilde{T} s' \right) \right] + \kappa \nabla \cdot \left( \tilde{\rho} \nabla \frac{p'}{\tilde{\rho}} \right) - \kappa \nabla \cdot \left[ \tilde{\rho} \nabla \left( \tilde{h}_{p,T} \mathcal{Z} \right) \right] \\ =& \kappa \nabla \cdot \left[ \tilde{\rho} \tilde{T} \nabla s' \right] - \frac{\kappa g}{c_p^{(h)}} \frac{\partial}{\partial z} (\tilde{\rho} s') + \kappa \nabla \cdot \left( \tilde{\rho} \nabla \frac{p'}{\tilde{\rho}} \right) \\ & - \kappa \nabla \cdot \left[ \tilde{\rho} \nabla \left( \tilde{h}_{p,T} \mathcal{Z} \right) \right], \end{aligned} \tag{4.204}$$

where we have used $\nabla \tilde{T} = \mathrm{d}_z \tilde{T} \hat{\mathbf{e}}_z = -g / c_p^{(h)} \hat{\mathbf{e}}_z + \mathcal{O}(\delta \Delta T / L)$ and we have defined

$$\mathcal{Z} = \xi' + \frac{r_m R \tilde{T} \tilde{\xi}}{\tilde{h}_{p,T}} \left[ \frac{\xi'}{\tilde{\xi}} - \left(1 + \frac{\xi'}{\tilde{\xi}}\right) \ln \left(1 + \frac{\xi'}{\tilde{\xi}}\right) \right]. \tag{4.205}$$

On the other hand the Navier-Stokes equation by the use of (4.130) can be cast in the form

$$\begin{aligned} \frac{\partial \mathbf{u}}{\partial t} + \frac{1}{\tilde{\rho}} \nabla \cdot (\tilde{\rho} \mathbf{u} \mathbf{u}) =& -\nabla \left( \frac{p'}{\tilde{\rho}} \right) + \frac{s'}{c_p^{(h)}} g \hat{\mathbf{e}}_z + \tilde{\chi} \xi' g \hat{\mathbf{e}}_z + \frac{1}{\tilde{\rho}} \nabla \cdot (2 \tilde{\rho} \nu \mathbf{G}^s) \\ & + \frac{1}{\tilde{\rho}} \nabla \left[ \left( \nu_b - \frac{2}{3} \nu \right) \tilde{\rho} \nabla \cdot \mathbf{u} \right], \end{aligned} \tag{4.206}$$

hence multiplying the latter equation by $\tilde{\rho}$, taking its divergence and utilizing the constraint $\nabla \cdot (\tilde{\rho} \mathbf{u}) = 0$ we get

$$\nabla\cdot\left(\tilde{\rho}\nabla\frac{p'}{\tilde{\rho}}\right)=\frac{g}{c_p^{(h)}}\frac{\partial}{\partial z}\left(\tilde{\rho}s'\right)+g\frac{\partial}{\partial z}\left(\tilde{\rho}\tilde{\chi}\xi'\right)+\nabla\cdot\left[\nabla\cdot\left(2\tilde{\rho}\nu\mathbf{G}^s-\tilde{\rho}\mathbf{uu}\right)\right]$$
$$+\nabla^2\left[\left(\nu_b-\frac{2}{3}\nu\right)\tilde{\rho}\nabla\cdot\mathbf{u}\right]. \tag{4.207}$$

Substitution of the formula (4.207) into (4.204) allows to express the thermal diffusion solely in terms of the entropy $s'$ and the mass fraction $\xi'$. Therefore we can now write down the final set of dynamical equations formulated in terms of the entropy and the mass fraction fluctuations in the following form

$$\frac{\partial\mathbf{u}}{\partial t}+(\mathbf{u}\cdot\nabla)\,\mathbf{u}=-\nabla\left(\frac{p'}{\tilde{\rho}}\right)+\frac{s'}{c_p^{(h)}}g\hat{\mathbf{e}}_z+\tilde{\chi}\xi'g\hat{\mathbf{e}}_z+\nu\nabla^2\mathbf{u}+\left(\frac{\nu}{3}+\nu_b\right)\nabla\left(\nabla\cdot\mathbf{u}\right)$$
$$+\frac{2}{\tilde{\rho}}\nabla\left(\tilde{\rho}\nu\right)\cdot\mathbf{G}^s+\frac{1}{\tilde{\rho}}\nabla\left(\tilde{\rho}\nu_b-\frac{2}{3}\tilde{\rho}\nu\right)\nabla\cdot\mathbf{u}, \tag{4.208a}$$

$$\nabla\cdot\left(\tilde{\rho}\mathbf{u}\right)=0, \tag{4.208b}$$

$$\tilde{\rho}\left(\frac{\partial\xi'}{\partial t}+\mathbf{u}\cdot\nabla\xi'\right)+\tilde{\rho}u_z\frac{\mathrm{d}\tilde{\xi}}{\mathrm{d}z}=\nabla\cdot\left(K\nabla\xi'\right), \tag{4.208c}$$

$$\tilde{\rho}\tilde{T}\left(\frac{\partial s'}{\partial t}+\mathbf{u}\cdot\nabla s'\right)-\tilde{\rho}c_p^{(h)}u_z\Delta_S+\tilde{\rho}\tilde{h}_{p,T}u_z\frac{\mathrm{d}\tilde{\xi}}{\mathrm{d}z}=\kappa\nabla\cdot\left(\tilde{\rho}\tilde{T}\nabla s'\right)+\Xi+\mathcal{J}+Q'. \tag{4.208d}$$

In the above the term

$$\Xi=-\kappa\nabla\cdot\left[\tilde{\rho}\nabla\left(\tilde{h}_{p,T}\mathcal{Z}\right)\right]+\kappa g\frac{\partial}{\partial z}\left(\tilde{\chi}\tilde{\rho}\xi'\right)+K\Upsilon\left(\nabla\xi'\right)^2$$
$$-2\tilde{j}_\xi\Upsilon\frac{\partial\xi'}{\partial z}-\frac{\tilde{j}_\xi^2}{K}\frac{r_mR^{(h)}\tilde{T}}{\tilde{\xi}+\xi'}\frac{\xi'}{\tilde{\xi}}, \tag{4.209}$$

(where $\mathcal{Z}$ is given in (4.205)), depends on the mass fraction fluctuation only, i.e. no other type of thermodynamic fluctuations such as $\rho'$, $p'$, $T'$ nor $s'$ contributes to the above expression for $\Xi$, and

$$\mathcal{J}=\kappa\nabla\cdot\left[\nabla\cdot\left(2\tilde{\rho}\nu\mathbf{G}^s-\tilde{\rho}\mathbf{uu}\right)\right]+\kappa\nabla^2\left[\left(\nu_b-\frac{2}{3}\nu\right)\tilde{\rho}\nabla\cdot\mathbf{u}\right]$$
$$+2\tilde{\rho}\nu\mathbf{G}^s:\mathbf{G}^s+\tilde{\rho}\left(\nu_b-\frac{2}{3}\nu\right)\left(\nabla\cdot\mathbf{u}\right)^2, \tag{4.210}$$

is the same as in the case of entropy formulation for a single-component fluid, cf. (3.107). The coefficients $\tilde{h}_{p,T}$ and $\tilde{\chi}$ are provided in (4.182a, 4.182d), with $T$, $p$, $\xi$ replaced by $\tilde{T}$, $\tilde{p}$, $\tilde{\xi}$; the parameter $\Upsilon$ at the leading order was given in (4.194). Once the dynamical equations (4.208a–4.208d) are solved and in particular the fluctuations of pressure $p'$, entropy $s'$ and the mass fraction $\xi'$ are determined, the density and temperature fluctuations can be found from

$$\frac{\rho'}{\tilde{\rho}} = -\frac{s'}{c_p^{(h)}} + \frac{p'}{\gamma^{(h)}\tilde{p}} - \tilde{\chi}\xi', \qquad \frac{T'}{\tilde{T}} = \frac{s'}{c_p^{(h)}} + \frac{\gamma^{(h)}-1}{\gamma^{(h)}}\frac{p'}{\tilde{p}} - \frac{\tilde{h}_{p,T}}{c_p^{(h)}\tilde{T}}\xi'. \quad (4.211)$$

## Review Exercises

**Exercise 1** In the case of a weak solution of two perfect gases the reference state takes the form provided in Sect. 4.5.1 (cf. (4.185) and (3.68a–3.68c)). Calculate the compositional Rayleigh number and the total heat flux for this reference state.
*Hint*: utilize (4.159) (why?), (4.135) and (4.133).

**Exercise 2** For the case of Exercise 1 demonstrate explicitly, that $\mathrm{d}\tilde{\mu}_c/\mathrm{d}z + g\tilde{\chi}$ is an $\mathcal{O}(1)$ quantity in terms of the small parameter $\delta$, which under the anelastic approximation is a unique feature of weak solutions.
*Hint*: utilize (4.188).

**Exercise 3** Under the assumptions $k_T = 0$, $k_p = 0$, $g = \text{const}$ and $\chi = \text{const}$ calculate the mean vertical molecular material flux $\langle \mathbf{j}_{\xi,\mathrm{mol}} \cdot \hat{\mathbf{e}}_z \rangle$, given the total material flux through the system $G_0$, the total viscous dissipation rate

$$D_\nu = 2\left\langle \mu \mathbf{G}^s : \mathbf{G}^s \right\rangle - \left\langle \left(\frac{2}{3}\mu - \mu_b\right)(\nabla \cdot \mathbf{u})^2 \right\rangle,$$

and the mean work of the thermal buoyancy force

$$W_{th} = \left\langle \frac{\tilde{\alpha}\tilde{T}g\tilde{\rho}}{\tilde{c}_{p,\xi}} u_z s' \right\rangle.$$

*Hint*: utilize the results of Sect. 4.4.1.

**Acknowledgements** The author wishes to thank Professor Chris Jones for many fruitfull discussions on the topic of compressible convection, which have greatly helped to improve the third chapter of the book. The support of the National Science Centre of Poland (grant no. 2017/26/E/ST3/00554) is gratefully acknowledged.

## References

Braginsky, S.I., and P.H. Roberts. 1995. Equations governing convection in Earth's core and the geodynamo. *Geophysical and Astrophysical Fluid Dynamics* 79: 1–97.

de Groot, S.R., and P. Mazur. 1984. *Non-equilibrium thermodynamics*. New York: Dover Publications.

Glansdorff, P., and I. Prigogine. 1971. *Thermodynamic theory of structure, stability and fluctuations*. London: Wiley.

Guminski, K. 1974. *Termodynamika*. Warsaw: Polish Scientific Publishers PWN.

Landau, L.D., and E.M. Lifshitz. 1980. *Statistical physics, Course of theoretical physics*, vol. 5. Oxford: Elsevier.

Landau, L.D., and E.M. Lifshitz. 1987. *Fluid mechanics, Course of theoretical physics*, vol. 6. Oxford: Elsevier.

# Index





K. A. Mizerski, *Foundations of Convection with Density Stratification*, GeoPlanet: Earth and Planetary Sciences, https://doi.org/10.1007/978-3-030-63054-6

Printed in the United States
by Baker & Taylor Publisher Services